CAMBRIDGE STUDIES IN
ADVANCED MATHEMATICS 112

PARTIAL DIFFERENTIAL EQUATIONS
FOR PROBABALISTS

This book deals with equations that have played a central role in the interplay between partial differential equations and probability theory. Most of this material has been treated elsewhere, but it is rarely presented in a manner that makes it readily accessible to people whose background is probability theory. Many results are given new proofs designed for readers with limited expertise in analysis.

The author covers the theory of linear, second order, partial differential equations of parabolic and elliptic types. Many of the techniques have antecedents in probability theory, although the book also covers a few purely analytic techniques. In particular, a chapter is devoted to the De Giorgi–Moser–Nash estimates, and the concluding chapter gives an introduction to the theory of pseudodifferential operators and their application to hypoellipticity, including the famous theorem of Lars Hörmander.

Cambridge Studies in Advanced Mathematics

All the titles listed below can be obtained from good booksellers or from Cambridge University Press. For a complete series listing visit www.cambridge.org/series/sSeries.asp?code=CSAM

PARTIAL DIFFERENTIAL EQUATIONS FOR PROBABALISTS

DANIEL W. STROOCK
Massachusetts Institute of Technology

CAMBRIDGE
UNIVERSITY PRESS

CAMBRIDGE UNIVERSITY PRESS
Cambridge, New York, Melbourne, Madrid, Cape Town, Singapore, São Paulo, Delhi

Cambridge University Press
32 Avenue of the Americas, New York, NY 10013-2473, USA

www.cambridge.org
Information on this title: www.cambridge.org/9780521886512

First published 2008

Printed in the United States of America

A catalog record for this publication is available from the British Library.

Library of Congress Cataloging in Publication Data
Stroock, Daniel W.
Partial differential equations for probabilists / Daniel W. Stroock.
p. cm. – (Cambridge studies in advanced mathematics)
Includes bibliographical references and index.
ISBN 978-0-521-88651-2 (hardback)
1. Differential equations, Partial. 2. Differential equations, Parabolic. 3. Differential
equations, Elliptic. 4. Probabilities. I. Title.
QA377.S845 2008
515′.353 – dc22 2007048751

ISBN 978-0-521-88651-2 hardback

This book is dedicated to the memory of my friend Eugene Fabes.

Contents

Preface

There are few benefits to growing old, especially if you are a mathematician. However, one of them is that, over the course of time, you accumulate a certain amount of baggage containing information in which, if you are lucky and they are polite, your younger colleagues may express some interest.

Having spent most of my career at the interface between probability and partial differential equations, it is hardly surprising that this is the item in my baggage about which I am asked most often. When I was a student, probabilists were still smitten by the abstract theory of Markov processes which grew out of the beautiful work of G. Hunt, E.B. Dynkin, R.M. Blumenthal, R.K. Getoor, P.A. Meyer, and a host of others. However, as time passed, it became increasingly apparent that the abstract theory would languish if it were not fed a steady diet of hard, analytic facts. As A.N. Kolmogorov showed a long time ago, ultimately partial differential equations are the engine which drives the machinery of Markov processes. Until you solve those equations, the abstract theory remains a collection of "if, then" statements waiting for someone to verify that they are not vacuous.

Unfortunately for probabilists, the verification usually involves ideas and techniques which they find unpalatable. The strength of probability theory is that it deals with probability measures, but this is also its weakness. Because they model a concrete idea, probability measures have enormous intuitive appeal, much greater than that of functions. They are particularly useful when comparing the relative size of quantities: A is larger than B because it is more likely. For example, it is completely obvious that a diffusion is less likely to go from x to y before leaving a set Γ than it is to go from x to y when it is allowed to leave Γ. On the other hand, probability theory provides little help when it comes to determining whether one can talk sensibly about "the probability of going from x to y," with or without leaving Γ. Indeed, in a continuous setting, such events usually will have probability 0, and so one needs to discuss their probabilities in terms of (preferably smooth) densities with respect to a reference measure like Lebesgue's. Proving that such a density exists, much less checking that it possesses desirable properties, is not something for which probability reasoning is particularly well suited. As a consequence, probabilists have tended to avoid addressing these questions themselves and have relied on the hard work of analysts.

My purpose in writing this book has been to provide probabilists with a few tools with which they can understand how to prove on their own some of the basic facts about the partial differential equations with which they deal. In so far as possible, I have tried to base my treatment on ideas which are already familiar to anyone who has worked in stochastic analysis. In fact, there is nothing in the first two chapters which requires more than a course on measure theory (including a little Fourier analysis) and a semester of graduate-level probability theory. In Chapter 1, I prove a general existence result for solutions to Kolmogorov's forward equation by a method which is essentially due to K. Itô, even though it may take all but my more sophisticated readers a little time to recognize it as such. In summary, the upshot of Chapter 1 is that solutions nearly always exist, at least as probability measures. Chapter 2 is an initial attempt to prove that the solutions produced in Chapter 1 can be used to generate smooth solutions to Kolmogorov's backward equation, at least when the initial data are themselves smooth. Again, the ideas here will be familiar to anyone who has worked with stochastic integral equations. In fact, after seeing some of the contortions which I have to make for not doing so, experts will undoubtedly feel that I have paid a high price for not using Brownian motion. Be that as it may, the conclusion drawn in Chapter 2 is that solutions to Kolmogorov's backward equation preserve the regularity properties their initial data possess.

All the results in Chapters 1 and 2 can be viewed as translations to a measure setting of results which are well known for flows generated by a vector field. It is not until Chapter 3 that I begin discussing properties which are not shared by deterministic flows. Namely, when the diffusion coefficients in Kolmogorov's equation are "elliptic" (i.e., the diffusion matrix is strictly positive definite), the associated flow not only preserves but even smooths the initial data. The classic example of this smoothing property is the standard heat equation which immediately transforms any reasonably bounded initial data into a smooth (in fact, analytic) function. Until quite recently, probabilists have been at a complete loss when it came to proving such results in any generality. However, thanks to P. Malliavin, we now know a method which allows us to transfer via integration by parts smoothness properties of the Gauss distribution to the measures produced by Itô's construction. Malliavin himself and his disciples, like me, implemented his ideas in a pathspace context. However, if one is satisfied with less than optimal conclusions, then there is no need to work in pathspace, and there are good reasons not to. Specifically, the price one pays for the awkward treatment of Itô's theory in Chapters 1 and 2 turns out to buy one a tremendous technical advantage in Chapter 3. Briefly stated, the advantage is that solutions to Itô's stochastic differential equations are not classically smooth functions of the driving Brownian motion, even though they are infinitely smooth in the sense of Sobolev. Thus, to carry out Malliavin's program in pathspace, one has to jump through all sorts of annoying technical hoops. On the other hand, if you do all your integration by parts before passing to the limit in Itô's procedure, you avoid all those difficulties, and that is the reason why I adopted the approach which I did in Chapters 1 and 2.

By the end of Chapter 3, I have derived the basic regularity properties for solutions to Kolmogorov's equations with smooth coefficients which are uniformly elliptic. In particular, I have shown that the transition probability function admits a smooth density which, together with its derivatives, satisfies Gaussian estimates, and I have used these to derive Weyl's Lemma (i.e., hypoellipticity) for solutions to the associated elliptic and parabolic equations. For many applications to probability theory, this is all that one needs to know. However, for other applications, it is important to have more refined results, and perhaps the most crucial such refinement is the estimation of the transition probability density from below. In Chapter 4, I develop quite sharp upper and lower bounds on the transition probability using a methodology which has essentially nothing to do with probability theory. Instead of probability theory, the origin of the ideas here come from the calculus of variations, and so the usual form of Kolmogorov's equations gets replaced by their divergence form counterparts. As long as the coefficients are sufficiently smooth, this replacement hardly affects the generality of the conclusions derived. However, it has enormous impact on the mathematics used to draw those conclusions. In essence, everything derived in the first three chapters relies on the minimum principle (i.e., non-negativity preservation). By writing the equation in divergence form, a second powerful mathematical tool is made manifest: the theory of self-adjoint operators and their spectral theory. In his brilliant article about parabolic equations, J. Nash showed how, in conjunction with the minimum principle, self-adjointness can be used to prove surprising estimates for solutions of parabolic equations written in divergence form. Not only are these estimates remarkably sharp, they make no demands on the regularity of the coefficients involved. As a result, they are entirely different from the results derived in Chapter 3, all of which rely heavily on the smoothness of the coefficients. Supplementing Nash's ideas with a few more recent ones, I derive in Chapter 4 very tight upper and lower bounds on the transition probability density. The techniques used are all quite elementary, but the details are intricate.

In Chapter 5, I return to probabilistic techniques to localize the results proved earlier. Here the demands on the reader's probabilistic background are much greater than those made earlier. In particular, the reader is assumed to know how to pass from a transition probability function to a Markov process on pathspace. I have summarized some of the ideas involved, but I doubt if my summary will be sufficient for the uninitiated. In any case, for those who have the requisite background, Chapter 5 not only provides a ubiquitous localization procedure but also a couple of important applications of the results to which it leads. In particular, the chapter closes with proofs of both Nash's Continuity Theorem and Moser's extension of De Giorgi's Harnack principle. Chapter 6 can be viewed as a further application of the localization procedure developed in Chapter 5. Now the goal is to work on a differentiable manifold and to show how one can lift everything to that venue. Besides a familiarity with probability theory, the reader of Chapter 6 is assumed to have some acquaintance with the basic ideas of Riemannian differential geometry.

The concluding chapter, Chapter 7, represents an abrupt departure from the ones preceding it. Whereas in Chapter 4 the minimum principle still plays a role, in Chapter 7 it completely disappears. All the techniques introduced in Chapter 7 derive from Fourier analysis. I begin with a brief resumé of basic facts about Sobolev spaces. This is followed by a short course on pseudodifferential operators, one in which I avoid all but the most essential ingredients. I then apply pseudodifferential operators to prove far-reaching extensions of Weyl's Lemma, first for general, scalar valued, elliptic operators and then, following J.J. Kohn, for an interesting class of second order, degenerate operators with real valued coefficients. The latter extension, which is due to L. Hörmander, is the one of greater interest to probabilists. Indeed, it has already played a major role in many applications and promises to continue doing so in the future.

Even though this book covers much of the material about partial differential equations needed by probabilists, it does not cover it all. Perhaps the most egregious omission is the powerful ideas introduced by M. Crandell and P.L. Lions under the name of "viscosity solutions." In a very precise sense, their theory allows one to describe a "probabilistic solution" without reference to probability theory. The advantage to this is that it removes the need to develop the whole pathspace apparatus in order to get at one's solution, an advantage that becomes decisive when dealing with non-linear equations like those which arise in optimal control and free boundary value problems. For the reader who wishes to learn about this important topic, I know of no better place to begin than the superb book [14] by L.C. Evans. Less serious is my decision to deal only with time independent coefficients. For the material in Chapters 1 and 2 it makes no difference, since as long as one is making no use of ellipticity, one can introduce time as a new coordinate. However, it does make some difference in the later chapters, but not enough difference to persuade me to burden the whole text with the additional notation and considerations which the inclusion of time dependence would have required.

Each chapter ends with a section entitled "Historical Notes and Commentary." The reader should approach these sections with a healthy level of skepticism. I have not spent much time tracking down sufficient historical evidence to make me confident that I have always given credit where credit is due and withheld it where it is not due. Thus, these sections should be read for what they are: a highly prejudiced, impressionistic account of what I think may be the truth.

Finally, a word about Eugene Fabes, to whom I have dedicated this book. Gene and I met as competitors, he working with Nestor Riviere to develop the analytic theory of parabolic equations with continuous coefficients and I working with S.R.S. Varadhan to develop the corresponding probabilistic theory. However, as anyone who knew him knows, Gene was not someone with whom you could maintain an adversarial relationship for long. After spending a semester together in Minnesota talking mathematics, sharing smoked fish, and drinking martinis, we became fast friends and eventually collaborated on two articles. To my great sorrow, Gene died too young. Just how much too young becomes increasingly clear with each passing year.

Daniel W. Stroock

CHAPTER 1

Kolmogorov's Forward, Basic Results

The primary purpose of this chapter is to present some basic existence and uniqueness results for solutions to second order, parabolic, partial differential equations. Because this book is addressed to probabilists, the treatment of these results will follow, in so far as possible, a line of reasoning which is suggested by thinking about these equations in a probabilistic context. For this reason, I begin by giving an explanation of why, under suitable conditions, Kolmogorov's forward equations for the transition probability function of a continuous path Markov process is second order and parabolic. Once I have done so, I will use this connection with Markov processes to see how solutions to these equations can be constructed using probabilistically natural ideas.

1.1 Kolmogorov's Forward Equation

Recall that a *transition probability function* on $\mathbb{R}^N$ is a measurable map $(t, x) \in [0, \infty) \times \mathbb{R}^N \longmapsto P(t, x) \in \mathbf{M}_1(\mathbb{R}^N)$, where $\mathbf{M}_1(\mathbb{R}^N)$ is the space of Borel probability measures on $\mathbb{R}^N$ with the topology of *weak convergence*,[1] which, for each $x \in \mathbb{R}^N$, satisfies $P(0, x, \{x\}) = 1$ and the *Chapman–Kolmogorov equation*[2]

$$P(s + t, x, \Gamma) = \int P(t, y, \Gamma)\, P(s, x, dy)$$

(1.1.1)

$$\text{for all } s, t \in [0, \infty) \text{ and } \Gamma \in \mathcal{B}_{\mathbb{R}^N}.$$

Kolmogorov's forward equation is the equation which describes, for a fixed $x \in \mathbb{R}^N$, the evolution of $t \in [0, \infty) \longmapsto P(t, x) \in \mathbf{M}_1(\mathbb{R}^N)$.

1.1.1. Derivation of Kolmogorov's Forward Equation: In order to derive Kolmogorov's forward equation, we will make the assumption that

$$(1.1.2) \qquad L\varphi(x) \equiv \lim_{h \searrow 0} \frac{1}{h} \int \big(\varphi(y) - \varphi(x)\big)\, P(h, x, dy)$$

---

[1] That is, the smallest topology for which the map $\mu \in \mathbf{M}_1(\mathbb{R}^N) \longmapsto \int \varphi\, d\mu \in \mathbb{R}$ is continuous whenever $\varphi \in C_{\mathrm{b}}(\mathbb{R}^N; \mathbb{R})$. See Chapter III of [53] for more information.
[2] $\mathcal{B}_{\mathbb{R}^N}$ denotes the Borel σ-algebra over $\mathbb{R}^N$.

exists for each $x \in \mathbb{R}^N$ and $\varphi \in C_c^\infty(\mathbb{R}^N; \mathbb{R})$, the space of infinitely differentiable, real-valued functions with compact support. Under mild additional conditions, one can combine (1.1.2) with (1.1.1) to conclude that

(1.1.3)
$$\frac{d}{dt} \int \varphi(y) \, P(t, x, dy) = \int L\varphi(y) \, P(t, x, dy) \quad \text{or, equivalently,}$$

$$\int \varphi(y) \, P(t, x, dy) = \varphi(x) + \int_0^t \left(\int L\varphi(y) \, P(\tau, x, dy) \right) d\tau$$

for $\varphi \in C_c^2(\mathbb{R}^N; \mathbb{R})$. This equation is called Kolmogorov's forward equation because it describes the evolution of $P(t, x, dy)$ in terms of its *forward variable* y, the variable giving the distribution of the process at time t, as opposed the the *backward variable* x which gives the initial position.

Thinking of $\mathbf{M}_1(\mathbb{R}^N)$ as a subset of $C_c^\infty(\mathbb{R}^N; \mathbb{R})^*$, the dual of $C_c^\infty(\mathbb{R}^N; \mathbb{R})$, one can rewrite (1.1.3) as

(1.1.4)
$$\frac{d}{dt} P(t, x) = L^\top P(t, x),$$

where $L^\top$ is the adjoint of L. Kolmogorov's idea was to recover $P(t, x)$ from (1.1.4) together with the initial condition $P(0, x) = \delta_x$, the unit point mass at x. Of course, in order for his idea to be of any value, one must know what sort of operator L can be. A general answer is given in § 2.1.1 of [55]. However, because this book is devoted to differential equations, we will not deal here with the general case but only with the case when L is a differential operator. For this reason, we add the assumption that L is *local*[3] in the sense that $L\varphi(x) = 0$ whenever φ vanishes in a neighborhood of x. Equivalently, in terms of $P(t, x)$, locality is the condition

(1.1.5)
$$\lim_{h \searrow 0} \frac{1}{h} P\big(h, x, B(x, r)\complement\big) = 0, \quad x \in \mathbb{R}^N \text{ and } r > 0.$$

LEMMA[4] 1.1.6.   *Let* $\{\mu_h : h \in (0, 1)\} \subseteq \mathbf{M}_1(\mathbb{R}^N)$, *and assume that*

$$A\varphi \equiv \lim_{h \searrow 0} \frac{1}{h} \int \big(\varphi(y) - \varphi(0)\big) \mu_h(dy)$$

exists for each $\varphi \in C_c^\infty(\mathbb{R}^N; \mathbb{R})$. *Then* A *is a linear functional on* $C_c^\infty(\mathbb{R}^N; \mathbb{R})$ $\oplus \mathbb{R}$ *which satisfies the* minimum principle

(1.1.7)
$$\varphi(0) = \min_{x \in \mathbb{R}^N} \varphi(x) \implies A\varphi \geq 0.$$

---

[3] Locality of L corresponds to path continuity of the associated Markov process.
[4] Readers who are familiar with Petrie's characterization of local operators may be surprised how simple it is to prove what, at first sight, might appear to be a more difficult result. Of course, the simplicity comes from the minimum principle, which allows one to control everything in terms of the action of A on quadratic functions.

Moreover, if

$$\lim_{h \searrow 0} \frac{1}{h} \mu_h\big(B(0,r)\complement\big) = 0 \quad \text{for all } r > 0,$$

then A is local. Finally, if A is a linear functional on $C_c^\infty(\mathbb{R}^N; \mathbb{R}) \oplus \mathbb{R}$, then A is local and satisfies the minimum principle if and only if there exists a non-negative, symmetric matrix[5] $a = ((a_{ij}))_{1 \le i,j \le N} \in \mathrm{Hom}(\mathbb{R}^N; \mathbb{R}^N)$ and a vector $b = (b_i)_{1 \le i \le N} \in \mathbb{R}^N$ such that

$$A\varphi = \frac{1}{2} \sum_{i,j=1}^N a_{ij} \partial_{x_i} \partial_{x_j} \varphi(0) + \sum_{i=1}^N b_i \partial_{x_i} \varphi(0) \quad \text{for all } \varphi \in C_c^\infty(\mathbb{R}^N; \mathbb{R}).$$

PROOF: The first assertion requires no comment. To prove the "if" part of the second assertion, suppose A is given in terms of a and b with the prescribed properties. Obviously, A is then local. In addition, if φ achieves its minimum value at 0, then the first derivatives of φ vanish at 0 and its Hessian is non-negative definite there. Thus, after writing $\sum_{i,j=1}^N a_{ij} \partial_{x_i} \partial_{x_j} \varphi(0)$ as the trace of $a(0)$ times the Hessian of φ at 0, the non-negativity of Af comes down to the fact that the product of two non-negative definite, symmetric matrices has a non-negative trace, a fact that can be seen by first writing one of them as the square of a symmetric matrix and then using the commutation invariance properties of the trace.

Finally, suppose that A is local and satisfies the minimum principle. To produce the required a and b, we begin by showing that $A\varphi = 0$ if φ vanishes to second order at 0. For this purpose, choose $\eta \in C_c^\infty(\mathbb{R}^N; [0,1])$ so that $\eta = 1$ on $B(0,1)$ and $\eta = 0$ off $B(0,2)$, and set $\varphi_R(x) = \eta(R^{-1}x)\varphi(x)$ for $R > 0$. Then, by locality, $A\varphi = A\varphi_R$ for all $R > 0$. In addition, by Taylor's Theorem, there exists a $C < \infty$ such that $|\varphi_R| \le CR\psi$ for $R \in (0,1]$, where $\psi(x) \equiv \eta(x)|x|^2$. Hence, by the minimum principle applied to $CR\psi \mp \varphi_R$, $|A\varphi| = |A\varphi_R| \le CRA\psi$ for arbitrarily small R's.

To complete the proof from here, set $\psi_i(x) = \eta(x)x_i$, $\psi_{ij} = \psi_i\psi_j$, $b_i = A\psi_i$, and $a_{ij} = A\psi_{ij}$. Given φ, consider

$$\tilde{\varphi} = \varphi(0) + \frac{1}{2} \sum_{i,j=1}^N \partial_{x_i} \partial_{x_j} \varphi(0) \psi_{ij} + \sum_{i=1}^N \partial_{x_i} \varphi(0) \psi_i.$$

By Taylor's Theorem, $\varphi - \tilde{\varphi}$ vanishes to second order at 0, and therefore $A\varphi = A\tilde{\varphi}$. At the same time, by the minimum principle applied to the constant functions $\pm\varphi(0)$, A kills the first term on the right. Hence,

$$A\varphi = \frac{1}{2} \sum_{i,j=1}^N \partial_{x_i} \partial_{x_j} \varphi(0) A\psi_{ij} + \sum_{i=1}^N \partial_{x_i} \varphi(0) A\psi_i,$$

---

[5] We will use $\mathrm{Hom}(\mathbb{R}^M; \mathbb{R}^N)$ to denote the vector space of linear transformation from $\mathbb{R}^M$ to $\mathbb{R}^N$.

and so it remains to check that a is non-negative definite. But, if $\xi \in \mathbb{R}^N$ and $\psi_\xi(x) \equiv \eta(x)^2 (\xi, x)^2_{\mathbb{R}^N} = \sum_{i,j=1}^N \xi_i \xi_j \psi_{ij}(x)$, then, by the minimum principle, $0 \le 2A\psi_\xi = (\xi, a\xi)_{\mathbb{R}^N}$. $\square$

Since the origin can be replaced in Lemma 1.1.6 by any point $x \in \mathbb{R}^N$, we now know that, when (1.1.5) holds, the operator L which appears in Kolmogorov's forward equation has the form

$$(1.1.8) \qquad L\varphi(x) = \frac{1}{2} \sum_{i,j=1}^N a_{ij}(x) \partial_{x_i} \partial_{x_j} \varphi(x) + \sum_{i=1}^N b_i(x) \partial_{x_i} \varphi(x),$$

where $a(x) = \big((a_{ij}(x)) \big)_{1 \le i,j \le N}$ is a non-negative definite, symmetric matrix for each $x \in \mathbb{R}^N$. In the probability literature, a is called the diffusion coefficient and b is called the drift coefficient.

1.1.2. Solving Kolmogorov's Forward Equation: In this section we will prove the following general existence result for solutions to Kolmogorov's forward equation. Throughout we will use the notation $\langle \varphi, \mu \rangle$ to denote the integral $\int \varphi \, d\mu$ of the function φ with respect to the measure μ.

THEOREM 1.1.9.   *Let* $a : \mathbb{R}^N \longrightarrow \mathrm{Hom}(\mathbb{R}^N; \mathbb{R}^N)$ *and* $b : \mathbb{R}^N \longrightarrow \mathbb{R}^N$ *be continuous functions with the properties that* $a(x) = \big((a_{ij}(x)) \big)_{1 \le i,j \le N}$ *is symmetric and non-negative definite for each* $x \in \mathbb{R}^N$ *and*

$$(1.1.10) \qquad \Lambda \equiv \sup_{x \in \mathbb{R}^N} \frac{\mathrm{Trace}\big(a(x)\big) + 2\big(x, b(x)\big)^+_{\mathbb{R}^N}}{1 + |x|^2} < \infty.$$

Then, for each $\nu \in \mathbf{M}_1(\mathbb{R}^N)$, *there is a continuous* $t \in [0,\infty) \longmapsto \mu(t) \in \mathbf{M}_1(\mathbb{R}^N)$ *which satisfies*

$$(1.1.11) \qquad \langle \varphi, \mu(t) \rangle - \langle \varphi, \nu \rangle = \int_0^t \langle L\varphi, \mu(\tau) \rangle \, d\tau,$$

for all $\varphi \in C_c^2(\mathbb{R}^N; \mathbb{C})$, *where* L *is the operator in (1.1.8). Moreover,*

$$(1.1.12) \qquad \int (1 + |y|^2) \, \mu(t, dy) \le e^{\Lambda t} \int (1 + |x|^2) \, \nu(dx), \quad t \ge 0.$$

Before giving the proof, it may be helpful to review the analogous result for ordinary differential equations. Indeed, when applied to the case when $a = 0$, our proof is exactly the same as the usual one there. Namely, in that case, except for the initial condition, there should be no randomness, and so, when we remove the randomness from the initial condition by taking $\nu = \delta_x$, we expect that $\mu_t = \delta_{X(t)}$, where $t \in [0, \infty) \longmapsto X(t) \in \mathbb{R}^N$ satisfies

$$\varphi\big(X(t)\big) - \varphi(x) = \int_0^t \big(b(X(\tau)), \nabla\varphi(X(\tau)) \big)_{\mathbb{R}^N} \, d\tau.$$

Equivalently, $t \rightsquigarrow X(t)$ is an integral curve of the vector field b starting at x. That is,

$$X(t) = x + \int_0^t b(X(\tau)) \, d\tau.$$

To show that such an integral curve exists, one can use the following Euler approximation scheme. For each $n \geq 0$, define $t \rightsquigarrow X_n(t)$ so that $X_n(0) = x$ and

$$X_n(t) = X_n(m2^{-n}) + (t - m2^{-n})b(X(m2^{-n})) \quad \text{for } m2^{-n} < t \leq (m+1)2^{-n}.$$

Clearly,

$$X_n(t) = x + \int_0^t b(X_n([\tau]_n)) \, d\tau,$$

where[6] $[\tau]_n = 2^{-n}[2^n\tau]$ is the largest dyadic number $m2^{-n}$ dominated by τ. Hence, if we can show that $\{X_n : n \geq 0\}$ is relatively compact in the space $C([0, \infty); \mathbb{R}^N)$, with the topology of uniform convergence on compacts, then we can take $t \rightsquigarrow X(t)$ to be any limit of the X_n's.

To simplify matters, assume for the moment that b is bounded. In that case, it is clear that $|X_n(t) - X_n(s)| \leq \|b\|_u |t - s|$, and so the Ascoli–Arzela Theorem guarantees the required compactness. To remove the boundedness assumption, choose a $\psi \in C_c^\infty(B(0, 2); [0, 1])$ so that $\psi = 1$ on $\overline{B(0, 1)}$ and, for each $k \geq 1$, replace b by b_k, where $b_k(x) = \psi(k^{-1}x)$. Next, let $t \rightsquigarrow X_k(t)$ be an integral curve of b_k starting at x, and observe that

$$\frac{d}{dt}|X_k(t)|^2 = 2(X_k(t), b_k(X_k(t)))_{\mathbb{R}^N} \leq \Lambda(1 + |X_k(t)|^2),$$

from which it is an easy step to the conclusion that $|X_k(t)| \leq R(T) \equiv (1 + |x|^2)e^{t\Lambda}$. But this means that, for each $T > 0$, $|X_k(t) - X_k(s)| \leq C(T)|t - s|$ for $s, t \in [0, T]$, where $C(T)$ is the maximum value of $|b|$ on the closed ball of radius $R(T)$ centered at the origin, and so we again can invoke the Ascoli–Arzela Theorem to see that $\{X_k : k \geq 1\}$ is relatively compact and therefore has a limit which is an integral curve of b.

In view of the preceding, it should be clear that our first task is to find an appropriate replacement for the Ascoli–Arzela Theorem. The one which we will choose is the following variant of Lévy's Continuity Theorem (cf. Exercise 3.1.19 in [53]), which states that if $\{\mu_n : n \geq 0\} \subseteq \mathbf{M}_1(\mathbb{R}^N)$ and $\hat{\mu}_n$ is the characteristic function (i.e., the Fourier transform) of μ_n, then $\mu = \lim_{n \to \infty} \mu_n$ exists in $\mathbf{M}_1(\mathbb{R}^N)$ if and only if $\hat{\mu}_n(\xi)$ converges for each ξ and uniformly in a neighborhood of 0, in which case $\mu_n \longrightarrow \mu$ in $\mathbf{M}_1(\mathbb{R}^N)$ where $\hat{\mu}(\xi) = \lim_{n \to \infty} \hat{\mu}_n(\xi)$.

In the following, and elsewhere, we say that $\{\varphi_k : k \geq 1\} \subseteq C_b(\mathbb{R}^N; \mathbb{C})$ converges to φ in $C_b(\mathbb{R}^N; \mathbb{C})$ and write $\varphi_k \longrightarrow \varphi$ in $C_b(\mathbb{R}^N; \mathbb{C})$ if $\sup_k \|\varphi_k\|_u$

---

[6] We use $[\tau]$ to denote the integer part of a number $\tau \in \mathbb{R}$

$< \infty$ and $\varphi_k(x) \longrightarrow \varphi(x)$ uniformly for x in compact subsets of $\mathbb{R}^N$. Also, we say that $\{\mu_k : k \geq 1\} \subseteq C(\mathbb{R}^M; \mathbf{M}_1(\mathbb{R}^N))$ *converges to* μ *in* $C(\mathbb{R}^M; \mathbf{M}_1(\mathbb{R}^N))$ and write $\mu_k \longrightarrow \mu$ in $C(\mathbb{R}^M; \mathbf{M}_1(\mathbb{R}^N))$ if, for each $\varphi \in C_{\mathrm{b}}(\mathbb{R}^N; \mathbb{C})$, $\langle \varphi, \mu_k(z) \rangle \longrightarrow \langle \varphi, \mu(z) \rangle$ uniformly for z in compact subsets of $\mathbb{R}^M$.

THEOREM 1.1.13. *If* $\mu_k \longrightarrow \mu$ *in* $C(\mathbb{R}^M; \mathbf{M}_1(\mathbb{R}^N))$, *then*

$$\langle \varphi_k, \mu_k(z_k) \rangle \longrightarrow \langle \varphi, \mu(z) \rangle$$

whenever $z_k \longrightarrow z$ *in* $\mathbb{R}^M$ *and* $\varphi_k \longrightarrow \varphi$ *in* $C_{\mathrm{b}}(\mathbb{R}^N; \mathbb{C})$. *Moreover, if* $\{\mu_n : n \geq 0\} \subseteq C(\mathbb{R}^M; \mathbf{M}_1(\mathbb{R}^N))$ *and* $f_n(z, \xi) = \widehat{\mu_n(z)}(\xi)$, *then* $\{\mu_n : n \geq 0\}$ *is relatively compact in* $C(\mathbb{R}^M; \mathbf{M}_1(\mathbb{R}^N))$ *if* $\{f_n : n \geq 0\}$ *is equicontinuous at each* $(z, \xi) \in \mathbb{R}^M \times \mathbb{R}^N$. *In particular,* $\{\mu_n : n \geq 0\}$ *is relatively compact if, for each* $\xi \in \mathbb{R}^N$, $\{f_n(\cdot, \xi) : n \geq 0\}$ *is equicontinuous at each* $z \in \mathbb{R}^N$ *and, for each* $r \in (0, \infty)$,

$$\lim_{R \to \infty} \sup_{n \geq 0} \sup_{|z| \leq r} \mu_n(z, \mathbb{R}^N \setminus B(0, R)) = 0.$$

PROOF: To prove the first assertion, suppose $\mu_k \longrightarrow \mu$ in $C(\mathbb{R}^M; \mathbf{M}_1(\mathbb{R}^N))$, $z_k \longrightarrow z$ in $\mathbb{R}^M$, and $\varphi_k \longrightarrow \varphi$ in $C_{\mathrm{b}}(\mathbb{R}^N; \mathbb{C})$. Then, for every $R > 0$,

$$\varlimsup_{k \to \infty} |\langle \varphi_k, \mu_k(z_k) \rangle - \langle \varphi, \mu(z) \rangle|$$

$$\leq \varlimsup_{k \to \infty} \left(|\langle \varphi - \varphi_k, \mu_k(z_k) \rangle| + |\langle \varphi, \mu_k(z_k) \rangle - \langle \varphi, \mu(z_k) \rangle| \right.$$

$$\left. + |\langle \varphi, \mu(z_k) \rangle - \langle \varphi, \mu(z) \rangle| \right)$$

$$\leq \varlimsup_{k \to \infty} \sup_{y \in B(0,R)} |\varphi_k(y) - \varphi(y)| + \sup_k \|\varphi_k\|_{\mathrm{u}} \varlimsup_{k \to \infty} \mu_k(z_k, B(0, R)\complement)$$

$$\leq \sup_k \|\varphi_k\|_{\mathrm{u}} \mu(z, B(0, R)\complement)$$

since $\varlimsup_{k \to \infty} \mu_k(z_k, F) \leq \mu(z, F)$ for any closed $F \subseteq \mathbb{R}^N$. Hence, the required conclusion follows after one lets $R \to \infty$.

Turning to the second assertion, apply the Arzela–Ascoli Theorem to produce an $f \in C_{\mathrm{b}}(\mathbb{R}^M \times \mathbb{R}^N; \mathbb{C})$ and a subsequence $\{n_k : k \geq 0\}$ such that $f_{n_k} \longrightarrow f$ uniformly on compacts. By Lévy's Continuity Theorem, there is, for each $z \in \mathbb{R}^M$, a $\mu(z) \in \mathbf{M}_1(\mathbb{R}^N)$ for which $f(z, \cdot) = \widehat{\mu(z)}$. Moreover, if $z_k \longrightarrow z$ in $\mathbb{R}^M$, then, because $f_{n_k}(z_k, \cdot) \longrightarrow f(z, \cdot)$ uniformly on compact subsets of $\mathbb{R}^N$, another application of Lévy's Theorem shows that $\mu_{n_k}(z_k) \longrightarrow \mu(z)$ in $\mathbf{M}_1(\mathbb{R}^N)$, and from this it is clear that $\mu_{n_k} \longrightarrow \mu$ in $C(\mathbb{R}^M; \mathbf{M}_1(\mathbb{R}^N))$.

It remains to show that, under the conditions in the final assertion, $\{f_n : n \geq 0\}$ is equicontinuous at each (z, ξ). But, by assumption, for each

$\xi \in \mathbb{R}^N$, $\{f_n(\,\cdot\,,\xi) : n \geq 0\}$ is equicontinuous at every $z \in \mathbb{R}^M$. Thus, it suffices to show that if $\xi_k \longrightarrow \xi$ in $\mathbb{R}^N$, then, for each $r > 0$,

$$\lim_{k \to \infty} \sup_{n \geq 0} \sup_{|z| \leq r} \left| f_n(z, \xi_k) - f_n(z, \xi) \right| = 0.$$

To this end, note that, for any $R > 0$,

$$\left| f_n(z, \xi_k) - f_n(z, \xi) \right| \leq R|\xi_k - \xi| + 2\mu_n\big(z, B(0,R)\complement\big),$$

and therefore

$$\varlimsup_{k \to \infty} \sup_{n \geq 0} \sup_{|z| \leq r} \left| f_n(z, \xi_k) - f_n(z, \xi) \right| \leq 2 \sup_{n \geq 0} \sup_{|z| \leq r} \mu_n\big(z, B(0,R)\complement\big) \longrightarrow 0$$

as $R \to \infty$. $\quad\square$

Now that we have a suitable compactness criterion, the next step is to develop an Euler approximation scheme. To do so, we must decide what plays the role in $\mathbf{M}_1(\mathbb{R}^N)$ that linear translation plays in $\mathbb{R}^N$. A hint comes from the observation that if $t \rightsquigarrow X(t, x) = x + tb$ is a linear translation along the constant vector field b, then $X(s+t, x) = X(s, x) + X(t, 0)$. Equivalently, $\delta_{X(s+t,x)} = \delta_x \star \delta_{X(s,0)} \star \delta_{X(t,0)}$, where "$\star$" denotes convolution. Thus, "linear translation" in $\mathbf{M}_1(\mathbb{R}^N)$ should be a path $t \in [0, \infty) \longmapsto \mu(t) \in \mathbf{M}_1(\mathbb{R}^N)$ given by $\mu(t) = \nu \star \lambda(t)$, where $t \rightsquigarrow \lambda(t)$ satisfies $\lambda(0) = \delta_0$ and $\lambda(s + t) = \lambda(s) \star \lambda(t)$. That is, in the terminology of classical probability theory, $\mu(t) = \nu \star \lambda(t)$, where $\lambda(t)$ is an *infinitely divisible flow*. Moreover, because L is local and therefore the associated process has continuous paths, the only infinitely divisible laws which can appear here must be Gaussian (cf. §§,III.3 and III.4 in [53]). With these hints, we now take $Q(t, x) \in \mathbf{M}_1(\mathbb{R}^N)$ to be the normal distribution with mean $x + tb(x)$ and covariance $ta(x)$. Equivalently, if

(1.1.14)
$$\gamma(d\omega) \equiv (2\pi)^{-\frac{M}{2}} e^{-\frac{|\omega|^2}{2}} \, d\omega$$

is the standard normal distribution on $\mathbb{R}^M$ and $\sigma : \mathbb{R}^N \longrightarrow \mathrm{Hom}(\mathbb{R}^M; \mathbb{R}^N)$ is a square root[7] of a in the sense that $a(x) = \sigma(x)\sigma(x)^\top$, then $Q(t, x)$ is the distribution of $\omega \rightsquigarrow x + t^{\frac{1}{2}}\sigma(x)\omega + tb(x)$ under γ. To check that $Q(t, x)$ will play the role that $x + tb(x)$ played above, observe that if $\varphi \in C^2(\mathbb{R}^N; \mathbb{C})$ and φ together with its derivatives have at most exponential growth, then

$$\langle \varphi, Q(t, x) \rangle - \varphi(x) = \int_0^t \langle L^x \varphi, Q(\tau, x) \rangle \, d\tau,$$

(1.1.15)
$$\text{where } L^x \varphi(y) = \frac{1}{2} \sum_{i,j}^N a(x) \partial_{y_i} \partial_{y_j} \varphi(y) + \sum_{i=1}^N b_i(x) \partial_{y_i} \varphi(y).$$

---

[7] At the moment, it makes no difference which choice of square root one chooses. Thus, one might as well assume here that $\sigma(x) = a(x)^{\frac{1}{2}}$, the non-negative definite, symmetric square root $a(x)$. However, later on it will be useful to have kept our options open.

To verify (1.1.15), simply note that

$$\frac{d}{dt}\langle\varphi, Q(t,x)\rangle = \frac{d}{dt}\int \varphi\big(x+\sigma(x)\omega+tb(x)\omega\big)\,\gamma_t(d\omega),$$

where $\gamma_t(\omega) = g(t,\omega)\,d\omega$ with $g(t,\omega) \equiv (2\pi t)^{-\frac{M}{2}}e^{-\frac{|\omega|^2}{2t}}$ is the normal distribution on $\mathbb{R}^M$ with mean 0 and covariance tI, use $\partial_t g(t,\omega) = \frac{1}{2}\Delta g(t,\omega)$, and integrate twice by parts to move the Δ off of g. As a consequence of either (1.1.15) or direct computation, we have

(1.1.16) $$\int |y|^2\, Q(t,x,dy) = \big|x+tb(x)\big|^2 + t\,\mathrm{Trace}\big(a(x)\big).$$

Now, for each $n \geq 0$, define the Euler approximation $t \in [0,\infty) \longmapsto \mu_n(t) \in \mathbf{M}_1(\mathbb{R}^N)$ so that

(1.1.17)
$$\mu_n(0) = \nu \quad\text{and}\quad \mu_n(t) = \int Q\big(t-m2^{-n}, y\big)\,\mu_n(m2^{-n}, dy)$$
$$\text{for } m2^{-n} < t \leq (m+1)2^{-n}.$$

By (1.1.16), we know that

(1.1.18)
$$\int |y|^2\, \mu_n(t,dy) = \int \Big[\big|y+(t-m2^{-n})b(y)\big|^2$$
$$+\,(t-m2^{-n})\mathrm{Trace}\big(a(y)\big)\Big]\,\mu_n\big(m2^{-n}, dy\big)$$

for $m2^{-n} \leq t \leq (m+1)2^{-n}$.

LEMMA 1.1.19.   *Assume that*

(1.1.20) $$\lambda \equiv \sup_{x\in\mathbb{R}^N} \frac{\mathrm{Trace}\big(a(x)\big)+2|b(x)|^2}{1+|x|^2} < \infty.$$

Then

(1.1.21) $$\sup_{n\geq 0}\int (1+|y|^2)\,\mu_n(t,dy) \leq e^{(1+\lambda)t}\int (1+|x|^2)\,\nu(dx).$$

In particular, if $\int |x|^2\,\nu(dx) < \infty$, then $\{\mu_n : n \geq 0\}$ is a relatively compact subset of $C\big([0,\infty);\mathbf{M}_1(\mathbb{R}^N)\big)$ with the topology of uniform convergence on compacts.

PROOF: Suppose that $m2^{-n} \leq t \leq (m+1)2^{-n}$, and set $\tau = t - m2^{-n}$. First note that

$$\big|y+\tau b(y)\big|^2 + \tau\,\mathrm{Trace}\big(a(y)\big)$$
$$= |y|^2 + 2\tau\big(y, b(y)\big)_{\mathbb{R}^N} + \tau^2|b(y)|^2 + \tau\,\mathrm{Trace}\big(a(y)\big)$$
$$\leq |y|^2 + \tau\big[|y|^2 + 2|b(y)|^2 + \mathrm{Trace}\big(a(y)\big)\big] \leq |y|^2 + (1+\lambda)\tau(1+|y|^2),$$

and therefore, by (1.1.18),

$$\int (1 + |y|^2)\, \mu_n(t, dy) \le \left(1 + (1 + \lambda)\tau\right) \int (1 + |y|^2)\, \mu_n\left(m2^{-n}, dy\right).$$

Hence,

$$\int (1 + |y|^2)\, \mu_n(t, dy)$$

$$\le \left(1 + (1 + \lambda)2^{-n}\right)^m \left(1 + (1 + \lambda)\tau\right) \int (1 + |y|^2)\, \nu(dy)$$

$$\le e^{(1+\lambda)t} \int (1 + |x|^2)\, \nu(dx).$$

Next, set $f_n(t, \xi) = \widehat{[\mu_n(t)]}(\xi)$. Under the assumption that the second moment $S \equiv \int |x|^2 \nu(dx) < \infty$, we want to show that $\{f_n : n \ge 0\}$ is equicontinuous at each $(t, \xi) \in [0, \infty) \times \mathbb{R}^N$. Since, by (1.1.21),

$$\mu_n\left(t, \overline{B(0, R)}\complement\right) \le S(1 + R^2)^{-1} e^{(1+\lambda)t},$$

the last part of Theorem 1.1.13 says that it suffices to show that, for each $\xi \in \mathbb{R}^N$, $\{f_n(\cdot, \xi) : n \ge 0\}$ is equicontinuous at each $t \in [0, \infty)$. To this end, first observe that, for $m2^{-n} \le s < t \le (m+1)2^{-n}$,

$$\left| f_n(t, \xi) - f_n(s, \xi) \right| \le \int \left| [\widehat{Q(t, y)}](\xi) - [\widehat{Q(s, y)}](\xi) \right| \mu_n\left(m2^{-n}, dy\right)$$

and, by (1.1.15),

$$\left| [\widehat{Q(t, y)}](\xi) - [\widehat{Q(s, y)}](\xi) \right| = \left| \int_s^t \left(\int L^y e_\xi(y')\, Q(\tau, y, dy') \right) d\tau \right|$$

$$\le (t - s)\left(\tfrac{1}{2}\left(\xi, a(y)\xi\right)_{\mathbb{R}^N} + |\xi||b(y)| \right) \le \tfrac{1}{2}(1 + \lambda)(1 + |y|^2)(1 + |\xi|^2)(t - s),$$

where $e_\xi(y) \equiv e^{\sqrt{-1}\,\xi \cdot y}$. Hence, by (1.1.21),

$$\left| f_n(t, \xi) - f_n(s, \xi) \right| \le \frac{(1 + \lambda)(1 + |\xi|^2)}{2} e^{(1+\lambda)t} \int (1 + |x|^2)\, \nu(dx)(t - s),$$

first for $s < t$ in the same dyadic interval and then for all $s < t$. $\quad\square$

With Lemma 1.1.19, we can now prove Theorem 1.1.9 under the assumptions that a and b are bounded and that $\int |x|^2 \nu(dx) < \infty$. Indeed, because we know then that $\{\mu_n : n \ge 0\}$ is relatively compact in $C\left([0, \infty); \mathbf{M}_1(\mathbb{R}^N)\right)$, all that we have to do is show that every limit satisfies (1.1.11). For this purpose, first note that, by (1.1.15),

$$\langle \varphi, \mu_n(t) \rangle - \langle \varphi, \nu \rangle = \int_0^t \left(\int \langle L^y \varphi, Q(\tau - [\tau]_n, y) \rangle \mu_n([\tau]_n, dy) \right) d\tau$$

for any $\varphi \in C_b^2(\mathbb{R}^N; \mathbb{C})$. Next, observe that, as $n \to \infty$,

$$\langle L^y \varphi, Q(\tau - [\tau]_n, y) \rangle \longrightarrow L\varphi(y)$$

boundedly and uniformly for (τ, y) in compacts. Hence, if $\mu_{n_k} \longrightarrow \mu$ in $C([0, \infty); \mathbf{M}_1(\mathbb{R}^N))$, then, by Theorem 1.1.13,

$$\langle \varphi, \mu_{n_k}(t) \rangle \longrightarrow \langle \varphi, \mu(t) \rangle \quad \text{and}$$

$$\int_0^t \left(\int \langle L^y \varphi, Q(\tau - [\tau]_n, y) \rangle \mu_n([\tau]_n, dy) \right) d\tau \longrightarrow \int_0^t \langle L\varphi, \mu(\tau) \rangle d\tau.$$

Before moving on, we want to show that $\int |x|^2 \nu(dx) < \infty$ implies that (1.1.11) continues to hold for $\varphi \in C^2(\mathbb{R}^N; \mathbb{C})$ with bounded second order derivatives. Indeed, from (1.1.21), we know that

$$(*) \qquad \int (1 + |y|^2)\, \mu(t, dy) \leq e^{(1+\lambda)t} \int (1 + |y|^2)\, \nu(dy).$$

Now choose $\psi \in C_c^\infty(\mathbb{R}^N; [0, 1])$ so that $\psi = 1$ on $\overline{B(0, 1)}$ and $\psi = 0$ off of $B(0, 2)$, define ψ_R by $\psi_R(y) = \psi(R^{-1}y)$ for $R \geq 1$, and set $\varphi_R = \psi_R \varphi$. Observe that[8]

$$\frac{|\varphi(y)|}{1 + |y|^2} \vee \frac{|\nabla \varphi(y)|}{1 + |y|} \vee \|\nabla^2 \varphi(y)\|_{\text{H.S.}}$$

is bounded independent of $y \in \mathbb{R}^N$, and therefore so is $\frac{|L\varphi(y)|}{1+|y|^2}$. Thus, by (*), there is no problem about integrability of the expressions in (1.1.11). Moreover, because (1.1.11) holds for each φ_R, all that we have to do is check that

$$\langle \varphi, \mu(t) \rangle = \lim_{R \to \infty} \langle \varphi_R, \mu(t) \rangle$$

$$\int_0^t \langle L\varphi, \mu(\tau) \rangle\, d\tau = \lim_{R \to \infty} \int_0^t \langle L\varphi_R, \mu(\tau) \rangle\, d\tau.$$

The first of these is an immediate application of Lebesgue's Dominated Convergence Theorem. To prove the second, observe that

$$L\varphi_R(y) = \psi_R(y) L\varphi(y) + \big(\nabla \psi_R(y), a(y)\nabla \varphi\big)_{\mathbb{R}^N} + \varphi(y) L\psi_R(y).$$

Again the first term on the right causes no problem. To handle the other two terms, note that, because ψ_R is constant off of $\overline{B(0, 2R)} \setminus B(0, R)$ and because $\nabla \psi_R(y) = R^{-1}\nabla\psi(R^{-1}y)$ while $\nabla^2 \psi_R(y) = R^{-2}\nabla^2 \psi(R^{-1}y)$, one

---

[8] We use $\nabla^2 \varphi$ to denote the Hessian matrix of φ and $\|\sigma\|_{\text{H.S.}}$ to denote the Hilbert–Schmidt norm $\sqrt{\sum_{ij} \sigma_{ij}^2}$ of σ.

can easily check that they are dominated by a constant, which is independent of R, times $(1 + |y|^2)\mathbf{1}_{[R,2R]}(|y|)$. Hence, again Lebesgue's Dominated Convergence Theorem gives the desired result.

Knowing that (1.1.11) holds for $\varphi \in C^2(\mathbb{R}^N; \mathbb{C})$ with bounded second order derivatives, we can prove (1.1.12) by taking $\varphi(y) = 1 + |y|^2$ and thereby obtaining

$$
\int (1 + |y|^2)\,\mu(t, dy)
$$
$$
= \int (1 + |y|^2)\,\nu(dy) + \int_0^t \left(\int \Big[\mathrm{Trace}\big(a(y)\big) + 2\big(y, b(y)\big)_{\mathbb{R}^N} \Big] \mu(\tau, dy) \right) d\tau
$$
$$
\leq \int (1 + |y|^2)\,\nu(dy) + \Lambda \int_0^t \left(\int (1 + |y|^2)\,\mu(\tau, dy) \right) d\tau,
$$

from which (1.1.12) follows by Gronwall's inequality.

Continuing with the assumption that $\int |x|^2\,\nu(dx) < \infty$, we want to remove the boundedness assumption on a and b and replace it by (1.1.10). To do this, take ψ_R as above, set $a_k = \psi_k a$, $b_k = \psi_k b$, define L_k accordingly for a_k and b_k, and choose $t \rightsquigarrow \mu_k(t)$ so that (1.1.12) is satisfied and (1.1.11) holds when μ and L are replaced there by μ_k and L_k. Because of (1.1.12), the argument which we used earlier can be repeated to show that $\{\mu_k : k \geq 1\}$ is relatively compact in $C\big([0, \infty); \mathbf{M}_1(\mathbb{R}^N)\big)$. Moreover, if μ is any limit of $\{\mu_k : k \geq 1\}$, then (1.1.12) is satisfied and, just as we did above, one can check (1.1.11), first for $\varphi \in C_c^2(\mathbb{R}^N; \mathbb{C})$ and then for all $\varphi \in C^2(\mathbb{R}^N; \mathbb{C})$ with bounded second order derivatives.

Finally, to remove the second moment condition on ν, assume that it fails, and choose $r_k \nearrow \infty$ so that $\alpha_1 \equiv \nu\big(B(0, r_1)\big) > 0$ and $\alpha_k \equiv \nu\big(B(0, r_k) \setminus \overline{B(0, r_{k-1})}\big) > 0$ for each $k \geq 2$, and set $\nu_1 = \alpha_1^{-1}\nu \upharpoonright B(0, r_1)$ and $\nu_k = \alpha_k^{-1}\nu \upharpoonright B(0, r_k) \setminus \overline{B(0, r_{k-1})}$ when $k \geq 2$. Finally, choose $t \rightsquigarrow \mu_k(t)$ for L and ν_k, and define $\mu(t) = \sum_{k=1}^{\infty} \alpha_k \mu_k(t)$. It is an easy matter to check that this μ satisfies (1.1.11) for all $\varphi \in C_c^2(\mathbb{R}^N; \mathbb{C})$.

REMARK 1.1.22. In order to put the result in Theorem 1.1.9 into a partial differential equations context, it is best to think of $t \rightsquigarrow \mu(t)$ as a solution to $\partial_t \mu = L^\top \mu(t)$ in the sense of (Schwartz) distributions. Of course, when the coefficients of L are not smooth, one has to be a little careful about the meaning of $L^\top \mu(t)$. The reason why this causes no problem here is that, by virtue of the minimum principle (cf. § 2.4.1), the only distributions with which we need to deal are probability measures.

1.2 Transition Probability Functions

Although we have succeeded in solving Kolmogorov's forward equation in great generality, we have not yet produced a transition probability function. To be more precise, let $S(x)$ denote the set of maps $t \rightsquigarrow \mu(t)$ satisfying (1.1.11) with $\nu = \delta_x$. In order to construct a transition probability function, we must make a measurable "selection" $x \in \mathbb{R}^N \longmapsto P(\,\cdot\,,x) \in S(x)$ in such a way that the Chapman–Kolmogorov equation (1.1.1) holds. Thus, the situation is the same as that one encounters in the study of ordinary differential equations when trying to construct a flow on the basis of only an existence result for solutions. In the absence of an accompanying uniqueness result, how does one go about showing that there is a "selection" of solutions which fit together nicely into a flow?

It turns out that, under the hypotheses in Theorem 1.1.9, one can always make a selection $x \in \mathbb{R}^N \longmapsto P(\,\cdot\,,x) \in S(x)$ which forms a transition probability function. The underlying idea is to introduce enough spurious additional conditions to force uniqueness. In doing so, one has to take advantage of the fact that, for each $x \in \mathbb{R}^N$, $S(x)$ is a compact, convex subset of $\mathbf{M}_1(\mathbb{R}^N)$ and that, if $y \to x$ in $\mathbb{R}^N$, then $\overline{\lim}_{y \to x} S(y) \subseteq S(x)$ in the sense of Hausdorff convergence of compact sets. Because we will not be using this result, we will not discuss it further. The interested reader should see Chapter XII of [52] for more details.

Another strategy for the construction of transition probability functions is to see whether one can show that the subsequence $\{n_k : k \geq 0\}$ in Lemma 1.1.19 can be chosen independent of ν. To be more precise, let a and b be given as in Lemma 1.1.19, and, for each $n \geq 0$, let $t \rightsquigarrow P_n(t,x)$ be constructed by the prescription in (1.1.17) with $\nu = \delta_x$. Suppose that we could find one subsequence $\{n_k : k \geq 0\}$ and a $(t,x) \in [0,\infty) \times \mathbb{R} \longmapsto P(t,x) \in \mathbf{M}_1(\mathbb{R}^N)$ with the property that $P_{n_k} \longrightarrow P$ in $C\big([0,\infty) \times \mathbb{R}^N; \mathbf{M}_1(\mathbb{R}^N)\big)$. It would then follow that $(t,x) \rightsquigarrow P(t,x)$ has to be a continuous transition probability function. Indeed, the continuity would be obvious. As for the Chapman–Kolmogorov equation, note that, by construction, for any $n \geq 0$, $\varphi \in C_{\mathrm{b}}(\mathbb{R}^N; \mathbb{C})$, and $t \in [0,\infty)$,

$$\langle \varphi, P_n(s+t,x) \rangle = \int \langle \varphi, P_n(t,y) \rangle \, P_n(s,x,dy)$$

whenever $s = m2^{-n}$ for some $(m,n) \in \mathbb{N}^2$. Hence, after passing along the subsequence, we would have

$$\langle \varphi, P(s+t,x) \rangle = \int \langle \varphi, P(t,y) \rangle \, P(s,x,dy)$$

whenever $s = m2^{-n}$ for some $(m,n) \in \mathbb{N}^2$, which, by continuity with respect to s, would lead immediately to (1.1.1).

1.2.1. Lipschitz Continuity and Convergence: Our goal in this subsection is to prove the following theorem.

THEOREM 1.2.1.   Let $a : \mathbb{R}^N \longrightarrow \mathrm{Hom}(\mathbb{R}^N; \mathbb{R}^N)$ and $b : \mathbb{R}^N \longrightarrow \mathbb{R}^N$ be given, where, for each $x \in \mathbb{R}^N$, $a(x)$ is a symmetric, non-negative definite matrix, and define L accordingly, as in (1.1.8). Further, assume that there exists a square root $\sigma : \mathbb{R}^N \longrightarrow \mathrm{Hom}(\mathbb{R}^M; \mathbb{R}^N)$ of a such that

$$(1.2.2) \qquad \sup_{x \neq x'} \frac{\|\sigma(x) - \sigma(x')\|_{\mathrm{H.S.}} \vee |b(x) - b(x')|}{|x - x'|} < \infty.$$

Then there exists a continuous transition probability function $(t, x) \in [0, \infty)$ $\times \mathbb{R}^N \longmapsto P(t, x) \in \mathbf{M}_1(\mathbb{R}^N)$ to which the $\{P_n : n \geq 0\}$ of the preceding discussion converges in $C([0, \infty) \times \mathbb{R}^N; \mathbf{M}_1(\mathbb{R}^N))$. Furthermore, (1.1.20) holds with

$$\lambda \leq \mathrm{Trace}(a(0)) + \|\sigma\|_{\mathrm{Lip}}^2 + 2(|b(0)| + \|b\|_{\mathrm{Lip}})^2 < \infty,$$

and

(1.2.3) $$\sup_{n \geq 0} \int (1 + |y|^2) \, P_n(t, x, dy) \leq e^{(1+\lambda)t}(1 + |x|^2)$$

and (cf. (1.1.10))

(1.2.4) $$\int (1 + |y|^2) \, P(t, x, dy) \leq e^{\Lambda t}(1 + |x|^2).$$

In addition, for each $x \in \mathbb{R}^N$, $t \leadsto P(t, x)$ solves Kolmogorov's forward equation. In fact,

(1.2.5) $$\langle \varphi, P(t, x) \rangle = \varphi(x) + \int_0^t \langle L\varphi, P(\tau, x) \rangle \, d\tau$$

for any $\varphi \in C^2(\mathbb{R}^N; \mathbb{C})$ with bounded second order derivatives.

In order to prove this theorem, we must learn how to estimate the difference between $P_{n_1}(t, x)$ and $P_{n_2}(t, x)$ and show that this difference is small when n_1 and n_2 are large. The method which we will use to measure these differences is *coupling*. That is, suppose that $(\mu_1, \mu_2) \in \mathbf{M}_1(\mathbb{R}^N)^2$. Then a coupling of μ_1 to μ_2 is any $\tilde{\mu} \in \mathbf{M}_1(\mathbb{R}^N \times \mathbb{R}^N)$ for which μ_1 and μ_2 are its marginal distributions on $\mathbb{R}^N$, in the sense that $\mu_1(\Gamma) = \tilde{\mu}(\Gamma \times \mathbb{R}^N)$ and $\mu_2(\Gamma) = \tilde{\mu}(\mathbb{R}^N \times \Gamma)$. Given a coupling of μ_1 to μ_2, an estimate of their difference is given by $\int |y - y'|^2 \, \tilde{\mu}(dy \times dy')$. Indeed,

$$|\langle \varphi, \mu_2 \rangle - \langle \varphi, \mu_1 \rangle| \leq \mathrm{Lip}(\varphi) \left(\int |y - y'|^2 \, \tilde{\mu}(dy \times dy') \right)^{\frac{1}{2}}.$$

Equivalently, a coupling of μ_1 to μ_2 means that one has found a probability space $(\Omega, \mathcal{F}, \mathbb{P})$ on which there are random variables X_1 and X_2 for which μ_i is the distribution of X_i. Of course, it is only when the choice of $\tilde{\mu}$ is made judiciously that the method yields any information. For example, taking $\tilde{\mu} = \mu_1 \times \mu_2$ yields essentially no information.

LEMMA 1.2.6. *Define $F : [0, \infty) \times \mathbb{R}^N \times \mathbb{R}^M \longrightarrow \mathbb{R}^N$ by*

$$F(t, x, \omega) = x + t^{\frac{1}{2}} \sigma(x)\omega + tb(x).$$

Next, for $(x, x') \in (\mathbb{R}^N)^2$, define $\tilde{Q}_n(t, x, x') \in \mathbf{M}_1(\mathbb{R}^N \times \mathbb{R}^N)$ for $t \in [0, 2^{-n}]$ so that $\tilde{Q}_n(t, x, x')$ is the distribution of

$$\omega \in \mathbb{R}^M \longmapsto \begin{pmatrix} F(t, x, \omega) \\ F(t, x', \omega) \end{pmatrix} \in \mathbb{R}^N \times \mathbb{R}^N$$

under (cf. (1.1.14)) γ when $t \in [0, 2^{-n-1}]$ and of

$$(\omega_1, \omega_2) \in (\mathbb{R}^M)^2 \longmapsto \begin{pmatrix} F(2^{-n-1}, x, \omega_1) + F(t - 2^{-n-1}, x, \omega_2) - x \\ F(t - 2^{-n-1}, F(2^{-n-1}, x', \omega_1), \omega_2) \end{pmatrix}$$

under γ^2 when $t \in (2^{-n-1}, 2^{-n}]$. Finally, for $n \geq 0$ and $(t, x) \in [0, \infty) \times \mathbb{R}^N$, define $\tilde{\mu}_n(t, x) \in \mathbf{M}_1(\mathbb{R}^N \times \mathbb{R}^N)$ so that $\tilde{\mu}_n(0, x) = \delta_x \times \delta_x$ and

$$\tilde{\mu}_n(t, x) = \int \tilde{Q}_n(t - m2^{-n}, y, y') \, \tilde{\mu}_n(m2^{-n}, dy \times dy')$$

for $m2^{-n} \leq t \leq (m + 1)2^{-n}$. Then $\tilde{\mu}_n(t, x)$ is a coupling of $P_n(t, x)$ to $P_{n+1}(t, x)$ and there exists a $K < \infty$, depending only on the value of σ and b at the origin and their uniform Lipschitz norms, such that

$$\int |y - y'|^2 \, \tilde{\mu}_n(t, x, dy \times dy') \leq 2^{-n} e^{Kt} (1 + |x|^2).$$

PROOF: Once one remembers that the distribution of $(\omega_1, \omega_2) \rightsquigarrow c_1 \omega_1 + c_2 \omega_2$ under γ^2 is the same as the distribution of $\omega \rightsquigarrow (c_1^2 + c_2^2)^{\frac{1}{2}} \omega$ under γ, it is easy to verify that $\tilde{\mu}_n(t, x)$ is a coupling of $P_n(t, x)$ to $P_{n+1}(t, x)$.

To prove the asserted estimate, set

$$\tilde{q}_n(t, x, x') = \int |y' - y|^2 \, \tilde{Q}_n(t, x, x', dy \times dy') \quad \text{for } t \in [0, 2^{-n}].$$

Clearly,

$$u_n(t, x) \equiv \int |y' - y|^2 \, \tilde{\mu}_n(t, x, dy \times dy')$$

$$= \int \tilde{q}_n(t - m2^{-n}, y, y') \, \tilde{\mu}_n(m2^{-n}, x, dy \times dy')$$

for $m2^{-n} \leq t \leq (m + 1)2^{-n}$.

Set $C = \sup_{x' \neq x} |x' - x|^{-2} \left(2|b(x') - b(x)|^2 + \|\sigma(x') - \sigma(x)\|_{\text{H.S.}}^2 \right)$. When $t \in [0, 2^{-n-1}]$,

$$\tilde{q}_n(t, x, x') = \left| (x' - x) + t(b(x') - b(x)) \right|^2 + t\|\sigma(x') - \sigma(x)\|_{\text{H.S.}}^2$$
$$\leq \left(1 + (1 + C)t \right) |x' - x|^2,$$

and so

(*)
$$u_n(t,x) \le \big(1 + (1+C)\tau\big) u_n(m2^{-n}, x)$$
for $m2^{-n} \le t \le (2m+1)2^{-n-1}$ and $\tau = t - m2^{-n}$.

When $t \in [2^{-n-1}, 2^{-n}]$ and $\tau = t - 2^{-n-1}$, $\tilde{q}_n(t, x, x')$ equals

$$\int \Big[\big|(y'-y) + \tau\big(b(y') - b(x)\big)\big|^2$$
$$+ \tau \big\|\sigma(y') - \sigma(x)\big\|_{\text{H.S.}}^2 \Big] \tilde{Q}_n(2^{-n-1}, x, x', dy \times dy')$$

$$\le (1+\tau)\tilde{q}_n(2^{-n-1}, x, x')$$
$$+ \tau \int \Big[2|b(y') - b(x)|^2 + \big\|\sigma(y') - \sigma(x)\big\|_{\text{H.S.}}^2 \Big] \tilde{Q}_n(2^{-n-1}, x, x', dy \times dy')$$

$$\le (1+\tau)\tilde{q}_n(2^{-n-1}, x, x')$$
$$+ 2\tau \int \Big[2|b(y') - b(y)|^2 + \big\|\sigma(y') - \sigma(y)\big\|_{\text{H.S.}}^2 \Big] \tilde{Q}_n(2^{-n-1}, x, x', dy \times dy')$$
$$+ 2\tau \int \Big[2|b(y) - b(x)|^2 + \big\|\sigma(y) - \sigma(x)\big\|_{\text{H.S.}}^2 \Big] Q(2^{-n-1}, x, dy)$$

$$\le \big(1 + (1+2C)\tau\big)\tilde{q}_n(2^{-n-1}, x, x') + 2C\tau \int |y-x|^2 \, Q(2^{-n-1}, x, dy)$$
$$= \big(1 + (1+2C)\tau\big)\tilde{q}_n(2^{-n-1}, x, x') + 2C\tau 2^{-n-1}\big[|b(x)|^2 + \text{Trace}\big(a(x)\big)\big].$$

Since $|b(x)|^2 + \text{Trace}\big(a(x)\big) \le \lambda(1+|x|^2)$, we now can use (1.1.21) to arrive at

$$u_n(t,x) \le \big(1+(1+2C)\tau\big)u_n\big((2m+1)2^{-n-1}, x\big) + 2C\lambda e^{(1+\lambda)t}\tau 2^{-n-1}\big(1+|x|^2\big)$$

for $(2m+1)2^{-n-1} \le t \le (m+1)2^{-n}$ and $\tau = t - (2m+1)2^{-n-1}$.

Putting this together with (*), we conclude there is a $C' < \infty$, with the required dependence, such that

(**)
$$u_n(t,x) \le (1 + C'\tau)u_n(m2^{-n}, x) + C'e^{C't}\tau 2^{-n}\big(1 + |x|^2\big)$$

for $m \in \mathbb{N}$, $\tau \in [0, 2^{-n}]$, and $t = m2^{-n} + \tau$. Finally, working by induction on m and remembering that $u_n(0,x) = 0$, one obtains from (**) first that

$$u_n(m2^{-n}, x) \le 2^{-n}\Big(e^{C't}(1 + |x|^2)\big(1 + C'2^{-n}\big)^m - 1\Big) \quad \text{for } m \le 2^n t$$

and then that the asserted estimate holds with $K = 2C'$. $\square$

PROOF OF THEOREM 1.2.1: Given the preceding, we know that

$$\big|\langle \varphi, P_n(t,x)\rangle - \langle \varphi, P_m(t,x)\rangle\big| \le \sum_{\ell=m}^{n-1} \big|\langle \varphi, P_{\ell+1}(t,x)\rangle - \langle \varphi, P_\ell(t,x)\rangle\big|$$

$$\le e^{\frac{Kt}{2}}(1 + |x|^2)^{\frac{1}{2}} \text{Lip}(\varphi) \sum_{\ell=m}^{n-1} 2^{-\frac{\ell}{2}}$$

and therefore that

$$(1.2.7) \qquad \left| \langle \varphi, P_n(t,x) \rangle - \langle \varphi, P_m(t,x) \rangle \right| \le \frac{2^{-\frac{m}{2}}}{2^{\frac{1}{2}} - 1} e^{\frac{Kt}{2}} (1 + |x|^2) \mathrm{Lip}(\varphi)$$

for any $0 \le m < n$. Starting from (1.2.7), it is easy to check that the sequence $\{P_n : n \ge 0\}$ converges in $C\big([0,\infty) \times \mathbb{R}^N; \mathbf{M}_1(\mathbb{R}^N)\big)$. Indeed, by applying (1.2.7) to $\varphi = e_\xi$ and setting $f_n(t,x,\xi) = [\widehat{P_n(t,x)}](\xi)$, we see that $\{f_n : n \ge 0\}$ is Cauchy convergent in the metric of uniform convergence on compact subsets of $[0,\infty) \times \mathbb{R}^N \times \mathbb{R}^N$. Hence, not only does it converge uniformly on compacts to some $f \in C_b\big([0,\infty) \times \mathbb{R}^N \times \mathbb{R}^N; \mathbb{C}\big)$, but, by Theorem 1.1.13, there is a continuous $(t,x) \in [0,\infty) \times \mathbb{R}^N \longmapsto P(t,x) \in \mathbf{M}_1(\mathbb{R}^N)$ such that $f(t,x) = \widehat{P(t,x)}$ and to which P_n converges in $C\big([0,\infty) \times \mathbb{R}^N; \mathbf{M}_1(\mathbb{R}^N)\big)$. Furthermore, by the general considerations in the introduction to this section, $(t,x) \rightsquigarrow P(t,x)$ is a transition probability function. Also, (1.2.3) and (1.2.4) as well as (1.2.5) for $\varphi \in C^2(\mathbb{R}^N; \mathbb{C})$ with bounded second order derivatives are all applications of the analogous statements in § 1.1. $\square$

COROLLARY 1.2.8. *Let everything be as in Theorem 1.2.1. Given $\nu \in \mathbf{M}_1(\mathbb{R}^N)$, define $\{\mu_n : n \ge 0\}$ as in (1.1.17). Then $\mu_n(t) = \int P_n(t,x)\,\nu(dx)$ and $t \rightsquigarrow \mu_n(t)$ converges to*

$$t \rightsquigarrow \mu(t) \equiv \int P(t,x)\,\nu(dx) \quad in \ C\big([0,\infty); \mathbf{M}_1(\mathbb{R}^N)\big).$$

Furthermore, (1.1.11) holds for all $\varphi \in C_c^2(\mathbb{R}^N; \mathbb{C})$. Finally, $\mu(t)$ satisfies (1.1.12), and, when $\int |y|^2\,\nu(dy) < \infty$, (1.1.11) continues to hold for all $\varphi \in C^2(\mathbb{R}^N; \mathbb{C})$ with bounded second order derivatives.

PROOF: Because, as we have already noted, $\mu_n(t) = \int P_n(t,x)\,\nu(dx)$, everything follows immediately from Theorem 1.2.1 combined with Fubini's Theorem. $\square$

1.3 Some Important Extensions

In the sequel, it will be important for us to have available several extensions and refinements of the results in §§ 1.1 and 1.2.

1.3.1. Higher Moments: Here we want to show that the mean-square results which we have been proving admit higher moment analogs. A key step in proving these analogs is the development of higher moment estimates to replace

$$\int |y|^2\,Q(t,x,dy) = |x + tb(x)|^2 + t\,\mathrm{Trace}\big(a(x)\big).$$

When doing computations involving an operator L of the sort in (1.1.8), it is useful to have checked that

$$(1.3.1) \qquad L(f \circ \varphi) = \tfrac{1}{2}\big(\nabla\varphi, a\nabla\varphi\big)_{\mathbb{R}^N} f'' \circ \varphi + (L\varphi) f' \circ \varphi$$

for $f \in C^2(\mathbb{R}; \mathbb{C})$ and $\varphi \in C^2(\mathbb{R}^N; \mathbb{R})$.

LEMMA 1.3.2.   *For each $r \in [1, \infty)$, there exists a universal $\kappa_r < \infty$ such that*

$$\int (1 + |Ty|^2)^r \, Q(t, x, dy)$$

$$\leq (1 + |Tx|^2)^r + \kappa_r t \Big((1 + |Tx|^2)^r + |Tb(x)|^{2r} + \big(\text{Trace} \, Ta(x)T^\top \big)^r \Big)$$

for $(t, x) \in (0, 1] \times \mathbb{R}^N$ and $T \in \text{Hom}(\mathbb{R}^N, \mathbb{R}^{N'})$.

PROOF: Begin by observing that, without loss in generality, we may assume that $N' = N$ and $T = I$. Indeed, the distribution of $y \rightsquigarrow Ty$ under $Q(t, x, \cdot)$ is $Q'(t, Tx, \cdot)$, where $Q'(t, x', \cdot)$ is defined relative to $Tb(x)$ and $Ta(x)T^\top$. Thus, we will proceed under this assumption.

Given an $\mathbb{R}^N$-valued normal random variable X with mean 0 and covariance A, $|X|^2$ has the same distribution as $\sum_{i=1}^N \alpha_i Y_i^2$, where $\{\alpha_1, \ldots, \alpha_N\}$ are the eigenvalues of A and the Y_i's are independent, $\mathbb{R}$-valued standard normal variables. Thus, for any $p \in [2, \infty)$, $|X|^{2p}$ has the same distribution as $\left(\sum_{i=1}^N \alpha_i Y_i^2 \right)^p$. Therefore, if M_r is the rth moment of $|Y_1|$, then, by Minkowski's inequality,

$$\mathbb{E}\big[|X|^{2p}\big] = \mathbb{E}\left[\left(\sum_{i=1}^N \alpha_i Y_i^2 \right)^p \right] \leq \big(\text{Trace} \, A \big)^p M_{2p}$$

when $p \geq 1$, and

$$\mathbb{E}\big[|X|^{2p}\big] \leq \mathbb{E}\big[|X|^2\big]^p = \big(\text{Trace} \, A \big)^p$$

if $p \in (0, 1]$. That is,

$$\mathbb{E}\big[|X|^{2p}\big] \leq \big(\text{Trace} \, A \big)^p M_{p \vee 2} \quad \text{for all } p \in (0, \infty).$$

Applying this to $\omega \rightsquigarrow \sigma(x)\omega$, we see that

$$\int (1 + |y|^2)^p \, Q(t, x, dy) = \int \left(1 + |x + tb(x) + t^{\frac{1}{2}} \sigma(x)\omega|^2 \right)^p \gamma(d\omega)$$

$$\leq 2^{(p-1)^+} \left[1 + \int |x + tb(x) + t^{\frac{1}{2}} \sigma(x)\omega|^{2p} \gamma(d\omega) \right]$$

$$\leq 2^{(p-1)^+} \left[1 + 3^{(2p-1)^+} \Big(|x|^{2p} + |b(x)|^{2p} + \big(\text{Trace} \, a(x) \big)^p M_{p \vee 2} \Big) \right]$$

for any $p \in (0, \infty)$ and $t \in (0, 1]$. Hence, for each $p \in (0, \infty)$ there is a universal $C_p < \infty$ such that

$$(*) \quad \int (1 + |y|^2)^p \, Q(t, x, dy) \leq C_p \big[(1 + |x|^2)^p + |b(x)|^{2p} + \big(\text{Trace} \, a(x) \big)^p \big]$$

for $t \in (0, 1]$.

Now let $r \geq 1$ be given, and apply (1.3.1) with $f(\xi) = \xi^r$ and $\varphi(y) = 1 + |y|^2$ to see that $L^x (1 + |y|^2)^r$ equals

$$2r(r-1)\big(y, a(x)y\big)_{\mathbb{R}^N} (1 + |y|^2)^{r-2}$$
$$+ r\big(\mathrm{Trace}\, a(x)\big)(1 + |y|^2)^{r-1} + 2r\big(y, b(x)\big)_{\mathbb{R}^N} (1 + |y|^2)^{r-1}$$
$$\leq r\big[(2r-1)\big(\mathrm{Trace}\, a(x)\big) + |b(x)|\big](1 + |y|^2)^{r-1}.$$

Hence, by (1.1.15) applied to $\varphi(y) = (1 + |y|^2)^r$,

$$\frac{d}{dt} \int (1 + |y|^2)^r \, Q(t, x, dy)$$
$$\leq r\big[(2r-1)\big(\mathrm{Trace}\, a(x)\big) + |b(x)|\big] \int (1 + |y|^2)^{r-1} \, Q(t, x, dy).$$

Clearly the desired estimate follows when one combines this with (*) for $p = r - 1$ and then applies the Fundamental Theorem of Calculus. $\square$

Our first application of Lemma 1.3.2 is to the situation described in Lemma 1.1.19.

LEMMA 1.3.3.   *Under the hypotheses in Lemma 1.1.19, for each $r \in [1, \infty)$ there exists a $\lambda_r < \infty$, depending only on r and the λ in (1.1.20), such that*

$$\sup_{n \geq 0} \int |y|^{2r} \, \mu_n(t, dy) \leq e^{\lambda_r t} \int (1 + |x|^{2r}) \, \nu(dx).$$

PROOF: By Lemma 1.3.2 and (1.1.20),

$$\int (1 + |y|^2)^r \, Q(t, x, dy) \leq \big(1 + \kappa_r(1 + \lambda^r)t\big)\big(1 + |x|^2\big)^r$$

for $(t, x) \in [0, 1] \times \mathbb{R}^N$. Hence, we can find a λ_r, with the required dependence, such that

$$\int (1 + |y|^2)^r \, Q(t, x, dy) \leq (1 + \lambda_r t)(1 + |x|^2)^r \quad \text{for } (t, x) \in [0, 1] \times \mathbb{R}^N.$$

Starting from here, the rest of the proof differs in no way from the derivation of (1.1.21). $\square$

In the following theorem and elsewhere, a function $\varphi : \mathbb{R}^N \longrightarrow \mathbb{C}$ is said to be *slowly increasing* if it has at most polynomial growth at infinity in the sense that $|\varphi(x)| \leq C\big(1 + |x|^{2r}\big)$ for some $C < \infty$ and $r \in [0, \infty)$. Also, we will say that a set S of functions is *uniformly slowly increasing* if the choice of C and r can be made independent of $\varphi \in S$.

THEOREM 1.3.4. *Refer to the hypotheses and notation in Theorem 1.2.1. Then, for each $r \in (0, \infty)$ there exists $\lambda_r \in (1, \infty)$, depending only on r, $\|\sigma(0)\|_{\text{H.S.}}$, $|b(0)|$, and the Lipschitz norms of σ and b, such that*

$$\sup_{n \geq 0} \int |y|^{2r} \, P_n(t, x, dy) \leq e^{\lambda_r t}\big(1 + |x|^{2r}\big).$$

In particular,

$$\int |y|^{2r} \, P(t, x, dy) \leq e^{\lambda_r t}\big(1 + |x|^{2r}\big)$$

and

$$\langle \varphi_n, P_n(t_n, x_n) \rangle \longrightarrow \langle \varphi, P(t, x) \rangle$$

when $(t_n, x_n) \longrightarrow (t, x)$ and $\{\varphi_n : n \geq 0\} \subset C(\mathbb{R}^N; \mathbb{C})$ is a uniformly slowly increasing sequence which tends to φ uniformly on compacts; and (1.2.5) holds for any $\varphi \in C^2(\mathbb{R}^N; \mathbb{C})$ whose second derivatives are slowly increasing.

PROOF: In view of Lemma 1.3.3, only the final two assertions need comment. For this purpose, choose $\psi \in C_c^\infty\big(B(0, 2); [0, 1]\big)$ so that $\psi = 1$ on $\overline{B(0, 1)}$, and set $\psi_R(y) = \psi(R^{-1}y)$. Since we already know that $P_n \longrightarrow P$ in $C\big([0, \infty) \times \mathbb{R}^N; \mathbf{M}_1(\mathbb{R}^N)\big)$, we know that the desired convergence takes place when φ_n and φ are replaced by $\psi_R \varphi_n$ and $\psi_R \varphi$. At the same time, for any $r \geq 1$,

$$\lim_{R \to \infty} \sup_{n \geq 0} \int_{B(0, R)\complement} (1 + |y|^{2r}) \big(P_n(t_n, x, dy) + P(t, x, dy)\big) = 0.$$

Thus, because the φ_n's are uniformly slowly increasing, the convergence continues to hold for the original φ_n's and φ. The proof of (1.2.5) for φ's with slowly increasing second derivatives is another application of the same cut-off procedure. $\square$

1.3.2. Introduction of a Potential: In this subsection we will consider $L + V$, where L is given by (1.1.8) and $V \in C_b^1(\mathbb{R}^N; \mathbb{R})$. Because, at least when $L = \frac{1}{2}\Delta$, such operators arise as the Hamiltonian in Schrödinger mechanics, in which case V is the potential energy, V is called a *potential*. Our goal is to produce a map $(t, x) \leadsto P^V(t, x)$ which plays the same role for $L + V$ as $(t, x) \leadsto P(t, x)$ does for L.

We begin by observing that we cannot expect $P^V(t, x)$ will be a probability measure. Indeed, when V is constant, it should be clear that $P^V(t, x) = e^{tV} P(t, x)$. Thus, we will have to deal with $\mathbf{M}(\mathbb{R}^N)$, the space of finite, non-negative Borel measures on $\mathbb{R}^N$. Like $\mathbf{M}_1(\mathbb{R}^N)$, we will give $\mathbf{M}(\mathbb{R}^N)$ the topology corresponding to weak convergence. That is, a sequence $\{\nu_n : n \geq 0\} \subseteq \mathbf{M}(\mathbb{R}^N)$ converges to ν in $\mathbf{M}(\mathbb{R}^N)$ if $\langle \varphi, \nu_n \rangle \longrightarrow \langle \varphi, \nu \rangle$ for every $\varphi \in C_b(\mathbb{R}^N; \mathbb{C})$. Equivalently, $\nu_n \longrightarrow \nu$ if

$\nu(\mathbb{R}^N) = \lim_{n \to \infty} \nu_n(\mathbb{R}^N)$ and, when $\nu(\mathbb{R}^N) > 0$, $\bar{\nu}_n \longrightarrow \bar{\nu}$ in $\mathbf{M}_1(\mathbb{R}^N)$, where

$$\bar{\nu}_n = \begin{cases} \frac{\nu_n}{\nu_n(\mathbb{R}^N)} & \text{if } \nu_n(\mathbb{R}^N) > 0 \\ \delta_0 & \text{if } \nu_n(\mathbb{R}^N) = 0 \end{cases} \quad \text{and} \quad \bar{\nu} = \frac{\nu}{\nu(\mathbb{R}^N)}.$$

Now assume L is given by (1.1.8) where $a = \sigma\sigma^\top$, and assume that σ and b are uniformly Lipschitz continuous. To carry out our construction, define[9] $\big(t, (x,\xi)\big) \in [0, \infty) \times \mathbb{R}^N \times \mathbb{R} \longmapsto \bar{Q}\big(t, (x,\xi)\big) \in \mathbf{M}_1(\mathbb{R}^N \times \mathbb{R})$ to be the distribution under (cf. (1.1.14)) γ of

$$\omega \in \mathbb{R}^M \longmapsto \begin{pmatrix} x \\ \xi \end{pmatrix} + t^{\frac{1}{2}} \begin{pmatrix} \sigma(x) \\ 0 \end{pmatrix} \omega + \begin{pmatrix} b(x) \\ V(x) \end{pmatrix} \in \mathbb{R}^N \times \mathbb{R},$$

and, for $n \geq 0$, define $\bar{P}_n\big(t, (x,\xi)\big)$ so that $\bar{P}_n\big(0, (x,\xi)\big) = \delta_{(x,\xi)}$ and

$$\bar{P}_n\big(t, (x,\xi)\big) = \int \bar{Q}\big(t - m2^{-n}, (y, \eta)\big) \, \bar{P}\big(m2^{-m}, (x,\xi), dy \times d\eta\big)$$

for $m2^{-n} < t \leq (m+1)2^{-n}$. Then Theorem 1.2.1 applies to $\{\bar{P}_n\big(t, (x,\xi)\big) : n \geq 0\}$ and says that there is a continuous $(t, (x,\xi)) \rightsquigarrow \bar{P}\big(t, (x,\xi)\big)$ to which the $\bar{P}_n\big(t, (x,\xi)\big)$'s converge in $C\big([0, \infty) \times \mathbb{R}^N \times \mathbb{R}; \mathbf{M}_1(\mathbb{R}^N \times \mathbb{R})\big)$. Moreover, it is easy to check that

$$(1.3.5) \qquad \int \psi(y, \eta) \bar{P}_n\big(t, (x,\xi), dy \times d\eta\big) = \int \psi(y, \xi + \eta) \bar{P}_n\big(t, (x,0), dy \times d\eta\big)$$

$$\text{and} \quad \bar{P}_n\big(t, (x,0), \mathbb{R}^N \times [-t\|V^-\|_{\mathrm{u}}, t\|V^+\|_{\mathrm{u}}]\complement\big) = 0,$$

and therefore that $(t, (x,\xi)) \rightsquigarrow \bar{P}\big(t, (x,\xi)\big)$ inherits the same properties.

THEOREM 1.3.6. *Suppose that σ and b are uniformly Lipschitz continuous, and let L be given by (1.1.8) with $a = \sigma\sigma^\top$. Next, given $V \in C_{\mathrm{b}}^1(\mathbb{R}^N; \mathbb{R})$, refer to the preceding, and define $(t, x) \in [0, \infty) \times \mathbb{R}^N \longmapsto P^V(t, x) \in \mathbf{M}(\mathbb{R}^N)$ so that*

$$P^V(t, x, \Gamma) = \int_{\Gamma \times \mathbb{R}} e^\eta \, \bar{P}\big(t, (x,0), dy \times d\eta\big).$$

Then, $(t, x) \rightsquigarrow P^V(t, x) \in \mathbf{M}(\mathbb{R}^N)$ is continuous,

$$e^{-t\|V^-\|_{\mathrm{u}}} \leq P^V(t, x, \mathbb{R}^N) \leq e^{t\|V^+\|_{\mathrm{u}}}, \quad \text{and for all } r \geq 1$$

$$\int |y|^{2r} \, P^V(t, x, dy) \leq e^{t(\lambda_r + \|V\|_{\mathrm{u}})} \big(1 + |x|^{2r}\big).$$

---

[9] The "bar" here is simply a device to distinguish the "barred" quantities from "unbarred" ones. It does not indicate complex conjugate.

Moreover,

$$P^V(s+t,x) = \int P^V(t,y)\,P^V(s,x,dy),$$

and for all $\varphi \in C^2(\mathbb{R}^N;\mathbb{C})$ *with slowly increasing second order derivatives,*

$$\langle \varphi, P^V(t,x)\rangle - \varphi(x) = \int_0^t \langle (L+V)\varphi, P^V(\tau,x)\rangle\,d\tau.$$

PROOF: We need only address the last part of the statement. To this end, first observe that, by the preceding discussion,

$$e^\xi P^V(t,x,\Gamma) = \int_{\Gamma\times\mathbb{R}} e^\eta\,\bar{P}\big(t,(x,\xi),dy\times d\eta\big).$$

Hence, by the Chapman–Kolmogorov equation for $\big(t,(x,\xi)\big)\rightsquigarrow\bar{P}\big(t,(x,\xi)\big)$, $P^V(s+t,x,\Gamma)$ is equal to

$$\int \left(\int_{\Gamma\times\mathbb{R}} e^{\xi''}\bar{P}\big(t,(x',\xi'),dx''\times d\xi''\big)\right)\bar{P}\big(s,(x,0),dx'\times d\xi'\big)$$
$$= \int e^{\xi'}\left(\int_{\Gamma\times\mathbb{R}} e^{\xi''}\bar{P}\big(t,(x',0),dx''\times d\xi''\big)\right)\bar{P}\big(s,(x,0),dx'\times d\xi'\big)$$
$$= \int P^V(t,y,\Gamma)\,P^V(s,x,dy).$$

Finally, given $\varphi \in C^2(\mathbb{R}^N;\mathbb{C})$ with bounded second order derivatives, set $\bar{\varphi}(x,\xi) = \varphi(x)e^\xi$. Then, taking into account the second line of (1.3.5), we know that

$$\frac{d}{dt}\langle \varphi, P^V(t,x)\rangle = \frac{d}{dt}\langle \bar{\varphi}, \bar{P}\big(t,(x,0))\big)\rangle$$
$$= \langle \bar{L}\bar{\varphi}, \bar{P}\big(t,(x,0))\big)\rangle = \langle (L+V)\varphi, P^V(t,x)\rangle,$$

where $\bar{L}$ is the operator corresponding to $(\xi,x)\rightsquigarrow\begin{pmatrix} a(x) & 0 \\ 0 & 0 \end{pmatrix}$, $(\xi,x)\rightsquigarrow\begin{pmatrix} b(x) \\ 0 \end{pmatrix}$. □

1.3.3. Lower Triangular Systems: In Chapters 2 and 3, we will need to know how to deal with coefficients which, although they are smooth, have unbounded derivatives. Even in the setting of ordinary differential equations, such coefficients can cause problems. For example, consider the $\mathbb{R}$-valued equation $\dot{X}(t) = X(t)^2$ with $X(0) = 1$. Obviously, the one and only solution is $X(t) = (1-t)^{-1}$, which explodes as $t \nearrow 1$. On the other hand, the $\mathbb{R}^2$-valued equation

$$\frac{d}{dt}\begin{pmatrix} X_1(t) \\ X_2(t) \end{pmatrix} = \begin{pmatrix} X_1(t) \\ X_1(t)^2 + X_2(t) \end{pmatrix} \quad \text{with} \quad \begin{pmatrix} X_1(0) \\ X_2(0) \end{pmatrix} = \begin{pmatrix} 1 \\ 1 \end{pmatrix}$$

has no such problems. Namely, its solution is

$$\begin{pmatrix} X_1(t) \\ X_2(t) \end{pmatrix} = \begin{pmatrix} e^t \\ e^{2t} \end{pmatrix}.$$

Of course, the point is that this is a *lower triangular system* and the quadratic growth of the coefficients occurs only in the second line and is only in the direction of the first coordinate.

With this example in mind, we will look at coefficients

(1.3.7)

$$x = \begin{pmatrix} x_1 \\ x_2 \end{pmatrix} \in \mathbb{R}^{N_1} \times \mathbb{R}^{N_2}$$

$$\longmapsto \boldsymbol{\sigma}(x) = \begin{pmatrix} \sigma_1(x_1) \\ \sigma_2(x) \end{pmatrix} \in \mathrm{Hom}(\mathbb{R}^M; \mathbb{R}^{N_1} \times \mathbb{R}^{N_2})$$

$$x = \begin{pmatrix} x_1 \\ x_2 \end{pmatrix} \in \mathbb{R}^{N_1} \times \mathbb{R}^{N_2}$$

$$\longmapsto \mathbf{b}(x) = \begin{pmatrix} b_1(x_1) \\ b_2(x) \end{pmatrix} \in \mathbb{R}^{N_1} \times \mathbb{R}^{N_2},$$

where, for some $r_1 \geq 1$ and $C_1 < \infty$,

(1.3.8) $$\|\sigma_2(x)\|_{\mathrm{H.S.}}^2 + 2|b_2(x)|^2 \leq C_1\big(1 + |x_1|^{2r_1} + |x_2|^2\big).$$

The replacement for Lemma 1.3.3 in this setting is the following.

LEMMA 1.3.9.   *Let $\boldsymbol{\sigma}$ and $\mathbf{b}$ have the form in (1.3.7), and assume that (1.3.8) holds. If*

(1.3.10) $$\sup_{n \geq 0} \int |y_1|^{2rr_1}\, \mu_n(t, dy) \leq A_r e^{A_r t} \int \big(1 + |x_1|^{2rr_1}\big)\, \nu(dx)$$

for some $r \geq 1$ and $A_r < \infty$, then

$$\sup_{n \geq 0} \int |y|^{2r}\, \mu_n(t, dy) \leq A e^{At} \int \big(1 + |x|^{2rr_1}\big)\, \nu(dx)$$

for an $A < \infty$ which depends only on r, A_r, and the r_1 and C_1 in (1.3.8).

PROOF: Using Lemma 1.3.2, with T equal to orthogonal projection onto the second coordinates, and (1.3.8), one sees that, when $t \in [0, 1]$,

$$\int (1 + |y_2|^2)^r\, Q(t, x, dy) \leq (1 + C't)(1 + |x_2|^2)^r + C't|x_1|^{2rr_1},$$

for a $C' < \infty$ depending only on r and C_1. Hence, for $m2^{-n} \leq t \leq (m+1)2^{-n}$ and $\tau = t - m2^{-n}$, (1.3.10) says that

$$\int (1 + |y_2|^2)^r\, \mu_n(t, dy) \leq (1 + C'\tau) \int (1 + |y_2|^2)^r\, \mu_n(m2^{-n}, dy)$$

$$+ A_r C' e^{A_r t} \tau \int \big(1 + |x_1|^{2rr_1}\big)\, \nu(dx),$$

from which the desired result follows by the same reasoning as was used at the analogous place at the end of the proof of Lemma 1.2.6.   □

We next want to development a replacement for Lemma 1.2.6.

LEMMA 1.3.11.   *Let* σ *and* **b** *be as in Lemma 1.3.9, and assume that* $(1.3.8)$, $(1.3.10)$ *with* $r = r_1$, *and*

$$(1.3.12) \quad \begin{aligned} &\|\sigma_1(x_1)\|_{\mathrm{H.S.}}^2 + 2|b_1(x_1)|^2 \leq C\big(1 + |x_1|^{2r_1}\big) \\ &\|\sigma_2(x') - \sigma_2(x)\|_{\mathrm{H.S.}}^2 + 2|b_2(x') - b_2(x)|^2 \\ &\qquad \leq C\big[\big(1 + |x'|^{r_1-1} + |x|^{r_1-1}\big)|x_1' - x_1|^2 + |x_2' - x_2|^2\big] \end{aligned}$$

all hold. If $(t, x) \leadsto \tilde{\mu}_n(t, x)$ *is defined for* σ *and* **b** *as in Lemma 1.2.6, and if*

$$(1.3.13) \quad \int |y_1' - y_1|^2\, \tilde{\mu}_n(t, x, dy \times dy') \leq 2^{-n\rho} C e^{Ct}\big(1 + |x_1|^{2r_1}\big)$$

for some $\rho \in (0, 1]$, *then there is a* $C' < \infty$, *depending only on* C, r_1, A_{r_1}, *and* ρ, *such that*

$$\int |y' - y|^2\, \tilde{\mu}_n(t, x, dy \times dy') \leq 2^{-\frac{\rho n}{2}} C' e^{C't}\big(1 + |x|^{2r_1^2}\big).$$

PROOF: Obviously, it is sufficient to prove the asserted estimate with $|y' - y|^2$ replaced by $|y_2' - y_2|^2$ in the integrand on the left-hand side. Thus, set

$$u_n(t, x) = \int |y_2' - y_2|^2\, \tilde{\mu}_n(t, x, dy \times dy').$$

Then,

$$u_n(t, x) = \int \tilde{q}_n\big(t - [t]_n, y, y'\big)\, \tilde{\mu}_n\big([t]_n, x, dy \times dy'\big),$$

where, for $t \in [0, 2^{-n}]$,

$$\tilde{q}_n(t, x, x') \equiv \int |y_2' - y_2|^2\, \tilde{Q}_n(t, x, x', dy \times dy')$$

and $\tilde{Q}_n(t, x, x')$ is defined as in Lemma 1.2.6, only now for σ and **b**.

Proceeding as in the proof of Lemma 1.2.6, we see that when $t \leq 2^{-n-1}$, $\tilde{q}_n(t, x, x')$ is equal to

$$\big|(x_2' - x_2) + t\big(b_2(x') - b_2(x)\big)\big|^2 + t\|\sigma_2(x') - \sigma_2(x)\|_{\mathrm{H.S.}}^2$$

and is therefore dominated by[10]

$$\begin{aligned} &\big(1 + (1 + 2C)t\big)|x_2' - x_2|^2 + 2Ct\big(1 + |x|^{r_1-1} + |x'|^{r_1-1}\big)|x_1' - x_1|^2 \\ &\qquad \leq (1 + C't)|x_2' - x_2|^2 + C't\big(1 + |x|^{2r_1} + |x'|^{2r_1}\big)^{\frac{1}{2}}|x_1' - x_1|. \end{aligned}$$

---

[10] In this proof, we will adopt the convention that C' stands for a constant, with the required dependence, which can change from line to line.

Thus, if $m2^{-n} \le t \le (2m+1)2^{-n-1}$ and $\tau = t - m2^{-n}$, then $u_n(t,x)$ is dominated by

$$(1 + C'\tau)u_n(m2^{-n}, x)$$
$$+ \tau C' \int \left(1 + |y|^{2r_1} + |y'|^{2r_1}\right)^{\frac{1}{2}} |y_1' - y_1|\, \tilde{\mu}_n(m2^{-n}, x, dy \times dy')$$
$$\le (1 + C'\tau)u_n(m2^{-n}, x)$$
$$+ \tau C' \left(\int \left(1 + |y|^{2r_1} + |y'|^{2r_1}\right) \tilde{\mu}_n(m2^{-n}, x, dy \times dy') \right)^{\frac{1}{2}}$$
$$\times \left(\int |y_1' - y_1|^2\, \tilde{\mu}_n(m2^{-n}, x, dy \times dy') \right)^{\frac{1}{2}},$$

and because

$$\int \left(1 + |y|^{2r_1} + |y'|^{2r_1}\right) \tilde{\mu}_n(m2^{-n}, x, dy \times dy')$$
$$= 1 + \int |y|^{2r_1} \left(P_n(m2^{-n}, x, dy) + P_{n+1}(m2^{-n}, x, dy)\right),$$

this, together with our hypotheses and Lemma 1.3.9, shows that

$$(*) \qquad u_n(t,x) \le (1 + C'\tau)u_n(m2^{-n}, x) + \tau 2^{-\frac{\rho n}{2}} C' e^{C't}(1 + |x|^{2r_1^2})$$
$$\text{for } t = m2^{-n} + \tau \text{ with } \tau \le 2^{-n-1}.$$

If $t = 2^{-n-1} + \tau$ for some $\tau \in [0, 2^{-n-1}]$, then, just as in the proof of Lemma 1.2.6,

$$\tilde{q}_n(t, x, x') = \int \left[\left| (y_2' - y_2) + \tau\left(b_2(y') - b_2(x)\right) \right|^2 \right.$$
$$\left. + \tau \left\| \sigma_2(y') - \sigma_2(x) \right\|_{\text{H.S.}}^2 \right] \tilde{Q}_n(2^{-n-1}, x, x', dy \times dy')$$
$$\le (1 + \tau)\tilde{q}_n(2^{-n-1}, x, x')$$
$$+ 2\tau \int \left[2\left| b_2(y') - b_2(y) \right|^2 \right.$$
$$\left. + \left\| \sigma_2(y') - \sigma_2(y) \right\|_{\text{H.S.}}^2 \right] \tilde{Q}_n(2^{-n-1}, x, x', dy \times dy')$$
$$+ 2\tau \int \left[2\left| b_2(y) - b_2(x) \right|^2 + \left\| \sigma_2(y) - \sigma_2(x) \right\|_{\text{H.S.}}^2 \right] Q(2^{-n-1}, x, dy)$$
$$\le \left(1 + (1 + 2C)\tau\right)\tilde{q}_n(2^{-n-1}, x, x')$$
$$+ 2C\tau \int \left(1 + |y'|^{2r_1} + |y|^{2r_1}\right)^{\frac{1}{2}} |y_1' - y_1|\, \tilde{Q}_n(2^{-n-1}, x, x', dy \times dy')$$
$$+ 2C\tau \int \left(1 + |y|^{2r_1} + |x|^{2r_1}\right)^{\frac{1}{2}} |y_1 - x_1|\, Q(2^{-n-1}, x, dy)$$
$$+ 2C\tau \int |y_2 - x_2|^2\, Q(2^{-n-1}, x, dy),$$

and so, if $t = (2m+1)2^{-n-1} + \tau$ with $\tau \leq 2^{-n-1}$, then $u_n(t,x)$ is dominated by

$$(1 + C'\tau)u_n\big((2m+1)2^{-n-1}, x\big)$$

$$+ C'\tau \int \big(1 + |y'|^{2r_1} + |y|^{2r_1}\big)^{\frac{1}{2}} |y_1' - y_1| \, \tilde{\mu}_n\big((2m+1)2^{-n-1}, x, dy \times dy'\big)$$

$$+ C'\tau \int \left(\int \big(1 + |y|^{2r_1} + |\xi|^{2r_1}\big)^{\frac{1}{2}} \right.$$

$$\left. \times |y_1 - \xi_1| \, Q(2^{-n-1}, \xi, dy') \right) P_n(m2^{-n}, x, d\xi)$$

$$+ C'\tau \int \left(\int |y_2 - \xi_2|^2 \, Q(2^{-n-1}, \xi, dy) \right) P_n(m2^{-n}, x, d\xi).$$

Using the same reasoning as in the derivation of $(*)$, one can dominate the second term by $\tau 2^{-\frac{\rho n}{2}} C' e^{C't} (1 + |x|^{2r_1^2})$. At the same time, the inner integral in third term is dominated by

$$\left(\int \big(1 + |y|^{2r_1} + |\xi|^{2r_1}\big) Q(2^{-n-1}, \xi, dy) \right)^{\frac{1}{2}} \left(\int |y_1 - \xi_1|^2 Q(2^{-n-1}, \xi, dy) \right)^{\frac{1}{2}},$$

and so the third term is dominated by a constant times

$$\tau \left(1 + \int |y|^{2r_1} \big[P_n\big((2m+1)2^{-n-1}, x, dy\big) + P_n\big(m2^{-n}, x, dy\big) \big] \right)^{\frac{1}{2}}$$

$$\times \left(2^{-n-1} \int \big(|b_1(y_1)|^2 + \|\sigma_1(y_1)\|_{\text{H.S.}}^2 \big) P_n(m2^{-n}, x, dy) \right)^{\frac{1}{2}}$$

$$\leq \tau 2^{-\frac{n+1}{2}} C' e^{C't} (1 + |x|^{2r_1^2}).$$

Similarly, but with less effort, one can show that $2^{-n-1} C' e^{C't} (1 + |x|^{2r_1^2})$ dominates the fourth term. Hence,

$$u_n(t,x) \leq (1 + C'\tau)u_n\big((2m+1)2^{-n-1}, x\big) + \tau 2^{-\frac{\rho n}{2}} C' e^{C't} \big(1 + |x|^{2r_1^2}\big)$$

for $t = (2m+1)2^{-n-1} + \tau$ with $\tau \leq 2^{-n-1}$. After combining this with $(*)$, one gets the desired estimate by the same argument as we used at the end of the proof of Lemma 1.2.6. $\square$

We are now ready to prove the result for which we have been preparing. In its statement, $\ell \geq 1$, $N = \sum_{k=1}^{\ell} N_k$, where $\{N_1, \ldots, N_\ell\} \subset \mathbb{Z}^+$, $\sigma_k \in C^1\big(\mathbb{R}^{N_1} \times \cdots \times \mathbb{R}^{N_k}; \text{Hom}(\mathbb{R}^M; \mathbb{R}^{N_k})\big)$, $b_k \in C^1\big(\mathbb{R}^{N_1} \times \cdots \times \mathbb{R}^{N_k}; \mathbb{R}^{N_k}\big)$, and

$$\boldsymbol{\sigma}(x) = \begin{pmatrix} \sigma_1(x_1) \\ \sigma_2(x_1, x_2) \\ \vdots \\ \sigma_\ell(x_1, \ldots, x_\ell) \end{pmatrix} \quad \text{and} \quad \mathbf{b}(x) = \begin{pmatrix} b_1(x_1) \\ b_2(x_1, x_2) \\ \vdots \\ b_\ell(x_1, \ldots, x_\ell) \end{pmatrix}$$

for $\mathbb{R}^N \ni x = (x_1, \ldots, x_\ell) \in \mathbb{R}^{N_1} \times \cdots \times \mathbb{R}^{N_\ell}$. Finally, $(t,x) \in [0,\infty) \times \mathbb{R}^N \longmapsto P_n(t,x) \in \mathbf{M}_1(\mathbb{R}^N)$ is defined, as in §1.2.1, relative to $\boldsymbol{\sigma}$ and $\mathbf{b}$.

Theorem 1.3.14.   *Referring to the preceding, assume that*

$$\sum_{k=1}^{\ell}\sum_{i=1}^{N_k}\left(\left\|\partial_{(x_k)_i}\sigma_k\right\|_{\mathrm{u}}+\left\|\partial_{(x_k)_i}b_k\right\|_{\mathrm{u}}\right)<\infty$$

and that, for $2\le k\le\ell$ and $1\le j<k$, $\partial_{(x_j)_i}\sigma_k$ and $\partial_{(x_j)_i}b_k$ are slowly increasing. Then there exists an $\mathbf{r}\ge 1$ such that

$$\sup_{n\ge 0}\sup_{x\in\mathbb{R}^N}(1+|x|^{r\mathbf{r}})^{-1}\int|y|^r\,P_n(t,x,dy)<\infty\quad\text{for all }r>0.$$

Furthermore, there exists a continuous transition probability function (t,x) $\rightsquigarrow P(t,x)$ to which $\{P_n:n\ge 0\}$ converges in $C\big([0,\infty)\times\mathbb{R}^N;\mathbf{M}_1(\mathbb{R}^N)\big)$. In particular,

$$\sup_{x\in\mathbb{R}^N}(1+|x|^{r\mathbf{r}})^{-1}\int|y|^r\,P(t,x,dy)<\infty\quad\text{for all }r\in[1,\infty)$$

and

$$\langle\varphi_n,P_n(t_n,x_n)\rangle\longrightarrow\langle\varphi,P(t,x)\rangle$$

if $(t_n,x_n)\longrightarrow(t,x)$ and $\{\varphi_n:n\ge 0\}\subseteq C(\mathbb{R}^N;\mathbb{C})$ is uniformly slowly increasing and $\varphi_n\longrightarrow\varphi$ uniformly on compacts. Finally, if L is defined as in (1.1.8) with $a=\sigma\sigma^{\top}$ and $b=\mathbf{b}$, then (1.2.5) holds for any $\varphi\in C^2(\mathbb{R}^N:\mathbb{C})$ with slowly increasing second derivatives.

Proof: The proof is done by induction on $\ell\ge 1$. The case when $\ell=1$ is covered by Theorem 1.3.4, and assuming the result for $\ell-1$ for some $\ell\ge 2$, it follows for ℓ when one combines Lemma 1.3.9 with Lemma 1.3.11 and applies them with the roles of σ_1, b_1, σ_2, and b_2 there being played here by

$$\begin{pmatrix}\sigma_1\\\vdots\\\sigma_{\ell-1}\end{pmatrix},\begin{pmatrix}b_1\\\vdots\\b_{\ell-1}\end{pmatrix},\sigma_\ell,\text{ and }b_\ell.$$ The details are left to the reader. □

1.4 Historical Notes and Some Commentary

Kolmogorov's equations appear in [29], which can be considered the birthplace of modern Markov process theory. In view of the fact that his motivation was probabilistic, it is interesting that Kolmogorov's own treatment of his equations is more or less purely analytic. It is also interesting that he seems to have been unaware of the much earlier work [35] on parabolic equations by E.E. Levi, who initiated the parametrix method which, after several iterations, led to the construction by W. Pogorzelski [48] and D. Aronson [3] of fundamental solutions for second order parabolic equations with uniformly elliptic, Hölder continuous coefficients. Even though it appeared in 1964, A. Friedman's exposition [19] remains an excellent source for this line of research, as does the more encyclopedic treatment in [34].

Anyone familiar with the work [24] of K. Itô will recognize that the presentation given in this chapter is a poor man's adaptation of his ideas. In particular, it was Itô who interpreted Kolmogorov's forward equation as the equation for an integral curve in $\mathbf{M}_1(\mathbb{R}^N)$. Further, he realized that, just as a vector field in $\mathbb{R}^N$ can be integrated by concatenating straight lines, so too solutions of Kolmogorov's forward equation can be built by concatenating infinitely divisible flows, which are the analog in $\mathbf{M}_1(\mathbb{R}^N)$ of straight lines. See [55] for a more detailed account of these matters.

From a technical standpoint, the main advantage of adopting Itô's approach is that it allows one to work entirely in the realm of probability measures. There are two reasons why this is important. First, it provides one with an easy and powerful way to handle questions of compactness. In the traditional analytic approach to solving equations like Kolmogorov's, one looks for solutions which are functions and one attempts to construct them by an approximation procedure which requires one to check a compactness criterion for an appropriate space of functions. In Itô's approach, both the solution and the approximates are probability measures, for which compactness is relatively trivial. The second advantage of Itô's approach is that it makes no use of *ellipticity* (i.e., strict positive definiteness) of the diffusion matrix a. Thus, his theory handles degenerate coefficients (i.e., ones for which a can degenerate) just as well as it does elliptic ones. Of course, this advantage becomes something of a liability when it comes to proving regularity results which are true only under suitable non-degeneracy conditions, but even so I think its virtues outweigh its flaws.

CHAPTER 2

Non-Elliptic Regularity Results

Let L be an operator of the form in (1.1.8). In the preceding chapter, we studied Kolmogorov's forward equation $\partial_t \mu(t) = L^\top \mu(t)$ and showed that, in great generality, solutions exist in the sense that, for each initial $\nu \in \mathbf{M}_1(\mathbb{R}^N)$, there is a continuous $t \rightsquigarrow \mu(t) \in \mathbf{M}_1(\mathbb{R}^N)$ such that (1.1.11) holds for all $\varphi \in C_c^2(\mathbb{R}^N; \mathbb{C})$. Moreover, in the generality that we have been working, this is the best sense in which one can hope to have solutions. For example, if $a \equiv 0$ and $b(x) \equiv b$, then the only solution, in the sense of Schwartz distributions, to the corresponding forward equation with initial value ν is $t \rightsquigarrow \mu(t)$, where

$$\langle \varphi, \mu(t) \rangle = \int \varphi(x + tb)\, \nu(dx).$$

In particular, if $\nu = \delta_x$, then $\mu(t) = \delta_{x+tb}$.

Now suppose that $(t, x) \rightsquigarrow P(t, x)$ is a continuous transition probability which solves Kolmogorov's forward equation in the sense that (1.2.5) is well defined and holds for all $\varphi \in C^2(\mathbb{R}^N; \mathbb{C})$ with slowly increasing second order derivatives. For example, by Theorems 1.2.1, such a $P(t, x)$ will exist when $a = \sigma\sigma^\top$ and σ and b are uniformly Lipschitz continuous. Although we know that $(t, x) \rightsquigarrow P(t, x)$ will, in general, be no better than continuous, when tested against smooth functions, it may nonetheless possess smoothness properties. For instance, in the preceding example, $\langle \varphi, P(t, x) \rangle = \varphi(x + bt)$, and so $(t, x) \rightsquigarrow \langle \varphi, P(t, x) \rangle$ will be just as smooth as φ. The reason why we are interested in this possibility is that it provides solutions to *Kolmogorov's backward equation*

$$(2.0.1) \qquad \partial_t u = Lu \quad \text{in } (0, \infty) \times \mathbb{R}^N \quad \text{with} \quad \lim_{t \searrow 0} u(t) = \varphi.$$

To understand this, let $\varphi \in C_b(\mathbb{R}^N; \mathbb{C})$ be given, and set

$$(2.0.2) \qquad u_\varphi(t, x) \equiv \langle \varphi, P(t, x) \rangle.$$

Further, assume that $u_\varphi(t, \cdot)$ has slowly increasing second order derivatives for each $t \in (0, \infty)$. Then, by the Chapman–Kolmogorov equation and

(1.2.5),

$$u_\varphi(t+h,x) - u_\varphi(t,x) = \langle u_\varphi(t), P(h,x) \rangle - u_\varphi(t,x)$$
$$= \int_0^h \langle Lu_\varphi(t), P(\tau,x) \rangle \, d\tau,$$

and so u_φ is differentiable with respect to $t \in (0,\infty)$ and $\partial_t u_\varphi = Lu_\varphi$. Hence, since $u_\varphi(t,x) \longrightarrow \varphi(x)$ as $t \searrow 0$, it follows that u_φ is a solution to (2.0.1).

In this chapter, we will find conditions under which $P(t,x)$ has the smoothness property required for the preceding line of reasoning to work. However, before doing so, we will explain how the existence of solutions to one of Kolmogorov's equation can be used to prove the uniqueness of solutions to his other equation. See, for example Corollaries 2.1.4, 2.1.6, and 2.1.7 below.

2.1 The Existence-Uniqueness Duality

For anyone familiar with operator theory, what is going on here is a simple manifestation of the general duality principle that existence for an equation implies uniqueness for the adjoint equation.

2.1.1. The Basic Duality Result: Let L be the operator defined from $a : \mathbb{R}^N \longrightarrow \mathrm{Hom}(\mathbb{R}^N; \mathbb{R}^N)$ and $b : \mathbb{R}^N \longrightarrow \mathbb{R}^N$, as in (1.1.8). Before stating our main result, we will need to have the contents of the following lemma.

LEMMA 2.1.1. *Assume that a and b satisfy the condition in (1.1.20), and let $t \rightsquigarrow \mu(t)$ be any continuous solution to (1.1.11) for $\varphi \in C_c^2(\mathbb{R}^N; \mathbb{C})$. Then, there is a non-decreasing function $r \in [0,\infty) \longmapsto \lambda_r \in [0,\infty)$, depending only on r and the λ in (1.1.20), such that $\lambda_0 = 0$ and*

$$(2.1.2) \qquad \int |y|^{2r} \, \mu(t,dy) \le e^{\lambda_r t} \int (1+|x|^{2r}) \, \nu(dx).$$

In particular, if $\int |x|^{2r} \, \nu(dx) < \infty$, then (1.1.11) continues to hold for any $\varphi \in C^2(\mathbb{R}^N; \mathbb{C})$ satisfying

$$\sup_{x \in \mathbb{R}^N} \frac{|\varphi(x)| + (\nabla\varphi(x), a(x)\nabla\varphi(x))_{\mathbb{R}^N}^{\frac{1}{2}} + |L\varphi(x)|}{1 + |x|^{2r}} < \infty.$$

PROOF: We begin by proving the last statement under the assumption that $t \rightsquigarrow \int |y|^{2r} \, \mu(t,dy)$ is bounded on compact intervals. To this end, choose $\psi \in C_c^\infty(\mathbb{R}^N; [0,1])$ so that $\psi = 1$ on $\overline{B(0,1)}$ and $\psi = 0$ off of $B(0,2)$, and set $\psi_R(x) = \psi(R^{-1}x)$ and $\varphi_R = \psi_R \varphi$ for $R \ge 1$. Clearly, by (1.1.11) and Lebesgue's Dominated Convergence Theorem,

$$\langle \varphi, \mu(t) \rangle - \langle \varphi, \nu \rangle = \lim_{R \to \infty} \int_0^t \langle L\varphi_R, \mu(\tau) \rangle \, d\tau,$$

and so we need only show that the limit on the right equals $\int_0^t \langle L\varphi, \mu(\tau) \rangle \, d\tau$. But

$$L\varphi_R(x) = \psi_R(x)L\varphi + \left(\nabla \varphi(x), a(x)\nabla \psi_R(x) \right)_{\mathbb{R}^N} + \varphi(x)L\psi_R(x),$$

and so, again by Lebesgue's Dominated Convergence Theorem, it suffices to show that

$$\int_0^t \left(\int \left(\left(\nabla \varphi(y), a(y)\nabla \psi_R(y) \right)_{\mathbb{R}^N} + \varphi(x)L\psi_R(y) \right) \mu(\tau, dy) \right) d\tau \longrightarrow 0.$$

For this purpose, first note that, because $a(y)$ is symmetric and non-negative definite,

$$\left| \left(\nabla \varphi(y), a(y)\nabla \psi_R(y) \right)_{\mathbb{R}^N} \right|^2$$
$$\leq \left(\nabla \varphi(y), a(y)\nabla \phi(y) \right)_{\mathbb{R}^N} \left(\nabla \psi_R(y), a(y)\nabla \psi_R(y) \right)_{\mathbb{R}^N}.$$

Because

$$\left(\nabla \psi_R(y), a(y)\nabla \psi_R(y) \right)_{\mathbb{R}^N}$$
$$\leq \|\nabla \psi\|_u^2 \sup_{R \leq |y'| \leq 2R} \frac{\text{Trace}\left(a(y') \right)}{R^2} \mathbf{1}_{B(0,R)\complement}(y),$$

we now see that there is a $C < \infty$ such that

$$\left| \left(\nabla \varphi(y), a(y)\nabla \psi_R(y) \right)_{\mathbb{R}^N} \right| \leq C(1 + |y|^{2r})\mathbf{1}_{B(0,R)\complement}(y).$$

Similarly, one can show $|L\psi_R(y)|$ is also bounded by a constant, independent of $R \geq 1$, times $(1 + |y|^{2r})\mathbf{1}_{B(0,R)\complement}(y)$. Thus, the desired conclusion follows after yet another application of Lebesgue's Dominated Convergence Theorem.

Returning to the proof of (2.1.2), consider the functions

$$\varphi_{r,\epsilon}(x) = \left(\frac{1 + |x|^2}{1 + \epsilon|x|^2} \right)^r \quad \text{for } r \in (0, \infty) \text{ and } \epsilon \in (0, 1).$$

Then

$$\nabla \varphi_{r,\epsilon}(x) = \frac{2r(1 - \epsilon)\varphi_{r,\epsilon}(x)}{(1 + \epsilon|x|^2)(1 + |x|^2)^{\frac{1}{2}}} \frac{x}{(1 + |x|^2)^{\frac{1}{2}}}$$

and

$$\nabla^2 \varphi_{r,\epsilon}(x) = \frac{2r(1 - \epsilon)\varphi_{r,\epsilon}(x)}{(1 + |x|^2)(1 + \epsilon|x|^2)} \left[I + 2 \left(\frac{(r + 1)(1 - \epsilon)}{1 + \epsilon|x|^2} - 2 \right) \frac{x \otimes x}{1 + |x|^2} \right].$$

In particular, the preceding result (applied with the r there equal to 0) shows that

$$\langle \varphi_{r,\epsilon}, \mu(t) \rangle \leq \langle \varphi_{r,\epsilon}, \nu \rangle + \lambda_r \int_0^t \langle \varphi_{r,\epsilon}, \mu(\tau) \rangle \, d\tau$$

for some $\lambda_r < \infty$ which is independent of ϵ and depends only on the λ in (1.1.20). Hence, by Fatou's Lemma and Gronwall's inequality,

$$\int (1 + |y|^{2r})\, \mu(t, dy) = \lim_{\epsilon \searrow 0} \langle \varphi_{r,\epsilon}, \mu(t) \rangle \leq e^{\lambda_r t} \int (1 + |x|^2)^r \, \nu(dx). \quad \square$$

THEOREM 2.1.3. *Assume that a and b satisfy the growth condition in (1.1.20), and suppose that $t \in [0, \infty) \longrightarrow \mu(t) \in \mathbf{M}_1(\mathbb{R}^N)$ is a continuous solution to (1.1.11) for all $\varphi \in C_c^2(\mathbb{R}^N; \mathbb{C})$. If $\int |x|^{2r} \nu(dx) < \infty$ for some $r \in [0, \infty)$ and if, for some $T \in (0, \infty)$, $u \in C^{1,2}([0, T] \times \mathbb{R}^N; \mathbb{C})$ is a solution to $\partial_t u = Lu$ which satisfies*

$$\sup_{t \in [0,T] \times \mathbb{R}^N} \frac{|u(t, x)| + (\nabla u(t, x), a(x) \nabla u(t, x))_{\mathbb{R}^N}^{\frac{1}{2}}}{1 + |x|^{2r}} < \infty,$$

then

$$\langle u(0), \mu(T) \rangle = \langle u(T), \nu \rangle.$$

PROOF: By Lemma 2.1.1 we know that (2.1.2) holds. Now define ψ_R, $R \geq 1$, as in the preceding proof, and set $u_R = \psi_R u$. Then, by Lebesgue's Dominated Convergence Theorem,

$$\langle u(0), \mu(T) \rangle - \langle u(T), \nu \rangle = \lim_{R \to \infty} \left(\langle u_R(0), \mu(T) \rangle - \langle u_R(T), \nu \rangle \right)$$

$$= \lim_{R \to \infty} \int_0^T \frac{d}{dt} \langle u_R(T - t), \mu(t) \rangle \, dt$$

$$= \lim_{R \to \infty} \int_0^T \langle Lu_R(T - t) - \psi_R Lu(T - t), \mu(t) \rangle \, dt$$

$$= \lim_{R \to \infty} \int_0^T \langle u(T - t) L\psi_R + (\nabla u(T - t), a\nabla \psi_R)_{\mathbb{R}^N}, \mu(t) \rangle \, dt.$$

By the argument used in the preceding proof, both $u(T - t, x) L\psi_R(x)$ and $\left| (\nabla u(T - t, x), a(x) \nabla \psi_R(x))_{\mathbb{R}^N} \right|$ are bounded by a constant, independent of $t \in [0, T]$ and $R \geq 1$, times $(1 + |x|^{2r}) \mathbf{1}_{B(0,R)\complement}(x)$. Hence, by Lebesgue's Dominated Convergence Theorem,

$$\lim_{R \to \infty} \int_0^T \langle u(T - t) L\psi_R + (\nabla u(t), a\nabla \psi_R)_{\mathbb{R}^N}, \mu(t) \rangle \, dt = 0. \quad \square$$

2.1.2. Uniqueness of Solutions to Kolmogorov's Equations: Our first application of Theorem 2.1.3 shows that the existence of solutions to Kolmogorov's forward equation guarantees uniqueness for the solutions of his backward equation.

COROLLARY 2.1.4. *Let a and b be as in Theorem 2.1.3, define L from a and b by (1.1.8), and assume that $(t, x) \rightsquigarrow P(t, x)$ is a continuous transition probability function for which (1.2.5) holds whenever $\varphi \in C_c^\infty(\mathbb{R}^N; \mathbb{R})$. Then, for each slowly increasing $\varphi \in C(\mathbb{R}^N; \mathbb{C})$ there is at most one solution $u \in C^{1,2}((0, \infty) \times \mathbb{R}^N; \mathbb{C})$ to (2.0.1) which satisfies the growth condition*

$$\sup_{t \in [0,T] \times \mathbb{R}^N} \frac{|u(t, x)| + (\nabla u(t, x), a(x) \nabla u(t, x))_{\mathbb{R}^N}^{\frac{1}{2}}}{1 + |x|^{2r}} < \infty, \quad T \in (0, \infty),$$

for some $r \in [0, \infty)$. In fact, $u(t, x) = \langle \varphi, P(t, x) \rangle$.

PROOF: By applying the preceding to $(t,x) \in [0,T] \times \mathbb{R}^N \longmapsto u(s+t,x) \in \mathbb{C}$ for $0 < s < T$ and $0 \leq t \leq T - s$, we see that $u(s+T,x) = \langle u(s), P(T,x) \rangle$, and so the result follows when we let $s \searrow 0$. □

REMARK 2.1.5. By considering the Markov process whose transition probability function is $(t,x) \rightsquigarrow P(t,x)$ and using Doob's Stopping Time Theorem, one can dispense in the preceding with the growth condition on $(\nabla u, a \nabla u)_{\mathbb{R}^N}$.

Finally, we apply Theorem 2.1.3 to show that uniqueness for solutions to (1.1.11) is implied by the existence of solutions to (2.0.1).

COROLLARY 2.1.6. *Again let a and b be as in Theorem 2.1.3, and assume that for each $\varphi \in C_c^2(\mathbb{R}^N; \mathbb{R})$ there is a solution $u \in C^{1,2}((0,\infty) \times \mathbb{R}^N; \mathbb{R})$ to (2.0.1) satisfying*

$$\sup_{(t,x) \in [0,T] \times \mathbb{R}^N} \frac{\|\nabla^2 u(t,x)\|_{\text{H.S.}}}{1 + |x|^{2r}} < \infty$$

for some $r \in [0,\infty)$ and all $T \in (0,\infty)$. Then, for each $\nu \in \mathbf{M}_1(\mathbb{R}^N)$ with moments of all orders, there is at most one $t \rightsquigarrow \mu(t)$ satisfying (1.1.11) for all $\varphi \in C_c^2(\mathbb{R}^N; \mathbb{R})$. In particular, if $t \rightsquigarrow P(t,x)$ is the solution with $\nu = \delta_x$ and $\{P_n(t,x) : n \geq 0\}$ is defined relative to a and b as in § 1.2, then $(t,x) \rightsquigarrow P(t,x)$ is a continuous probability function and $\{P_n : n \geq 0\}$ converges to P in the sense that

$$\langle \varphi_n, P_n(t_n, x_n) \rangle \longrightarrow \langle \varphi, P(t,x) \rangle$$

whenever $(t_n, x_n) \to (t,x)$ in $[0,\infty) \times \mathbb{R}^N$ and $\{\varphi_n : n \geq 0\} \subseteq C(\mathbb{R}^N; \mathbb{C})$ is a uniformly slowly increasing sequence which converges to φ uniformly on compacts.

PROOF: Using the same argument as we used in the proof of Corollary 2.1.4, one sees that

$$\langle u(T), \nu \rangle = \lim_{s \searrow 0} \langle u(s), \mu(T) \rangle = \langle \varphi, \mu(T) \rangle.$$

Thus $\mu(T)$ is uniquely determined for each $T \in [0,\infty)$. To prove the asserted convergence result, suppose that $(t_n, x_n) \longrightarrow (t,x)$, and note that, starting from (1.1.21), we can use the argument given in the proof of the last part of Lemma 1.1.19 to check that $\{P_n(t_n, x_n) : n \geq 1\}$ is relatively compact in $C([0,\infty) \times \mathbb{R}^N; \mathbb{R}^N)$. Thus, since every limit will be a solution to (1.2.5) and, as we just showed, there is only one such solution, it follows that $\{P_n(t_n, x_n) : n \geq 0\}$ converges in $\mathbf{M}_1(\mathbb{R}^N)$ to the unique $P(t,x)$ satisfying (1.2.5). Hence, we now know $P_n \longrightarrow P$ in $C([0,\infty) \times \mathbb{R}^N; \mathbf{M}_1(\mathbb{R}^N))$. In particular, this means that $(t,x) \rightsquigarrow P(t,x)$ is continuous. To see that it is a transition probability function, one can use the remark given in introduction to § 1.2. Alternatively, one can argue by uniqueness. That is, observe that both

$$t \rightsquigarrow P(s+t, x) \quad \text{and} \quad t \rightsquigarrow \int P(t, y) P(s, x, dy)$$

satisfy (1.1.11) with $\nu = P(s, x)$ and therefore, by uniqueness, must be equal. Finally, to complete the proof, suppose the $\{\varphi_n : n \geq 0\}$ and φ are as the final assertion. To prove the asserted convergence, note that, by Lemma 1.3.3, $\sup_{n \geq 0} \int |y|^{2r} P_n(t, x, dy) < \infty$ for each $r \geq 1$. Hence, exactly the same argument as was used to prove Theorem 1.3.4 applies here. $\square$

Before moving on, we mention a somewhat technical extension of Corollary 2.1.6, one which is useful in borderline situations.

COROLLARY 2.1.7.   *In the setting of Corollary 2.1.6, all the conclusions there continue to hold if, for each $\varphi \in C_c^\infty(\mathbb{R}^N; \mathbb{R})$ and $T > 0$, there exists a sequence $\{u_k : k \geq 1\} \subseteq C^{1,2}([0, T] \times \mathbb{R}^N : \mathbb{R})$ with the properties that $u_k(0) \longrightarrow \varphi$ uniformly on compacts, and there exists an $r_T \geq 0$ such that*

$$\sup_{k \geq 0} \sup_{(t,x) \in [0,T]} \frac{|\partial_t u_k(t,x)| \vee |\nabla u_k(t,x)| \vee \|\nabla^2 u_k(t,x)\|_{\text{H.S.}}}{1 + |x|^{2r_T}} < \infty$$

and

$$\lim_{k \to \infty} \sup_{(t,x) \in [0,T] \times \mathbb{R}^N} \frac{|(\partial_t - L) u_k(t,x)|}{1 + |x|^{2r_T}} = 0.$$

PROOF: By looking at its proof, one realizes that everything in Corollary 2.1.6 will follow once we prove the uniqueness statement. For this purpose, choose $\psi \in C_c^\infty(B(0, 2); [0, 1])$ so that $\psi = 1$ on $\overline{B(0, 1)}$, and set $v_k(t, x) = \psi(k^{-1}x) u_k(t, x)$, $\varphi_k = v_k(0)$, and $g_k = (L - \partial_t) v_k$. Then the argument used in the proof of Corollary 2.1.4 shows that

$$\langle \varphi_k, \mu(T) \rangle - \langle v_k(T), \nu \rangle = \int_0^T \langle g_k(T-t), \mu(t) \rangle \, dt.$$

At the same time, it is easy to check that

$$\lim_{k \to \infty} \sup_{(t,x) \in [0,T] \times \mathbb{R}^N} \frac{|\varphi_k(x) - \varphi(x)| \vee |g_k(t,x)|}{1 + |x|^{2(r_T + 1)}} = 0.$$

Hence, by (2.1.2) and Lebesgue's Dominated Convergence Theorem, $\langle \varphi, \mu(T) \rangle = \lim_{k \to \infty} \langle v_k(T), \nu \rangle$. $\square$

2.2 Smoothness in the Backward Variable

Let $\sigma : \mathbb{R}^N \longrightarrow \text{Hom}(\mathbb{R}^M; \mathbb{R}^N)$ and $b : \mathbb{R}^N \longrightarrow \mathbb{R}^N$ be smooth functions, and assume that their first derivatives are bounded and that their higher order derivatives are slowly increasing. Let $(t, x) \rightsquigarrow P(t, x)$ be the corresponding transition probability function constructed in § 1.2. In this section we will show that $x \rightsquigarrow \langle \varphi, P(t, x) \rangle$ is ℓ-times continuously differentiable whenever φ is a function which is ℓ-times differentiable with derivatives that are slowly increasing.

2.2.1. Differentiating the Backward Variable: Let σ and b be as above, and define $\{P_n : n \geq 0\}$ accordingly, as in § 1.2. That is, $P_n(0, x) = \delta_x$ and, for $m2^{-n} \leq t \leq (m+1)2^{-n}$,

$$P_n(t, x) = \int Q(t - m2^{-n}, y) \, P_n(m2^{-n}, x, dy),$$

where $Q(t, x)$ is the distribution of

$$\omega \in \mathbb{R}^M \longmapsto F(t, x, \omega) \equiv x + t^{\frac{1}{2}} \sigma(x) \omega + tb(x) \in \mathbb{R}^N$$

under γ.

Next, set $\Omega = (R^M)^{\mathbb{Z}^+}$ and $\Gamma = \gamma^{\mathbb{Z}^+}$. Define $X_n : [0, \infty) \times \mathbb{R}^N \times \Omega \longrightarrow \mathbb{R}^N$ so that

(2.2.1)
$$X_n(0, x, \boldsymbol{\omega}) = x \ \& \ X_n(t, x, \boldsymbol{\omega}) = F\big(t - m2^{-n}, X_n(m2^{-n}, \boldsymbol{\omega}), \omega_{m+1}\big)$$
$$\text{for } m2^{-n} \leq t \leq (m+1)2^{-n} \text{ and } \boldsymbol{\omega} = (\omega_1, \dots, \omega_m, \dots) \in \Omega.$$

Notice that, for $t \leq m2^{-n}$, $\boldsymbol{\omega} \rightsquigarrow X_n(t, x, \boldsymbol{\omega})$ depends only on $(\omega_1, \dots, \omega_m)$ and is therefore independent of $\sigma(\{\omega_\ell : \ell \geq m + 1\})$ under Γ.

LEMMA 2.2.2.   *For each $n \geq 0$ and $(t, x) \in [0, \infty) \times \mathbb{R}^N$, $P_n(t, x)$ is the distribution of $\boldsymbol{\omega} \rightsquigarrow X_n(t, x, \boldsymbol{\omega})$ under Γ. That is, for any slowly increasing $\varphi \in C(\mathbb{R}^N; \mathbb{C})$,*

$$\langle \varphi, P_n(t, x) \rangle = \int \varphi \circ X_n(t, x, \boldsymbol{\omega}) \, \Gamma(d\boldsymbol{\omega}).$$

PROOF: Obviously, there is nothing to do when $t = 0$. Now assume the result for $t \in [0, m2^{-n}]$, and let $t = m2^{-n} + \tau$, where $\tau \in [0, 2^{-n}]$. Then, by the preceding remark about independence,

$$\langle \varphi, P_n(t, x) \rangle = \int \langle \varphi, Q(\tau, y) \rangle \, P_n(m2^{-n}, x, dy)$$

$$= \int \langle \varphi, Q\big(\tau, X_n(m2^{-n}, x, \boldsymbol{\omega})\big) \rangle \, \Gamma(d\boldsymbol{\omega})$$

$$= \int \left(\int \varphi \circ F\big(\tau, X_n(m2^{-n}, x, \boldsymbol{\omega}), \omega_{m+1}\big) \, \gamma(d\omega_{m+1}) \right) \Gamma(d\boldsymbol{\omega})$$

$$= \int \varphi \circ X_n(t, x, \boldsymbol{\omega}) \, \Gamma(d\boldsymbol{\omega}).$$

Hence, by induction, on $m \geq 0$, we are done. $\square$

Before taking the next step, we need to introduce some notation. Given $\ell \geq 1$ and $\Phi \in C^\ell(\mathbb{R}^N; \mathbb{R}^{N'})$, we use $\nabla^\ell \Phi$ to denote the mapping from $\mathbb{R}^N$ into $\mathrm{Hom}\big((\mathbb{R}^N)^{\otimes \ell}; \mathbb{R}^{N'}\big)$ determined by

$$\nabla^\ell \Phi(x)\xi_1 \otimes \cdots \otimes \xi_\ell = \frac{\partial^\ell}{\partial_{t_1} \cdots \partial_{t_\ell}} \Phi\left(x + \sum_{k=1}^\ell t_k \xi_k\right)\Bigg|_{t_1 = \cdots = t_\ell = 0}$$

for $(\xi_1, \ldots, \xi_\ell) \in (\mathbb{R}^N)^\ell$. Using the chain rule and induction on ℓ, one can check that if $\Phi \in C^\ell(\mathbb{R}^N; \mathbb{R}^{N'})$ and $\Psi \in C^\ell(\mathbb{R}^{N'}; \mathbb{R}^{N''})$, then

$$(2.2.3) \qquad \nabla^\ell(\Psi \circ \Phi) = \sum_{k=1}^\ell \sum_{\beta \in \mathfrak{B}(k,\ell)} c_\beta \big((\nabla^k \Phi) \circ \Phi\big) \nabla^{\otimes \beta} \Psi,$$

where $\mathfrak{B}(k, \ell) = \Big\{ \beta \in (\mathbb{Z}^+)^k : \sum_{j=1}^k \beta_j = \ell \Big\}$, $\nabla^{\otimes \beta} \Phi = \nabla^{\beta_1} \Phi \otimes \cdots \otimes \nabla^{\beta_k} \Phi$, and the coefficients c_β are given by the prescription: $c_\beta = 1$ if $\beta \in \mathfrak{B}(1, \ell) \cup \mathfrak{B}(\ell, \ell)$ and, for $\beta \in \mathfrak{B}(k, \ell + 1)$ with $2 \leq k \leq \ell$,

$$c_\beta = \delta_{\beta_1, 1} c_{(\beta_2, \ldots, \beta_{\ell+1})} + \sum_{\alpha \in \mathcal{P}(\beta)} c_\alpha,$$

where $\mathcal{P}(\beta)$ (the "parents of β") is the set of $\alpha \in \mathfrak{B}(k, \ell)$ with the property that $\beta_j = \delta_{i,j} + \alpha_j$, $1 \leq j \leq k$, for some $1 \leq i \leq k$.

Using (2.2.1), (2.2.3), and induction on $m \geq 0$, one sees that, for each $(t, \boldsymbol{\omega}) \in [m2^{-n}, (m+1)2^{-n}] \times \Omega$, $x \leadsto X_n(t, x, \boldsymbol{\omega})$ is ℓ-times continuously differentiable. In fact, if $X_n^\ell(t, x, \boldsymbol{\omega}) \equiv \nabla^\ell X_n(t, x, \boldsymbol{\omega})$ (the differentiation being with respect to the x-variables only), then

$$X_n^\ell(t, x, \boldsymbol{\omega})$$

$$(2.2.4) \quad = \sum_{k=1}^\ell \sum_{\beta \in \mathfrak{B}(k, \ell)} c_\beta \nabla^k F\big(\tau, X_n(m2^{-n}, x, \boldsymbol{\omega}), \omega_{m+1}\big) X_n^{\otimes \beta}(m2^{-n}, x, \boldsymbol{\omega})$$

$$\text{for } m2^{-n} \leq t \leq (m+1)2^{-n} \text{ and } \tau = t - m2^{-n},$$

where $X_n^{\otimes \beta} \equiv X_n^{\beta_1} \otimes \cdots \otimes X_n^{\beta_k}$. In addition, under the conditions we have imposed on σ and b, we know that there exist $C < \infty$ and $r \geq 1$ for which

$$|F(\tau, x, \omega)| \vee \max_{1 \leq k \leq \ell} \|\nabla^k F(\tau, x, \omega)\|_{\text{H.S.}} \leq C(1 + |x|)^r (1 + |\omega|)$$

when $(\tau, x, \omega) \in [0, 1] \times \mathbb{R}^N \times \mathbb{R}^M$, and so one can use (2.2.4) to check that, for each $n \in \mathbb{N}$, $r \geq 1$, $t \in (0, \infty)$, and $R > 0$,

$$\int \sup_{|x| \leq R} \left(|X_n(t, x, \boldsymbol{\omega})| \vee \max_{1 \leq k \leq \ell} \|X_n^k(t, x, \boldsymbol{\omega})\|_{\text{H.S.}}\right)^{2r} \Gamma(d\boldsymbol{\omega}) < \infty.$$

LEMMA 2.2.5.   *Given a slowly increasing* $\varphi \in C(\mathbb{R}^N; \mathbb{C})$, *set*

$$u_{\varphi,n}(t, x) = \int \varphi(y) \, P_n(t, x, dy).$$

If φ *is* ℓ-*times differentiable and its derivatives are slowly increasing, then, for each* $n \geq 0$ *and* $t \in [0, \infty)$, $u_{\varphi,n}(t, \cdot)$ *is* ℓ-*times continuously differentiable and*

$$\nabla^\ell u_{\varphi,n}(t, x) = \sum_{k=1}^{\ell} \sum_{\beta \in \mathfrak{B}(k,\ell)} c_\beta \int \nabla^k \varphi\big(X_n(t, x, \boldsymbol{\omega})\big) X_n^{\otimes \beta}(t, x, \boldsymbol{\omega}) \, \Gamma(d\boldsymbol{\omega}).$$

PROOF: The result is an immediate consequence of (2.2.3) and an application of the preceding integrability estimate to justify differentiation under the integral sign. □

2.2.2. The Distribution of Derivatives: Based on the result in Lemma 2.2.5, we now want to prove that (cf. (2.0.2)) $u_\varphi(t, \cdot)$ is just as smooth as φ and the coefficients σ and b are. For this purpose, we will use the contents of § 1.3.3 to control what happens when we let $n \to \infty$ in Lemma 2.2.5.

Given $\ell \geq 1$, set $E_\ell = \mathrm{Hom}\big((\mathbb{R}^N)^{\otimes \ell}; \mathbb{R}^N\big)$, $\mathbf{E}^{(\ell)} = E_1 \times \cdots \times E_\ell$, and

$$\|\mathbf{J}\|_{\mathbf{E}^{(\ell)}} = \left(\sum_{k=1}^{\ell} \|J_k\|_{\mathrm{H.S.}}^2 \right)^{\frac{1}{2}} \quad \text{and} \quad \mathbf{J}^\beta = J_1^{\beta_1} \otimes \cdots \otimes J_\ell^{\beta_\ell}$$

for $\mathbf{J} = (J_1, \ldots, J_\ell) \in \mathbf{E}^{(\ell)}$ and $\beta \in (\mathbb{Z}^+)^\ell$. Next, define $\sigma_\ell : \mathbb{R}^N \times \mathbf{E}^{(\ell)} \longrightarrow$ $\mathrm{Hom}(\mathbb{R}^M; E_\ell)$ and $b_\ell : \mathbb{R}^N \times \mathbf{E}^{(\ell)} \longrightarrow E_\ell$ so that

$$\sigma_\ell(x, \mathbf{J})\omega = \sum_{k=1}^{\ell} \sum_{\beta \in \mathfrak{B}(k,\ell)} c_\beta \nabla^k \big(\sigma(x)\omega\big) \mathbf{J}^{\otimes \beta}$$

and

$$b_\ell(x, \mathbf{J}) = \sum_{k=1}^{\ell} \sum_{\beta \in \mathfrak{B}(k,\ell)} c_\beta \nabla^k b(x) \mathbf{J}^{\otimes \beta}$$

for $x \in \mathbb{R}^N$ and $\mathbf{J} = (J_1, \ldots, J_\ell) \in \mathbf{E}^{(\ell)}$. Finally, determine $\boldsymbol{\sigma}^{(\ell)} : \mathbb{R}^N \times \mathbf{E}^{(\ell)}$ $\longrightarrow \mathrm{Hom}(\mathbb{R}^M; \mathbb{R}^N \times \mathbf{E}^{(\ell)})$ and $\mathbf{b}^{(\ell)} : \mathbb{R}^N \times \mathbf{E}^{(\ell)} \longrightarrow \mathbb{R}^N \times \mathbf{E}^{(\ell)}$ by

$$\boldsymbol{\sigma}^{(\ell)}(x, \mathbf{J}) = \begin{pmatrix} \sigma(x) \\ \sigma_1(x, J_1) \\ \vdots \\ \sigma_\ell(x, \mathbf{J}) \end{pmatrix} \quad \text{and} \quad \mathbf{b}^{(\ell)}(x, \mathbf{J}) = \begin{pmatrix} b(x) \\ b_1(x, J_1) \\ \vdots \\ b_\ell(x, \mathbf{J}) \end{pmatrix}.$$

We next define $P_n^{(\ell)}\big(t,(x,\mathbf{J})\big)$ in terms of $\boldsymbol{\sigma}^{(\ell)}$ and $\mathbf{b}^{(\ell)}$ by the usual prescription. That is, $P_n^{(\ell)}\big(0,(x,\mathbf{J})\big) = \delta_{(x,\mathbf{J})}$ and

$$P^{(\ell)}\big(t,(x,\mathbf{J})\big) = \int Q^{(\ell)}\big(t - [t]_n, (x',\mathbf{J}')\big)\, P_n^{(\ell)}\big([t]_n, (x,\mathbf{J}), dx' \times d\mathbf{J}')\big),$$

where $Q^{(\ell)}\big(t,(x,\mathbf{J})\big)$ is the distribution of

$$\omega \rightsquigarrow \begin{pmatrix} x \\ \mathbf{J} \end{pmatrix} + t^{\frac{1}{2}}\boldsymbol{\sigma}^{(\ell)}(x,\mathbf{J})\omega + t\mathbf{b}^{(\ell)}(x,\mathbf{J}) \quad \text{under } \gamma.$$

By repeating the argument used to prove Lemma 2.2.2, we see that, for any slowly increasing $\varphi \in C\big(\mathbb{R}^N \times \mathbf{E}^{(\ell)}; \mathbb{C}\big)$,

(2.2.6)
$$\int \varphi\big(X_n(t,x,\boldsymbol{\omega}), X_n^1(t,x,\boldsymbol{\omega}), \cdots, X^\ell(t,x,\boldsymbol{\omega})\big)\, \Gamma(d\omega)$$
$$= \big\langle \varphi, P^{(\ell)}\big(t,(x,\mathbf{J}_0)\big)\big\rangle, \quad \text{where } \mathbf{J}_0 \equiv (I, 0, \ldots, 0).$$

THEOREM 2.2.7. *Assume that σ and b are smooth functions whose first derivatives are bounded and whose higher order derivatives are slowly increasing. Set*

$$C_1 = \sup_{x \in \mathbb{R}^N}\big(\|\nabla b(x)\|_{\mathrm{H.S.}}^2 + \|\nabla\sigma(x)\|_{\mathrm{H.S.}}^2\big),$$

and, for $k \geq 2$, take $r_k \in [0,\infty)$ so that

$$C_k = \sup_{x \in \mathbb{R}^N}\frac{\|\nabla^k b(x)\|_{\mathrm{H.S.}}^2 + \|\nabla^k\sigma(x)\|_{\mathrm{H.S.}}^2}{1 + |x|^{2r_k}} < \infty.$$

Then, for each $\ell \geq 1$, there is an $\mathbf{r}^{(\ell)} \in [1,\infty)$ and a continuous map $r \in [1,\infty) \longmapsto C_r^{(\ell)} \in (0,\infty)$, depending only on ℓ, $\|\sigma(0)\|_{\mathrm{H.S.}}$, $|b(0)|$, $\max_{1 \leq k \leq \ell} r_k$, and $\max_{1 \leq k \leq \ell} C_k$, such that

$$\sup_{n \geq 0}\int \big(|x'|^{2r} + \|\mathbf{J}'\|_{\mathbf{E}^{(\ell)}}^{2r}\big)P_n^{(\ell)}\big(t,(x,\mathbf{J}), dx' \times d\mathbf{J}'\big)$$
$$\leq C_r^{(\ell)}e^{C_r^{(\ell)}t}\big(1 + |x|^{2r\mathbf{r}^{(\ell)}} + \|\mathbf{J}\|_{\mathbf{E}^{(\ell)}}^{2r\mathbf{r}^{(\ell)}}\big).$$

Furthermore, there is a continuous transition probability function $\big(t,(x,\mathbf{J})\big)$ $\rightsquigarrow P^{(\ell)}\big(t,(x,\mathbf{J})\big)$ to which $\{P_n^{(\ell)} : n \geq 0\}$ converges in $C\big([0,\infty) \times \mathbb{R}^N \times \mathbf{E}^{(\ell)}; \mathbf{M}_1(\mathbb{R}^N \times \mathbf{E}^{(\ell)})\big)$. In particular,

$$\int \big(|x'|^{2r} + \|\mathbf{J}'\|_{\mathbf{E}^{(\ell)}}^{2r}\big)P^{(\ell)}\big(t,(x,\mathbf{J}), dx' \times d\mathbf{J}'\big)$$
$$\leq C_r^{(\ell)}e^{C_r^{(\ell)}t}\big(1 + |x|^{2r\mathbf{r}^{(\ell)}} + \|\mathbf{J}\|_{\mathbf{E}^{(\ell)}}^{2r\mathbf{r}^{(\ell)}}\big)$$

and

$$\langle \varphi_n, P_n^{(\ell)}(t_n, (x_n, \mathbf{J}_n)) \rangle \longrightarrow \langle \varphi, P^{(\ell)}(t, (x, \mathbf{J})) \rangle$$

if $\big(t_n, (x_n, \mathbf{J}_n)\big) \longrightarrow \big(t, (x, \mathbf{J})\big)$ in $[0, \infty) \times \mathbb{R}^N \times \mathbf{E}^{(\ell)}$ and $\{\varphi_n : n \geq 0\} \subseteq C\big(\mathbb{R}^N \times \mathbf{E}^{(\ell)}; \mathbb{C}\big)$ is a uniformly slowly increasing sequence which tends to φ uniformly on compacts. Finally, if σ, b, and their derivatives of order through ℓ are bounded, then, for each $r \in [0, \infty)$, there is a continuous $r \in [0, \infty) \longmapsto C_r^{(\ell)} \in (0, \infty)$, depending only on the bounds on σ, b, and their derivatives through order ℓ, such that (cf. (2.2.6))

$$\sup_{n \geq 0} \int \big(|x' - x|^{2r} + \|\mathbf{J}'\|_{\mathbf{E}^{(\ell)}}^{2r}\big) P_n^{(\ell)}\big(t, (x, \mathbf{J}_0), dx' \times d\mathbf{J}'\big) \leq C_r^{(\ell)} e^{C_r^{(\ell)} t}$$

and

$$\int \big(|x' - x|^{2r} + \|\mathbf{J}'\|_{\mathbf{E}^{(\ell)}}^{2r}\big) P^{(\ell)}\big(t, (x, \mathbf{J}_0), dx' \times d\mathbf{J}'\big) \leq C_r^{(\ell)} e^{C_r^{(\ell)} t}.$$

PROOF: Because, for each $k \geq 1$, σ_k is linear as a function of J_k, all but the final assertion are simple applications of Theorem 1.3.14.

Now assume that σ, b, and their derivatives through order ℓ are bounded. One way to prove the asserted estimates is to trace through the proof given and check that, under these boundedness conditions, the right-hand side can be made independent of x. A second way is to begin by observing that these assumptions are translation invariant. Next, check that, for each $x \in \mathbb{R}^N$, $P^{(\ell)}\big(t, (x, \mathbf{J}_0)\big)$ is equal to $P_x^{(\ell)}\big(t, (0, \mathbf{J}_0)\big)$, where $P_x^{(\ell)}$ is defined relative to the translated coefficients $\sigma(x + \cdot)$ and $b(x + \cdot)$. Hence, the asserted estimate follows from the previous one applied to $P_x^{(\ell)}\big(t, (0, \mathbf{J}_0)\big)$. $\square$

COROLLARY 2.2.8.   *Refer to the conditions and notation in Theorem 2.2.7. If $\varphi \in C^\ell(\mathbb{R}^N; \mathbb{C})$ satisfies*

$$\sum_{k=0}^{\ell} \sup_{x \in \mathbb{R}^N} \frac{\|\nabla^k \varphi(x)\|_{\text{H.S.}}}{1 + |x|^r} < \infty$$

for some $r \in [1, \infty)$ and u_φ is given by (2.0.2), then, for each $t \geq 0$, $u_\varphi(t) \in C^\ell(\mathbb{R}^N; \mathbb{C})$,

$$\nabla^\ell u_\varphi(t, x) = \sum_{k=1}^{\ell} \sum_{\beta \in \mathfrak{B}(k, \ell)} c_\beta \int \nabla^k \varphi(y) \mathbf{J}^{\otimes \beta} \, \mathbf{P}^{(\ell)}\big(t, (x, 0), dy \times d\mathbf{J}\big),$$

and so

$$\|\nabla^\ell u_\varphi(t, x)\|_{\text{H.S.}} \leq \kappa_\ell e^{\kappa_\ell t} (1 + |x|)^{(r+\ell)\mathbf{r}^{(\ell)}} \sum_{k=0}^{\ell} \sup_{y \in \mathbb{R}^N} \frac{\|\nabla^k \varphi(y)\|_{\text{H.S.}}}{1 + |y|^r},$$

where $C_\ell(\varphi)$ depends only on ℓ, r, C, $\|\sigma(0)\|_{\mathrm{H.S.}}$, $|b(0)|$, $\max_{1\le k\le \ell} C_k$, and $\max_{1\le k\le \ell} r_k$. In particular, if $\ell = 2m$, where $m \ge 1$, then $u_\varphi \in C^{m,2m}\big([0,\infty)\times \mathbb{R}^N;\mathbb{C}\big)$ and $\partial_t^m u_\varphi = L^m u_\varphi$. Finally, if σ, b, and their derivatives through order ℓ are all bounded, then

$$\|\nabla^\ell u_\varphi(t)\|_{\mathrm{u}} \le \kappa_\ell e^{\kappa_\ell t}\sum_{k=0}^{\ell} \|\nabla^k \varphi\|_{\mathrm{u}}.$$

PROOF: The differentiability of u_φ and the formula for $\nabla^\ell u_\varphi(t,x)$ follow immediately from Lemma 2.2.5, (2.2.6), and Theorem 2.2.7. Given the formula for $\nabla^\ell u_\varphi(t,x)$, the stated estimates follow from the estimates in Theorem 2.2.7 and Schwarz's inequality.

Turning to the derivatives of u_φ with respect to t, we can apply the argument used in the derivation of (2.0.1). Namely, knowing that $u_\varphi(t, \cdot)$ is twice continuously differentiable with slowly increasing second order derivatives, that argument shows that $\partial_t u_\varphi = L u_\varphi$. Assuming the result for $m-1$, use the Chapman–Kolmogorov equation to see that

$$\partial_t^{m-1} u_\varphi(t+h, x) = \partial_t^{m-1}\int u_\varphi(t,y)\, P(h,x,dy)$$

$$= \int L^{m-1} u_\varphi(t,y)\, P(h,x,dy),$$

and apply the reasoning for (2.0.1), with $L^{m-1} u_\varphi(t)$ replacing φ, to get the desired conclusion. $\square$

2.2.3. Uniqueness of Solutions to Kolmogorov's Equation: By combining Corollaries 2.1.6 and 2.2.8, we arrive at the following uniqueness statement for solutions to Kolmogorov's forward equation. See Theorem Theorem 2.4.6 below for an important extension.

THEOREM 2.2.9. *Let L be given by (1.1.8), where $a = \sigma\sigma^\top$ and σ and b are twice continuously differentiable functions whose first derivatives are bounded and whose second derivatives are slowly increasing. Then the hypotheses of Corollary 2.1.7 are satisfied and therefore all the conclusions drawn in Corollary 2.1.6 hold. In particular, there is a unique continuous transition probability function $(t,x)\rightsquigarrow P(t,x)$ satisfying (1.2.5) and, for each ν with moments of all orders, $t\rightsquigarrow \mu(t) = \int P(t,x)\,\nu(dx)$ is the one and only solution to (1.1.11).*

PROOF: We must construct a sequence $\{u_k : k \ge 1\}$ of the sort described in Corollary 2.1.7. To this end, choose $\rho \in C_{\mathrm{c}}^\infty\big(B(0,1);[0,\infty)\big)$ with integral 1, and set $\rho_k(x) = k^N \rho(kx)$, $\sigma_k = \rho_k \star \sigma$, and $b_k = \rho_k \star b$. Then, for each k, σ_k and b_k are smooth and have slowly increasing derivatives of all orders. Moreover,

$$\sup_{k\ge 1}\sup_{x\in\mathbb{R}^N} \frac{\|\sigma_k\|_{\mathrm{H.S.}}^2 + \|b_k(x)\|_{\mathrm{H.S.}}^2}{1+|x|^2} < \infty$$

and

$$\sup_{k \geq 1} \sup_{x \in \mathbb{R}^N} \frac{\|\nabla^2 \sigma_k\|_{\text{H.S.}}^2 + \|\nabla^2 b_k(x)\|_{\text{H.S.}}^2}{1 + |x|^{2r}} < \infty$$

for some $r \geq 0$. Now let $(t, x) \rightsquigarrow P^k(t, x)$ denote the transition probability function constructed in § 1.2 for $a_k = \sigma_k \sigma_k^\top$ and b_k, and set $u_k(t, x) = \langle \varphi, P^k(t, x) \rangle$, where φ is a given element of $C_c^\infty(\mathbb{R}^N; \mathbb{R})$. For each k, u_k is a smooth solution to $\partial_t u = L_k u$ with $u(0) = \varphi$, where L_k is defined by (1.1.8) with $a = a_k$ and $b = b_k$. Moreover, $\|u_k\|_u \leq \|\varphi\|_u$ and, for each $T > 0$, $\{\nabla u_k(t, \cdot) : k \geq 1\}$ and $\{\nabla^2 u_k(t, \cdot) : k \geq 1\}$ are uniformly slowly increasing sequences. Hence, for some $r' \geq 0$,

$$\frac{|(\partial_t - L)u_k(t, x)|}{1 + |x|^{r'}} = \frac{|(L_k - L)u_k(t, x)|}{1 + |x|^{r'}} \longrightarrow 0$$

uniformly for $(t, x) \in [0, T] \times \mathbb{R}^N$. □

2.3 Square Roots

One of the most severe weaknesses of the theory developed in §§ 1.2 and 2.2 is that it relies on our being able to take a good square root of the matrix a. Indeed, when a can degenerate and we need its square root to be smooth, this flaw is fatal: In general, there is simply no way to take a smooth square root of a, even when a is analytic. Nonetheless, as we are about to see, there is no problem when a is non-degenerate and, if all that one requires is Lipschitz continuity, there is no problem as long as a is twice differentiable.

2.3.1. The Non-Degenerate Case: In the following, when a is a non-negative definite and symmetric matrix, $a^{\frac{1}{2}}$ denotes the non-negative definite, symmetric square root of a.

LEMMA 2.3.1. *Assume that $a : \mathbb{R}^N \longrightarrow \text{Hom}(\mathbb{R}^N; \mathbb{R}^N)$ is an ℓ-times continuously differentiable, symmetric, non-negative definite matrix-valued function for some $\ell \geq 1$. Further, assume that $a(x)$ is strictly positive definite, and let $\lambda_{\min}(x)$ denote its smallest eigenvalue. Then $a^{\frac{1}{2}}$ is ℓ-times continuously differentiable at x and there is a universal constant $C_\ell < \infty$ such that*

$$\|\nabla^\ell a^{\frac{1}{2}}(x)\|_{\text{H.S.}} \leq C_\ell \lambda_{\min}(x)^{\frac{1}{2}} \sum_{k=1}^\ell \left(\frac{\|a(x)\|_{\text{H.S.}}^{(k)}}{\lambda_{\min}(x)} \right)^k,$$

where $\|a(x)\|_{\text{H.S.}}^{(\ell)} = \max_{1 \leq k \leq \ell} \|\nabla^k a(x)\|_{\text{H.S.}}$.

PROOF: Given $R \in (1, \infty)$, let $A(R)$ be the space of symmetric $N \times N$-matrices a satisfying $R^{-1}I < a < RI$, and define $\Phi_R : A(R) \longrightarrow A(R^{\frac{1}{2}})$ by

$$\Phi_R(a) = R^{\frac{1}{2}} \sum_{m=0}^\infty \binom{\frac{1}{2}}{m} (R^{-1}a - I)^m,$$

where $\binom{\frac{1}{2}}{m}$ is the coefficient of ξ^m in the Taylor's expansion of $x \rightsquigarrow (1+\xi)^{\frac{1}{2}}$ at 0. By testing its action on the eigenvectors of a, one sees that $a^{\frac{1}{2}} = \Phi_R(a)$. Hence, by (2.2.3), in order to prove the result, it suffices to check that there exists a $B_\ell < \infty$ such that, for any symmetric matrices $E_1, \ldots, E_\ell \in \mathrm{Hom}(\mathbb{R}^N; \mathbb{R}^N)$ with $\|E_k\|_{\mathrm{H.S.}} = 1$,

$$\left\| \frac{\partial^\ell}{\partial_{E_1} \cdots \partial_{E_\ell}} \Phi_R(a) \right\| \le B_\ell \lambda^{\frac{1}{2}-\ell},$$

where λ is the smallest eigenvalue of a. But clearly

$$\frac{\partial^\ell}{\partial_{E_1} \cdots \partial_{E_\ell}} \Phi_R(a) = R^{\frac{1}{2}} \sum_{m=\ell}^{\infty} \binom{\frac{1}{2}}{m} D^{(m)}(a),$$

where $D^{(m)}(a)$ is the sum of $\frac{m!}{(m-\ell)!}$ terms, each of which is the product of m factors: ℓ being E_k's and the other $(m-\ell)$ being $\frac{a}{R} - I$. In particular,

$$\|D^{(m)}(a)\|_{\mathrm{H.S.}} \le \left\| \frac{a}{R} - I \right\|_{\mathrm{op}}^{m-\ell} \prod_{k=1}^{m-\ell} \|E_k\|_{\mathrm{H.S.}} = \left(1 - \frac{\lambda}{R}\right)^{m-\ell},$$

Hence

$$\left\| \frac{\partial^\ell}{\partial_{E_1} \cdots \partial_{E_\ell}} \Phi_R(a) \right\|_{\mathrm{H.S.}} \le R^{\frac{1}{2}} \sum_{m=\ell}^{\infty} \left| \binom{\frac{1}{2}}{m} \right| \frac{m!}{(m-\ell)!} \left(1 - \frac{\lambda}{R}\right)^{m-\ell}$$

$$= (-1)^{\ell+1} \frac{d^\ell}{d\xi^\ell} \left(R^{\frac{1}{2}} \sum_{m=0}^{\infty} \binom{\frac{1}{2}}{m} \left(\frac{\xi}{R} - 1\right)^m \right) \Bigg|_{\xi=\lambda}$$

$$= (-1)^{\ell+1} \frac{d^\ell}{d\xi^\ell} \xi^{\frac{1}{2}} \Bigg|_{\xi=\lambda} = \ell! \left| \binom{\frac{1}{2}}{\ell} \right| \lambda^{\frac{1}{2}-\ell}. \quad \square$$

2.3.2. The Degenerate Case: Obviously, Lemma 2.3.1 is very satisfactory as long as a is bounded below by a positive multiple of the identity. When a can degenerate, the following is essentially the best which one can say in general.

The basic fact which allows us to deal with degenerate a's is an elementary observation about non-negative functions. Namely, if $f \in C^2(\mathbb{R}; [0, \infty))$, then

(2.3.2) $$|f'(t)| \le \sqrt{2\|f''\|_{\mathrm{u}} f(t)}, \quad t \in \mathbb{R}.$$

To see this, use Taylor's theorem to write $0 \le f(t+h) \le f(t) + hf'(t) + \frac{h^2}{2}\|f''\|_{\mathrm{u}}$ for all $h \in \mathbb{R}$. Hence $|f'(t)| \le h^{-1} f(t) + \frac{h}{2}\|f''\|_{\mathrm{u}}$ for all $h > 0$, and so (2.3.2) results when one minimizes with respect to h.

LEMMA 2.3.3. *Assume that $a : \mathbb{R}^N \longrightarrow \mathrm{Hom}(\mathbb{R}^N; \mathbb{R}^N)$ is a twice contin-uously differentiable, non-negative definite, symmetric matrix valued func-tion. If*

$$\Lambda \equiv \sup\{\|\partial_e^2 a(x)\|_{\mathrm{op}} : x \in \mathbb{R}^N \text{ and } e \in \mathbb{S}^{N-1}\} < \infty,$$

then

$$\left\|a^{\frac{1}{2}}(y) - a^{\frac{1}{2}}(x)\right\|_{\mathrm{H.S.}} \leq N\sqrt{2\Lambda}\,|y - x|.$$

PROOF: First observe that it suffices to handle a's which are uniformly positive definite. Indeed, given the result in that case, we can prove the result in general by replacing a with $a + \epsilon I$ and then letting $\epsilon \searrow 0$. Thus, we will, from now on, assume that $a(x) \geq \epsilon I$ for some $\epsilon > 0$ and all $x \in \mathbb{R}^N$. In particular, by Lemma 2.3.1, this means that $a^{\frac{1}{2}}$ is twice continuously differentiable and that the required estimate will follow once we show that

$$(*) \qquad\qquad \left\|\partial_e a^{\frac{1}{2}}(x)\right\|_{\mathrm{H.S.}} \leq N\sqrt{2\Lambda}$$

for all $x \in \mathbb{R}^N$ and $e \in \mathbb{S}^{N-1}$.

To prove $(*)$, let x be given, and choose an orthonormal basis $(e_1, \ldots, e_N)$ with respect to which $a(x)$ is diagonal. Then, from $a = a^{\frac{1}{2}} a^{\frac{1}{2}}$ and Leibniz's rule, one obtains

$$\partial_e a_{ij} = \partial_e a_{ij}^{\frac{1}{2}}(x)\left(\sqrt{a_{ii}(x)} + \sqrt{a_{jj}(x)}\right),$$

where $a_{ij} = \left(e_i, ae_j\right)_{\mathbb{R}^N}$. Hence, because $\sqrt{\alpha} + \sqrt{\beta} \geq \sqrt{\alpha + \beta}$ for all $\alpha, \beta \geq 0$,

$$\left|\partial_e a_{ij}^{\frac{1}{2}}(x)\right| \leq \frac{|\partial_e a_{ij}(x)|}{\sqrt{a_{ii}(x) + a_{jj}(x)}}.$$

To complete the proof of $(*)$, set

$$f_{\pm}(t) = \left(e_i \pm e_j, a(x + te)e_i \pm e_j\right)_{\mathbb{R}^N} \geq 0,$$

note that

$$\left|\partial_e a_{ij}(x)\right| = \frac{|f'_+(0) - f'_-(0)|}{4} \leq \frac{|f'_+(0)| + |f'_-(0)|}{4},$$

and apply (2.3.2) to get

$$(2.3.4) \qquad\qquad \left|\partial_e a_{ij}(x)\right| \leq \sqrt{2\Lambda\left(a_{ii}(x) + a_{jj}(x)\right)},$$

which, in conjunction with the preceding, leads first to

$$\left|\partial_e a_{ij}^{\frac{1}{2}}(x)\right| \leq \sqrt{2\Lambda}$$

and thence to (*).   □

As we mentioned earlier, even if a is real analytic as a function of x, it will not always be possible to find a smooth σ such that $a = \sigma\sigma^\top$. The reasons for this have their origins in classical algebraic geometry. Indeed, D. Hilbert showed that it is not possible to express every non-negative polynomial as a finite sum of squares of polynomials. After combining this fact with Taylor's theorem, one realizes that it rules out the existence of a smooth choice of σ. Of course, the problem arises only at the places where a degenerates: Away from degeneracies, as the proof of Lemma 2.3.1 shows, the entries of $a^{\frac{1}{2}}$ are analytic functions of the entries of a.

In spite of the preceding, it should be recognized that, although Lipschitz continuity is the best one can do in general, there are circumstances in which one can do better, especially if one is willing to consider square roots other than the non-negative, symmetric one. For example, consider $a : \mathbb{R}^2 \longrightarrow \mathrm{Hom}(\mathbb{R}^2; \mathbb{R}^2)$ given by $a(x)_{ij} = x_i x_j$. Because $a(x)^2 = |x|^2 a(x)$, $a^{\frac{1}{2}}(x) = |x|^{-1} a(x)$, which is not smooth at the origin. On the other hand, $\sigma(x) = \begin{pmatrix} x_1 \\ x_2 \end{pmatrix}$ is smooth everywhere, and $a(x) = \sigma(x)\sigma(x)^\top$.

2.4 Oleinik's Approach

As the preceding section makes clear, basing our theory on properties of σ instead of a creates a serious weakness in degenerate situations. Thus, it is important to know that there is a way of getting estimates of the sort in Corollary 2.2.8 without relying on the existence of a smooth σ. The key ideas here are due to O. Oleinik.

2.4.1. The Weak Minimum Principle: In this subsection, $a : \mathbb{R}^N \longrightarrow \mathrm{Hom}(\mathbb{R}^N; \mathbb{R}^N)$ and $b : \mathbb{R}^N \longrightarrow \mathbb{R}^N$ are bounded, measurable functions; $a(x)$ is symmetric and non-negative definite for each $x \in \mathbb{R}^N$; and L is defined from a and b as in (1.1.8).

LEMMA 2.4.1.   *If $u \in C^{1,2}((0,T] \times \mathbb{R}^N; \mathbb{R})$ is bounded below and satisfies*

$$(L - \partial_t)u \le 0 \text{ on } (0,T] \times \mathbb{R}^N \quad and \quad \lim_{t \searrow 0} \inf_{x \in B(0,R)} u(t,x) \ge 0 \text{ for } R > 0,$$

then $u \ge 0$ on $(0,T] \times \mathbb{R}^N$.

PROOF: We first show that if $(L - \partial_t)u < 0$ on $(0,T] \times \mathbb{R}^N$, then u cannot achieve a minimum there. To this end, suppose that u achieved a minimum at (t_0, x_0). Then, by the first and second derivative tests, $\partial_t u(t_0, x_0) \le 0$, $\nabla u(t_0, x_0) = 0$, and $\nabla^2 u(t_0, x_0)$ would be symmetric and non-negative definite. But this would lead to the contradiction that $(L - \partial_t)u(t_0, x_0) \ge 0$, since the product of two symmetric, non-negative matrices has a non-negative trace.

We now return to the original assumptions. For $\delta > 0$ and $\epsilon > 0$, set

$$u_{\epsilon,\delta}(t,x) = u(t,x) + \delta t + \epsilon e^t |x|^2.$$

Then

$$(L - \partial_t)u_{\epsilon,\delta}(t,x) \leq -\delta + \epsilon e^t \left[-|x|^2 + \text{Trace}\big(a(x)\big) + 2\big(x,b(x)\big)_{\mathbb{R}^N} \right]$$
$$\leq -\delta + \epsilon e^t \left[|b(x)|^2 + \text{Trace}\big(a(x)\big) \right].$$

Thus, for each $\delta > 0$, there exists an $\epsilon(\delta) > 0$ such that $(L - \partial_t)u_{\epsilon,\delta} < 0$ in $(0,T] \times \mathbb{R}^N$ when $\epsilon < \epsilon(\delta)$. In addition, for any $\delta > 0$ and $\epsilon > 0$, $u_{\epsilon,\delta}(t,x) \longrightarrow \infty$ uniformly in $t \in (0,T]$ as $|x| \to \infty$. Hence, if $u_{\epsilon,\delta}$ were to become negative somewhere in $(0,T] \times \mathbb{R}^N$, then there would exist a sequence $\{(t_n, x_n) : n \geq 1\} \subseteq (0,T] \times \mathbb{R}^N$ which converges to some $(t_0, x_0) \in [0,T] \times \mathbb{R}^N$ and for which

$$0 > \inf\{u_{\epsilon,\delta}(t,x) : (t,x) \in (0,T] \times \mathbb{R}^N\} = \lim_{n\to\infty} u_{\epsilon,\delta}(t_n, x_n).$$

Moreover, by the second assumption made about u, $t_0 > 0$, and so $(t_0, x_0) \in (0,T] \times \mathbb{R}^N$ would be a point at which $u_{\epsilon,\delta}$ achieves its minimum value. In particular, because this cannot happen when $\epsilon < \epsilon(\delta)$, we have now shown that $u_{\epsilon,\delta} \geq 0$ for $\epsilon < \epsilon(\delta)$. Now let $\epsilon \searrow 0$ and then $\delta \searrow 0$. $\quad\square$

THEOREM 2.4.2.   *Assume that $u \in C^{1,2}\big((0,T] \times \mathbb{R}^N; \mathbb{R}\big) \cap C_b\big([0,T] \times \mathbb{R}^N; \mathbb{R}\big)$ satisfies*

$$\big(L - \partial_t + c(t)\big)u \geq -g(t) \quad on \ (0,T] \times \mathbb{R}^N,$$

where c and g are continuous $\mathbb{R}$-valued functions on $[0,T]$. Then

$$u(t,x) \leq \|u(0,\cdot)\|_u e^{C(t)} + \int_0^t g(\tau) e^{C(t)-C(\tau)} \, d\tau,$$

where $C(t) \equiv \int_0^t c(\tau) \, d\tau$.

PROOF: Set

$$v(t,x) = \|u(0,\cdot)\|_u - u(t,x)e^{-C(t)} - \int_0^t g(\tau) e^{-C(\tau)} \, d\tau,$$

and check that $(L - \partial_t)v \leq 0$ on $(0,T] \times \mathbb{R}^N$ and $\underline{\lim}_{t\searrow 0} \inf_{x \in \mathbb{R}^N} v(t,x) \geq 0$. Hence, by Lemma 2.4.1, $v \geq 0$, which is equivalent to the asserted estimate. $\quad\square$

2.4.2. Oleinik's Estimate: Let a, b, and L be as in the preceding section. An important role in Oleinik's argument is played by the following.

LEMMA 2.4.3.   *Given any symmetric $H \in \text{Hom}(\mathbb{R}^N; \mathbb{R}^N)$ and $e \in \mathbb{S}^{N-1}$,*

$$\left(\sum_{i,j=1}^N \partial_e a(x)_{ij} H_{ij} \right)^2 \leq 4N^2 \Lambda \text{Trace}\big(Ha(x)H\big),$$

where $\Lambda \equiv \sup\{\|\partial_e^2 a(x)\|_{op} : x \in \mathbb{R}^N \ \& \ e \in \mathbb{S}^{N-1}\}$.

PROOF: Let $x \in \mathbb{R}^N$ be given, choose an orthogonal transformation $\mathcal{O}$ so that $\mathcal{O}^\top a(x)\mathcal{O}$ is diagonal, and set $\bar{a}(y) = \mathcal{O}^\top a(y)\mathcal{O}$ and $\bar{H} = \mathcal{O}^\top H\mathcal{O}$. Then

$$\sum_{ij} \partial_e a_{ij}(x) H_{ij} = \sum_{ij} \partial_e \bar{a}_{ij}(x) \bar{H}_{ij},$$

and so, by Schwarz's inequality and (2.3.4),

$$\left(\sum_{ij} \partial_e a_{ij}(x) H_{ij} \right)^2 \leq N^2 \sum_{ij} \left(\partial_e \bar{a}_{ij}(x) \right)^2 \bar{H}_{ij}^2 \leq 4N^2 \Lambda \sum_{ij} \bar{a}(x)_{ii} \bar{H}_{ij}^2$$

$$= 4N^2 \Lambda \operatorname{Trace}\big(Ha(x)H\big). \quad \square$$

Recall the notation

$$\partial_x^\alpha f = \frac{\partial^{\|\alpha\|} f}{\partial_{x_1}^{\alpha_1} \cdots \partial_{x_N}^{\alpha_N}} \quad \text{for } \alpha \in \mathbb{N}^N,$$

where $\|\alpha\| \equiv \sum_1^N \alpha_i$. In addition, define

$$\|f\|_{\mathrm{u}}^{(\ell)} = \left(\sum_{\|\alpha\| \leq \ell} \|\partial^\alpha f\|_{\mathrm{u}}^2 \right)^{\frac{1}{2}}$$

for $\ell \geq 0$. Finally, write $\beta \leq \alpha$ when $\beta_k \leq \alpha_k$ for each $1 \leq k \leq N$.

THEOREM 2.4.4. *Assume that a and b have $\ell \geq 2$ bounded continuous derivatives, and let $c \in C_{\mathrm{b}}^\ell(\mathbb{R}; \mathbb{R})$ be given. Given $u \in C_{\mathrm{b}}^{0,\ell+2}([0,T] \times \mathbb{R}^N; \mathbb{R})$ with $\partial^\alpha u \in C^{1,0}([0,T] \times \mathbb{R}^N; \mathbb{R})$ for each $\|\alpha\| \leq \ell$, set*

$$g(t,x) = \big(\partial_t - L - c\big)u(t,x).$$

Then

$$\|u(t, \cdot)\|_{\mathrm{u}}^{(\ell)} \leq A_\ell \Big[\|u(0, \cdot)\|_{\mathrm{u}}^{(\ell)} + \sup_{\tau \in [0,t]} \|g(\tau, \cdot)\|_{\mathrm{u}}^{(\ell)} \Big] e^{B_\ell t},$$

where $A_\ell < \infty$ and $B_\ell < \infty$ can be chosen to depend only on N, ℓ, $\|a\|_{\mathrm{u}}^{(\ell)}$, $\|b\|_{\mathrm{u}}^{(\ell)}$, and $\|c\|_{\mathrm{u}}^{(\ell)}$.

PROOF: For $\|\alpha\| \leq \ell$, set $u_\alpha = \partial^\alpha u$ and $g_\alpha = \partial^\alpha g$, and use Leibniz's rule and induction to check that

$$\partial_t u_\alpha = L u_\alpha + \frac{1}{2} \sum_k{}' \sum_{i,j=1}^N \alpha_k (\partial_{x_k} a_{ij}) \partial_{x_i} \partial_{x_j} u_{\hat{\alpha}^k} + \sum_{\beta \leq \alpha} c_{\alpha,\beta} u_\beta = -g_\alpha,$$

where $\sum_k{}'$ denotes summation over $1 \leq k \leq N$ with $\alpha_k > 0$; the $c_{\alpha,\beta}$'s are linear combinations of a, b, c and their derivatives up to order $\|\alpha\|$; and, when $\alpha_k > 0$, $\hat\alpha^k$ is obtained from α by replacing α_k by $\alpha_k - 1$ and leaving the other coordinates unchanged. Now remember that $Lf^2 = 2fLf + (\nabla f, a\nabla f)_{\mathbb{R}^N}$, and use this to see that if $1 \le \ell' \le \ell$ and $w_{\ell'} \equiv \sum_{\|\alpha\|=\ell'} u_\alpha^2$, then

$$
\partial_t w_{\ell'} = L w_{\ell'} + \sum_{\|\alpha\|=\ell'} {\sum_k}' \sum_{i,j=1}^N \alpha_k u_\alpha (\partial_{x_k} a_{ij}) \partial_{x_i} \partial_{x_j} u_{\hat\alpha^k}
$$

$$
- \sum_{\|\alpha\|=\ell'} (\nabla u_\alpha, a \nabla u_\alpha)_{\mathbb{R}^N} + 2 \sum_{\|\alpha\|=\ell'} \left(u_\alpha g_\alpha + \sum_{\beta \le \alpha} c_{\alpha,\beta} u_\alpha u_\beta \right).
$$

At first sight, the preceding looks bad: There are $(\ell' + 1)$st order derivatives appearing in places where they should not be. Oleinik's crucial observation is that Lemma 2.4.3 allows us to eliminate them. Namely, set $\Lambda = \sup_{(x,e)\in\mathbb{R}^N\times\mathbb{S}^{N-1}} \|\partial_e^2 a(x)\|_{\mathrm{op}}$, and conclude that

$$
\left({\sum_k}' \sum_{i,j=1}^N \alpha_k u_\alpha (\partial_{x_k} a_{ij}) \partial_{x_i} \partial_{x_j} u_{\hat\alpha^k} \right)^2
$$

$$
\le u_\alpha^2 \left({\sum_k}' \alpha_k^2 \right) {\sum_k}' \left(\sum_{i,j=1}^N (\partial_{x_k} a_{ij}) \partial_{x_i} \partial_{x_j} u_{\hat\alpha^k} \right)^2
$$

$$
\le 4(N\ell')^2 \Lambda u_\alpha^2 {\sum_k}' \mathrm{Trace}\left(\nabla^2 u_{\hat\alpha^k} a \nabla^2 u_{\hat\alpha^k} \right).
$$

In particular, this shows that there is a $C_{\ell'} < \infty$, depending only on N, ℓ', and $\|a\|_u^{(2)}$, such that

$$
\left(\sum_{\|\alpha\|=\ell'} {\sum_k}' \sum_{i,j=1}^N \alpha_k u_\alpha (\partial_{x_k} a_{ij}) \partial_{x_i} \partial_{x_j} u_{\hat\alpha^k} \right)^2
$$

$$
\le 4 C_{\ell'} w_{\ell'} \sum_{\|\alpha\|=\ell'} (\nabla u_\alpha, a \nabla u_\alpha)_{\mathbb{R}^N},
$$

and therefore

$$
\sum_{\|\alpha\|=\ell'} {\sum_k}' \sum_{i,j=1}^N \alpha_k u_\alpha (\partial_{x_k} a_{ij}) \partial_{x_i} \partial_{x_j} u_{\hat\alpha^k} - \sum_{\|\alpha\|=\ell'} (\nabla u_\alpha, a \nabla u_\alpha)_{\mathbb{R}^N} \le C_{\ell'} w_{\ell'}.
$$

Using the preceding estimate in the equation for $\partial_t w_{\ell'}$, we arrive at

$$
(L - \partial_t + C_{\ell'}) w_{\ell'} \ge -2 \sum_{\|\alpha\|=\ell'} \sum_{\beta \le \alpha} c_{\alpha,\beta} u_\alpha u_\beta - 2 \sum_{\|\alpha\|=\ell'} u_\alpha g_\alpha.
$$

Note that

$$
2\left|\sum_{\|\alpha\|=\ell'}\sum_{\beta\leq\alpha}c_{\alpha,\beta}u_\alpha u_\beta\right| \leq 2\left(\sum_{\substack{\|\alpha\|=\|\beta\|=\ell'}}\|c_{\alpha,\beta}\|_{\mathrm{u}}^2\right)^{\frac{1}{2}}w_{\ell'}
$$
$$
+ \left(\sum_{\substack{\alpha=\ell'\\ \beta<\ell'}}\|c_{\alpha,\beta}\|_{\mathrm{u}}^2\right)^{\frac{1}{2}}\left(\|u(t,\,\cdot\,)\|_{\mathrm{u}}^{(\ell'-1)}\right)^2,
$$

whereas

$$
2\left|\sum_{\|\alpha\|=\ell'}u_\alpha g_\alpha\right| \leq w_{\ell'} + \left(\|g(t,\,\cdot\,)\|_{\mathrm{u}}^{(\ell')}\right)^2.
$$

Hence, after adjusting $C_{\ell'}$, we have

(**) $\qquad (L-\partial_t + C_{\ell'})w_{\ell'} \geq -C_{\ell'}\left(\|u(t,\,\cdot\,)\|_{\mathrm{u}}^{(\ell'-1)}\right)^2 - \left(\|g(t,\,\cdot\,)\|_{\mathrm{u}}^{(\ell')}\right)^2.$

Given (**), one proceeds by induction on ℓ'. By Theorem 2.4.2, we know that

$$
\|u(t,\,\cdot\,)\|_{\mathrm{u}} \leq e^{B_0 t}\left(\|u(0,\,\cdot\,)\|_{\mathrm{u}} + \sup_{\tau\in[0,t]}\|g(\tau,\,\cdot\,)\|_{\mathrm{u}}\right),
$$

where $B_0 = \|c\|_{\mathrm{u}}$. Next, one applies Theorem 2.4.2 to (**) to carry out each inductive step. $\square$

COROLLARY 2.4.5. *Under the conditions in Theorem 2.4.4, for each $\varphi \in C_{\mathrm{b}}^\ell(\mathbb{R}^N;\mathbb{R})$ there is a sequence $\{u_n : n \geq 1\} \subseteq C^{1,\infty}\big([0,\infty)\times\mathbb{R}^N;\mathbb{R}\big)$ with the properties that $\lim_{n\to\infty}\|u_n(0,\,\cdot\,) - \varphi\|_{\mathrm{u}} = 0$ and*

$$
\begin{aligned}
&\sup_{t\in[0,T]}\sup_{n\geq 1}\|u_n(t,\,\cdot\,)\|_{\mathrm{u}}^{(\ell)} < \infty\\
&\sup_{t\in[0,T]}\|(L-\partial_t)u_n(t,\,\cdot\,)\|_{\mathrm{u}} \longrightarrow 0
\end{aligned}
\qquad \text{for all } T\in(0,\infty).
$$

Moreover, there exists a unique $u_\varphi \in C_{\mathrm{b}}(\mathbb{R}^N;\mathbb{R})$ to which any such sequence converges uniformly on $[0,T]\times\mathbb{R}^N$ for each $T\in(0,\infty)$. In particular, $u_\varphi \in C^{1,\ell-1}\big([0,\infty)\times\mathbb{R}^N;\mathbb{R}\big)$ and, for each $T>0$ and $0\leq k<\ell$, $\nabla^k u_\varphi \upharpoonright [0,T]\times\mathbb{R}^N$ is bounded and uniformly Lipschitz continuous. Finally, if $\ell\geq 3$, then $u_\varphi \in C^{1,2}\big([0,\infty)\times\mathbb{R}^N;\mathbb{R}\big)$ and u_φ solves (2.0.1).

PROOF: Choose $\rho \in C_{\mathrm{c}}^\infty\big(B(0,1);[0,\infty)\big)$ with total integral 1; set $\rho_k(x) = k^N\rho(kx)$; take $a_k(x) = \rho_k \star a(x) + \frac{1}{k}I$, $b_k = \rho_k \star b$, and $\varphi_k = \rho_k \star \varphi$; and define L_k accordingly from a_k and b_k. Because, by Lemma 2.3.1, the positive definite, symmetric square root σ_k of a_k has bounded derivatives of all orders, Corollary 2.1.6 guarantees that L_k determines a transition probability function $(t,x) \rightsquigarrow P^k(t,x)$ and Corollary 2.2.8 says that $(t,x) \rightsquigarrow u_k(t,x) = \langle \varphi_k, P^k(t,x) \rangle$ is smooth and solves $\partial_t u = L_n u$. Furthermore, because φ_k, σ_k, b_k, and their derivatives are all bounded, the last part of Corollary 2.2.8 says that, for each $k \geq 1$, $\sup_{t \in [0,T]} \|u_k(t, \cdot)\|_u^{(\ell)} < \infty$. But, because the derivatives of a_k, b_k, and φ_k up to order ℓ are bounded independent of $k \geq 1$, Theorem 2.4.4 allows us to say that $\sup_{k \geq 1} \sup_{t \in [0,T]} \|u_k(t, \cdot)\|_u^{(\ell)} < \infty$. In particular, $g_k \equiv (L - \partial_t) u_k \longrightarrow 0$ uniformly on $[0,T] \times \mathbb{R}^N$. Finally, by Theorem 2.4.2, for $1 \leq k_1 < k_2$,

$$|u_{k_2}(t,x) - u_{k_1}(t,x)| \leq \|\varphi_{k_2} - \varphi_{k_1}\|_u + \int_0^t \|g_{k_2}(\tau) - g_{k_1}(\tau)\|_u \, d\tau,$$

and so there is a $u_\varphi \in C_b([0,\infty) \times \mathbb{R}^N; \mathbb{R})$ to which $\{u_k : k \geq 1\}$ converges uniformly on $[0,T] \times \mathbb{R}^N$ for each $T \in (0,\infty)$. □

As an immediate consequence of Corollaries 2.4.5 and 2.1.7, we have the following.

THEOREM 2.4.6.   *Assume that a and b are twice continuously differentiable and that they and their derivatives are bounded. Then all the conclusions in Theorem 2.4.6 and Corollary 2.1.6 hold.*

2.5 The Adjoint Semigroup

Up until now we have not given an operator-theoretic interpretation of our transition probabilities $(t,x) \rightsquigarrow P(t,x)$. However, for the purposes of this section, it will be important for us to do so now. Thus, define

$$\mathbf{P}_t \varphi(x) = \int \varphi(y) \, P(t,x,dy)$$

for Borel measurable $\varphi : \mathbb{R}^N \longrightarrow \mathbb{R}$ which are bounded below. Because $(t,x) \rightsquigarrow P(t,x)$ is continuous, it is clear that $(t,x) \rightsquigarrow \mathbf{P}_t \varphi$ is Borel measurable and that it is continuous when $\varphi \in C_b(\mathbb{R}^N; \mathbb{R})$. In fact, because $P(t,x)$ is a probability measure, $\varphi \geq 0 \implies \mathbf{P}_t \varphi \geq 0$, $\mathbf{P}_t \mathbf{1} = \mathbf{1}$, and so $\|\mathbf{P}_t \varphi\|_u \leq \|\varphi\|_u$. Moreover, the Chapman–Kolmogorov equation for $(t,x) \rightsquigarrow P(t,x)$ becomes the *semigroup property* $\mathbf{P}_{s+t} = \mathbf{P}_t \circ \mathbf{P}_t$ for the $\mathbf{P}_t$'s. Because it preserves non-negativity and the constant function $\mathbf{1}$, the family $\{\mathbf{P}_t : t \geq 0\}$ is said to be a *Markov semigroup*.

We now want to examine the $\mathbf{P}_t$'s as operators on $L^2(\mathbb{R}^N; \mathbb{R})$. In particular, we want to work under conditions which guarantee that $\mathbf{P}_t$ is a bounded operator there and to describe its adjoint $\mathbf{P}_t^\top$. Namely, we will assume that $\sigma : \mathbb{R}^N \longrightarrow \mathrm{Hom}(\mathbb{R}^M; \mathbb{R}^N)$ and $b : \mathbb{R}^N \longrightarrow \mathbb{R}^N$ are bounded smooth functions with bounded derivatives of all orders and will take L to be the operator given by (1.1.8) with $a = \sigma \sigma^\top$. Then the formal adjoint

$L^\top$ of L is the operator

$$
\begin{aligned}
L^\top \varphi &= \frac{1}{2} \sum_{i,j=1}^{N} \partial_{x_i} \partial_{x_j} \left(a_{ij} \varphi \right) - \sum_{i=1}^{N} \partial_{x_i} \left(b_i \varphi \right) \\
&= \frac{1}{2} \sum_{i,j=1}^{N} a_{ij} \partial_{x_i} \partial_{x_j} \varphi + \sum_{i=1}^{N} b_i^\top \partial_{x_i} \varphi + V^\top \varphi,
\end{aligned}
$$

(2.5.1)

where

(2.5.2) $\quad b_i^\top = -b_i + \sum_{j=1}^{N} \partial_{x_j} a_{ij} \quad \text{and} \quad V^\top = \frac{1}{2} \sum_{i,j=1}^{N} \partial_{x_i} \partial_{x_j} a_{ij} - \sum_{i=1}^{N} \partial_{x_i} b_i.$

That is, for $\varphi, \psi \in C^2(\mathbb{R}^N; \mathbb{C})$, one can easily check that

(2.5.3) $\qquad \left(\varphi, L\psi \right)_{L^2(\mathbb{R}^N)} = \left(L^\top \varphi, \psi \right)_{L^2(\mathbb{R}^N)}$

if either φ or ψ has compact support. Next, referring to §1.3.2, let $(t, x) \rightsquigarrow P^\top(t, x)$ be the transition function associated with $L^\top$. Our main goal is to show that the adjoint $\mathbf{P}_t^\top$ of $\mathbf{P}_t$ as an operator on $L^2(\mathbb{R}^N; \mathbb{R})$ is given by

(2.5.4) $\qquad \mathbf{P}_t^\top \psi(y) = \int \psi(x) \, P^\top(t, y, dx).$

2.5.1. The Adjoint Transition Function: Proceeding as in §1.3.2, we now define $\left(t, (y, \xi) \right) \in [0, \infty) \times (\mathbb{R}^N \times \mathbb{R}) \longmapsto \bar{P}_n^\top \left(t, (y, \xi) \right) \in \mathbf{M}_1(\mathbb{R}^N)$ relative to (cf. (2.5.2))

$$
\bar{\sigma}^\top(y, \xi) = \begin{pmatrix} \sigma(y) \\ 0 \end{pmatrix} \quad \text{and} \quad \bar{b}^\top(y, \xi) = \begin{pmatrix} b^\top(y) \\ V^\top(y) \end{pmatrix}.
$$

If $(t, y) \in [0, \infty) \times \mathbb{R}^N \longmapsto P^\top(t, y) \in \mathbf{M}(\mathbb{R}^N)$ is given by

(2.5.5) $\qquad P^\top(t, y, \Gamma) = \int_{\Gamma \times \mathbb{R}} e^\eta \, \bar{P}^\top \left(t, (y, \xi), dx \times d\eta \right),$

where $\bar{P}^\top \left(t, (y, \xi) \right) = \lim_{n \to \infty} \bar{P}_n^\top \left(t, (y, \xi) \right)$, then

$$
e^{-t \| (V^\top)^- \|_{\mathrm{u}}} \le P^\top(t, y, \mathbb{R}^N) \le e^{t \| (V^\top)^+ \|_{\mathrm{u}}},
$$

(2.5.6)

$$
P^\top(s + t, x) = \int P^\top(t, y) \, P^\top(s, x, dy), \quad \text{and}
$$

$$
\langle \psi, P^\top(t, x) \rangle - \psi(y) = \int_0^t \langle L^\top \psi, P^\top(\tau, x) \rangle \, d\tau
$$

for $\psi \in C^2(\mathbb{R}^N; \mathbb{C})$ with bound second order derivatives. Furthermore, if $\psi \in C_{\mathrm{b}}^2(\mathbb{R}^N; \mathbb{C})$ and

$$u_\psi^\top(t, y) = \langle \psi, P^\top(t, y) \rangle,$$

then $u_\psi^\top \in C^{1,2}([0,\infty) \times \mathbb{R}^N; \mathbb{C})$, $\partial_t u_\psi^\top = L^\top u_\psi^\top$, and, for each $T > 0$, $u_\psi^\top \upharpoonright [0,T] \times \mathbb{R}^N \in C_{\mathrm{b}}^{1,2}([0,T] \times \mathbb{R}^N; \mathbb{C})$. To check these last assertions, first note that

$$u_\psi^\top(t, y) = \langle \bar{\psi}, \bar{P}(t, (y, 0)) \rangle,$$

where $\bar{\psi}(y, \xi) = e^\xi \psi(y)$, remember (cf. (1.3.5))

$$\bar{P}^\top\big(t, (y, 0), \mathbb{R}^N \times [-t\|V^\top\|_{\mathrm{u}}, t\|V^\top\|_{\mathrm{u}}]\complement\big) = 0,$$

and apply Theorem 2.2.7 to check the required smoothness properties. Then, use (2.5.6) to justify

$$\partial_t u_\psi^\top(t, y) = \lim_{h \searrow 0} \frac{1}{h} \int_0^h \langle L^\top u^\psi(t), P^\top(\tau, y) \rangle \, d\tau = L^\top u_\psi^\top(t, y).$$

There is one more matter with which we must deal before we can prove (2.5.4). Indeed, as yet, we have not discussed the integrability properties of either $\mathbf{P}_t \varphi$ or $\mathbf{P}_t^\top \psi$. For this purpose, we will use the following simple lemma.

LEMMA 2.5.7.   For each $q \in (0, \infty)$ there exists a $C_q < \infty$ such that

$$\int \big(1 + |y - x|^2\big)^{\frac{q}{2}} P(t, x, dy) \vee \int \big(1 + |y - x|^2\big)^{\frac{q}{2}} P^\top(t, x, dy) \le C_q e^{C_q t}.$$

In particular, if $\varphi, \psi \in C_{\mathrm{c}}(\mathbb{R}^N; \mathbb{R})$, then

$$\lim_{|x| \to \infty} (1 + |x|^q) \sup_{\tau \in [0, t]} \big|\mathbf{P}_\tau \varphi(x)\big| \vee \big|\mathbf{P}_\tau^\top \psi(x)\big| = 0$$

for all $t \in [0, \infty)$ and $q \in (0, \infty)$.

PROOF: For any $q \in \mathbb{R}$, set

$$w_q(t) = \sup_{x \in \mathbb{R}^N} \int \big(1 + |y - x|^2\big)^{\frac{q}{2}} P(t, x, dy).$$

Obviously, $w_q(t) \le 1$ when $q \in (-\infty, 0]$. Moreover, because

$$\frac{d}{dt} \int \big(1 + |y - x|^2\big)^{\frac{q}{2}} P(t, x, dy)$$

$$= \int \left[\frac{q(q - 2)\big(y - x, a(y)(y - x)\big)_{\mathbb{R}^N}}{2(1 + |y - x|^2)} + \frac{q}{2} \mathrm{Trace}\big(a(y)\big) \right.$$

$$\left. + q\big(b(y), y - x\big)_{\mathbb{R}^N} \right] \big(1 + |y - x|^2\big)^{\frac{q}{2} - 1} P(t, x, dy),$$

we can find a $C < \infty$, depending only on the bounds on a and b, such that

$$w_q(t) \le 1 + C(1 + q^2) \int_0^t \left(w_{q-1}(\tau) + w_{q-2}(\tau) \right) d\tau.$$

Thus, proceeding by induction on $m \in \mathbb{N}$, one gets the required estimate on $P(t, x)$ for $m < q \le (m+1)$. Essentially the same argument works for $P^\top(t, x)$. The only change is that one picks up an extra factor of $e^{t\|V^\top\|_u}$.

Given the preceding, the second part follows easily. Namely, if φ vanishes outside of $B(0, R)$, then, for $q \in (0, \infty)$ and $|x| > R$,

$$|\mathbf{P}_t \varphi(x)| \le \frac{\|\varphi\|_u}{(|x| - R)^q} \int |y - x|^q \, P(t, x, dy),$$

and similarly for $|\mathbf{P}_t^\top \psi|$. $\square$

THEOREM 2.5.8. *If $(t, x) \leadsto P^\top(t, x)$ is given by (2.5.5) and $\mathbf{P}_t^\top$ denotes the adjoint of $\mathbf{P}_t$ as an operator on $L^2(\mathbb{R}^N; \mathbb{R})$, then (2.5.4) holds. In particular, for any $q \in [1, \infty)$,*

$$\|\mathbf{P}_t \varphi\|_{L^q(\mathbb{R}^N; \mathbb{R})} \le e^{\frac{t}{q}\|V^\top\|_u} \|\varphi\|_{L^q(\mathbb{R}^N; \mathbb{R})}.$$

PROOF: To prove the first assertion, let $\varphi, \psi \in C_c^\infty(\mathbb{R}^N; \mathbb{R})$ be given, and set $u(t) = \mathbf{P}_t \varphi$ and $v(t) = \mathbf{P}_t^\top \psi$. Then, by Lemma 2.5.7 and the last part of Corollary 2.2.8, we know that, for any $t > 0$ and $q \in [1, \infty)$,

$$\sup_{(\tau, x) \in [0, t] \times \mathbb{R}^N} \left[(1 + |x|)^q \left(|u(\tau, x)| \vee |v(\tau, x)| \right) \right] \vee |\nabla u(\tau, x)| < \infty.$$

Now choose $\eta \in C_c^\infty\big(B(0, 2); [0, 1]\big)$ so that $\eta = 1$ on $B(0, 1)$, and set $\eta_R(x) = \eta(R^{-1}x)$. Then, by (2.5.3), for any $R > 0$,

$$\frac{d}{d\tau} \int \eta_R(x) u(\tau, x) v(t - \tau, \xi) \, dx$$

$$= \int \eta_R(x) v(t - \tau) L u(\tau, x) \, dx - \int \eta_R(x) u(\tau, x) L^\top v(t - \tau, x) \, dx$$

$$= - \int v(t - \tau, x) \Big(\big(\nabla \eta_R(x), a(x) \nabla u(\tau, x) \big)_{\mathbb{R}^N} + u(\tau, x) L \eta_R(x) \Big) \, dx,$$

and the above estimates on u and v are more than sufficient to show that, as $R \to \infty$, the right-hand side tends to 0 uniformly in $\tau \in [0, t]$ and therefore that

$$\big(\mathbf{P}_t^\top \psi, \varphi \big)_{L^2(\mathbb{R}^N; \mathbb{R})} - \big(\psi, \mathbf{P}_t \varphi \big)_{L^2(\mathbb{R}^N; \mathbb{R})} = 0$$

first for $\varphi, \psi \in C_c^\infty(\mathbb{R}^N; \mathbb{R})$, and then, after mollifying and taking limits, for $\varphi \in C_c(\mathbb{R}^N; \mathbb{R})$. Hence, once we prove the estimate in the second assertion, we will be done.

To prove the second assertion, let $\varphi \in C_{\mathrm{c}}(\mathbb{R}^N; \mathbb{R})$ be given. Then, by Jensen's inequality, $|\mathbf{P}_t\varphi|^q \leq \mathbf{P}_t|\varphi|^q$, and so, for each $R > 0$,

$$\int \eta_R(x)|\mathbf{P}_t\varphi(x)|^q \, dx \leq \int \eta_R(x)\left[\mathbf{P}_t|\varphi|^q\right](x), \, dx$$

$$= \int \mathbf{P}_t^\top \eta_R(x)|\varphi(x)|^q \, dx \leq e^{t\|(V^\top)^+\|_u}\|\varphi\|_{L^q(\mathbb{R}^N;\mathbb{R})}^q. \quad \square$$

2.6 Historical Notes and Commentary

Except for those in § 2.4, the techniques in this chapter, like the ones in Chapter 1, are natural stochastic analogs of techniques which are familiar in the theory of ordinary differential equations. Basically, they simply confirm the principle, well known in the analysis of flows, that the dependence of the solution on its initial value will be as smooth as the vector field generating the flow. As such, they entirely miss possible smoothing provided by ellipticity when it is present. Thus, from the standpoint of aficionados in elliptic equations, the results in this chapter look rather pale, even though most experts in elliptic theory would be hard put to prove them. The idea of differentiating Itô's representation with respect to the starting point seems to have been introduced first by I.I. Gikhman (cf. [20]), although the first time I encountered it was in H.P. McKean's [39].

Besides their inability to produce elliptic regularity results, the techniques in § 2.2 suffer from their dependence on Itô's representation of solutions to Kolmogorov's forward equation. Specifically, his theory requires one to take the square root of the diffusion matrix. As was pointed out in § 2.3, taking the square root of a non-negative function which can vanish is a dicey business, one which will, in general, destroy regularity properties possessed by the original function. For this reason, the contents of § 2.4, which are taken from [46], are significant and may come as a surprise to practitioners of Itô's art. Indeed, I know no way to prove them using only Itô's theory.

CHAPTER 3

Preliminary Elliptic Regularity Results

So far, all our results depend only on the smoothness properties of the coefficients a and b. As a consequence, the regularity results which we obtained in Chapter 2 are modest: They simply say that the heat flow "does no harm." That is, if a, b, and φ are all sufficiently smooth, then the u_φ in (2.0.2) is smooth as well. They do not say that the heat flow "improves" data in the sense that $u_\varphi(t, \cdot)$ for $t > 0$ is smoother than φ. Of course, seeing as $u_\varphi(t, x) = \varphi(x + t\xi)$ when $a \equiv 0$ and $b \equiv \xi$, we should not have expected any such improvement in the absence of further conditions on a.

The canonical example of a heat flow which is smoothing is the classic heat flow, when $a \equiv I$ and $b \equiv 0$, in which case

$$u_\varphi(t, x) = (2\pi t)^{-\frac{N}{2}} \int e^{-\frac{|y-x|^2}{2t}} \varphi(y) \, dy.$$

Thus, one should suspect that the additional condition needed on a is a non-degeneracy assumption. More important, one can guess the role that non-degeneracy should play. Indeed, we already know (cf. Corollary 2.2.8) that

$$\nabla u_\varphi(t, x) = \int \nabla \varphi(y) J \, P^{(1)}\big(t, (x, I), dy \times dJ\big).$$

In order to use this formula to show that $u_\varphi(t, \cdot)$ may be differentiable even when φ itself is not, one has to remove the derivatives of φ from the right-hand side, and the only way to do so is to move them, via integration by parts, to $P^{(1)}\big(t, (x, I)\big)$. But before integrating by parts, we must check that $P^{(1)}\big(t, (x, I), dy \times dJ\big)$ can be differentiated as a function of y, and the only reason why this might be possible is that $P(t, x)$ is built out of the smooth measure γ. In other words, we are hoping to show that $P(t, x)$ inherits the smoothness of γ. But γ enters $P(t, x)$ only after multiplication by $\sigma = a^{\frac{1}{2}}$, and so, when a is degenerate the smoothness of γ might not get transferred. The purpose of this chapter is to show that the smoothness of γ does get transferred to $P(t, x)$ when a is non-degenerate.

3.1 Integration by Parts

As we just said, the basic tool which will allow us to capitalize on the non-degeneracy of a is integration by parts. In this section, we will develop the requisite integration by parts formulas on which our analysis will rest.

3.1.1. Gaussian Integration by Parts: We return to the notation introduced in § 2.2.1. Thus, $\Omega = (\mathbb{R}^M)^{\mathbb{Z}^+}$ and (cf. (1.1.14)) $\Gamma = \gamma^{\mathbb{Z}^+}$. Given a Hilbert space E, we use $C_0^\infty(\Omega; E)$ to denote the space of smooth, E-valued functions Φ on Ω which depend on only a finite number of coordinates and write $D_m\Phi$ to denote the gradient of Φ with respect to the mth coordinate. Given an $\mathbf{h} = (h_1, \ldots, h_m, \ldots) : \Omega \longrightarrow \Omega$, we will say that $\mathbf{h}$ is *adapted function* if h_1 is constant and, for $m \geq 2$, $\boldsymbol{\omega} \rightsquigarrow h_m(\boldsymbol{\omega})$ is a measurable function of $(\omega_1, \ldots, \omega_{m-1})$. Finally, given $\Phi \in C_0^\infty(\Omega; E)$ and a measurable $\mathbf{h} : \Omega \longrightarrow \Omega$, define $\mathbf{D_h}\Phi : \Omega \longrightarrow E$ by

$$\mathbf{D_h}\Phi(\boldsymbol{\omega}) \equiv \frac{d}{d\xi}\Phi(\boldsymbol{\omega} + \xi\mathbf{h})\big|_{\xi=0} = \sum_{m=1}^\infty \left(D_m\Phi(\boldsymbol{\omega}), h_m(\boldsymbol{\omega})\right)_{\mathbb{R}^M}.$$

LEMMA 3.1.1.   *Let $\Phi \in C_0^\infty(\Omega; E)$, and assume that $D_m\Phi = 0$ for $m \geq m_\Phi$. Next, let $\mathbf{h} : \Omega \longrightarrow \Omega$ be an adapted function. Finally, assume that*

$$\int \left(\sum_{m=1}^{m_\Phi} \left(|D_m\Phi(\boldsymbol{\omega})|^2 + |h_m(\boldsymbol{\omega})|^2\right)\right) \Gamma(d\boldsymbol{\omega}) < \infty.$$

Then

$$\int \mathbf{D_h}\Phi(\boldsymbol{\omega}) \, \Gamma(d\boldsymbol{\omega}) = \int \left(\sum_{m=1}^{m_\Phi} \left(h_m, \omega_m\right)_{\mathbb{R}^M}\right) \Phi(\boldsymbol{\omega}) \, \Gamma(d\boldsymbol{\omega}).$$

PROOF: Clearly, it suffices for us to prove that, for each $1 \leq m \leq m_\Phi$,

$$(*) \qquad \int \left(D_m\Phi(\boldsymbol{\omega}), h_m(\boldsymbol{\omega})\right)_{\mathbb{R}^M} \Gamma(d\boldsymbol{\omega}) = \int \left(h_m(\boldsymbol{\omega}), \omega_m\right)_{\mathbb{R}^M} \Phi(\boldsymbol{\omega}) \, \Gamma(d\boldsymbol{\omega}).$$

Moreover, by an obvious approximation argument, while proving (*) we may and will assume that Φ, $D_m\Phi$, and h_m are all bounded. Proceeding under these conditions, first note that the left-hand side of (*) equals

$$\int \left(\int \left(D_m\Phi(\boldsymbol{\omega}), h_m(\boldsymbol{\omega})\right)_{\mathbb{R}^M} \gamma(d\omega_m)\right) \Gamma(d\boldsymbol{\omega}).$$

Next, remember that h_m does not depend on ω_m, and use integration by parts to check that

$$\int \left(D_m\Phi(\boldsymbol{\omega}), h_m(\boldsymbol{\omega})\right)_{\mathbb{R}^M} \gamma(d\omega_m) = \int \left(h_m(\boldsymbol{\omega}), \omega_m\right)_{\mathbb{R}^M} \Phi(\boldsymbol{\omega}) \, \gamma(d\omega_m).$$

Finally, observe that the right-hand side of (*) equals

$$\int \left(\int \big(h_m(\boldsymbol{\omega}), \omega_m \big)_{\mathbb{R}^M} \Phi(\boldsymbol{\omega}) \, \gamma(d\omega_m) \right) \Gamma(d\boldsymbol{\omega}). \quad \square$$

3.1.2. A General Formula: Assume that $\sigma : \mathbb{R}^N \longrightarrow \mathrm{Hom}(\mathbb{R}^M; \mathbb{R}^N)$ and $b : \mathbb{R}^N \longrightarrow \mathbb{R}^N$ are smooth, slowly increasing functions; define $X_n : [0, \infty) \times \mathbb{R}^N \times \Omega \longrightarrow \mathbb{R}^N$ accordingly, as in (2.2.1); set $X_n^1(t, x, \boldsymbol{\omega}) = \nabla X_n(t, x, \boldsymbol{\omega})$; and assume that, for each $r \in [1, \infty)$ and $T \in (0, \infty)$, there exists a $C_r < \infty$ such that

$$\sup_{n \geq 0} \sup_{t \in [0,T]} \int \big(|X_n(t, x, \boldsymbol{\omega})|^2 + \|X_n^1(t, x, \boldsymbol{\omega})\|_{\mathrm{H.S.}}^2 \big)^r \Gamma(d\boldsymbol{\omega})$$

(3.1.2)

$$\leq C_r(1 + |x|^{2r}).$$

Next, referring to § 1.2 and § 2.2.1, define $(t, x) \in [0, \infty) \times \mathbb{R}^N \longmapsto P_n(t, x) \in \mathbf{M}_1(\mathbb{R}^N)$ and $\big(t, (x, J) \big) \in [0, \infty) \times (\mathbb{R}^N \times E_1) \longmapsto P_n^{(1)}\big(t, (x, J) \big) \in \mathbf{M}_1(\mathbb{R}^N \times E_1)$, as we did there. In particular, $P_n(t, x)$ and $P_n^{(1)}(t, (x, I))$ are the distributions of, respectively, $\boldsymbol{\omega} \rightsquigarrow X_n(t, x, \boldsymbol{\omega})$ and $\boldsymbol{\omega} \rightsquigarrow \big(X_n(t, x, \boldsymbol{\omega}), X_n^1(t, x, \boldsymbol{\omega}) \big)$ under Γ. We will assume that the $P_n^{(1)}$'s converge to a continuous $\big(t, (x, J) \big) \rightsquigarrow P^{(1)}\big(t, (x, J) \big) \big)$ in the sense that

$$\langle \varphi_n, P_n^{(1)}(t_n, x_n, J_n) \rangle \longrightarrow \langle \varphi, P^{(1)}\big(t, (x, J) \big) \rangle$$

if $(t_n, x_n, J_n) \to (t, x, J)$ in $[0, \infty) \times \mathbb{R}^N \times E_1$ and $\{\varphi_n : n \geq 0\} \subseteq C\big(\mathbb{R}^N \times E_1; \mathbb{C} \big)$ is a uniformly slowly increasing sequence which converges to φ uniformly on compacts. Finally, set $P(t, x, A) = P^{(1)}\big(t, (x, I), A \times E_1 \big) = \lim_{n \to \infty} P_n(t, x, A)$.

THEOREM 3.1.3.   *Suppose that* $\xi \in \mathbb{S}^{N-1}$ *has the property that* $(\xi, b)_{\mathbb{R}^N} \equiv 0$, *and let* $\varphi \in C^1(\mathbb{R}^N; \mathbb{C})$ *be a function whose derivatives are slowly increasing. If* $(\xi, x)_{\mathbb{R}^N} = 0$, *then*

$$\int_0^t \left(\int \left(\int \big(\nabla\varphi(x''), J''a(x')\xi \big)_{\mathbb{R}^N} \right. \right.$$

(3.1.4)

$$\left. \left. \times P^{(1)}\big(t - \tau, (x', I), dx'' \times dJ'' \big) \right) P(\tau, x, dx') \right) d\tau$$

$$= \int \varphi(y)\big(\xi, y \big)_{\mathbb{R}^N} P(t, x, dy),$$

where $a = \sigma\sigma^{\top}$.

PROOF: First note that it suffices to prove (3.1.4) for a dense set of $t \in (0, \infty)$. Hence, from now on, we will assume that $2^{n_0}t \in \mathbb{Z}^+$ for some $n_0 \in \mathbb{N}$. Next, let $n \geq n_0$ be given, and define $\mathbf{h}_n : \Omega \longrightarrow \Omega$ so that

$$h_{n,m}(\boldsymbol{\omega}) = 2^{-\frac{n}{2}}\sigma\big(X_n\big((m-1)2^{-n}, x, \boldsymbol{\omega}\big)\big)^\top \xi \quad \text{for } m \in \mathbb{Z}^+.$$

Obviously, $\mathbf{h}_n$ is adapted. In order to compute $\mathbf{D}_{\mathbf{h}_n}X_n(t, x, \boldsymbol{\omega})$, first note that

$$X_n(t, x, \boldsymbol{\omega}) = X_n\big(t - m2^{-n}, X_n(m2^{-n}, x, \boldsymbol{\omega}), S^m\boldsymbol{\omega}\big),$$

where $S^m : \Omega \longrightarrow \Omega$ is the shift map defined by

$$S^m(\omega_1, \ldots, \omega_j, \ldots) = (\omega_{m+1}, \ldots, \omega_{m+j}, \ldots).$$

Thus,

$$\begin{aligned}
&D_m X_n(t, x, \boldsymbol{\omega}) \\
&= 2^{-\frac{n}{2}} X_n^1\big(t - m2^{-n}, X_n(m2^{-n}, x, \boldsymbol{\omega}), S^m\boldsymbol{\omega}\big)\sigma\big(X_n\big((m-1)2^{-n}, x, \boldsymbol{\omega}\big)\big),
\end{aligned}$$

and so

$$\begin{aligned}
&\mathbf{D}_{\mathbf{h}_n}\varphi\big(X_n(t, x, \boldsymbol{\omega})\big) \\
&= 2^{-n}\sum_{m=1}^{2^n t}\Big(\nabla\varphi\big(X_n(t, x, \boldsymbol{\omega})\big), X_n^1\big(t - m2^{-n}, X_n(m2^{-n}, x, \boldsymbol{\omega}), S^m\boldsymbol{\omega}\big) \\
&\qquad\qquad \times a\big(X_n\big((m-1)2^{-n}, x, \boldsymbol{\omega}\big)\big)\xi\Big)_{\mathbb{R}^N}.
\end{aligned}$$

Next observe that our integrability assumptions are more than enough to guarantee that the difference between the right-hand side of the preceding and the same expression with $a\big(X_n\big((m-1)2^{-n}, x, \boldsymbol{\omega}\big)\big)$ replaced by $a\big(X_n(m2^{-n}, x, \boldsymbol{\omega})\big)$ goes to 0 in $L^1\big(\Gamma; \text{Hom}(\mathbb{R}^N; \mathbb{R}^N)\big)$ as $n \to \infty$. At the same time, because $X_n(t, x, \boldsymbol{\omega}) = X_n\big(t - m2^{-n}, X_n(m2^{-n}, x, \boldsymbol{\omega}, S^m\boldsymbol{\omega})\big)$,

$$\begin{aligned}
&\int \Big(\nabla\varphi\big(X_n(t, x, \boldsymbol{\omega})\big), X_n^1\big(t - m2^{-n}, X_n(m2^{-n}, x, \boldsymbol{\omega}), S^m\boldsymbol{\omega}\big) \\
&\qquad\qquad \times a\big(X_n(m2^{-n}, x, \boldsymbol{\omega})\big)\xi\Big)_{\mathbb{R}^N}\Gamma(d\boldsymbol{\omega}) \\
&= \int\bigg(\int\big(\nabla\varphi(x''), J''a\big(X_n(m2^{-n}, x, \boldsymbol{\omega})\big)\xi\big)_{\mathbb{R}^N} \\
&\qquad\qquad \times P_n^{(1)}\big(t - m2^{-n}, (X_n(m2^{-n}, x, \boldsymbol{\omega}), I), dx'' \times dJ''\big)\bigg)\Gamma(d\boldsymbol{\omega}) \\
&= \int\bigg(\int\big(\nabla\varphi(x''), J''a(x')\xi\big)_{\mathbb{R}^N} P_n^{(1)}\big(t - m2^{-n}, (x', I), dx'' \times dJ''\big)\bigg) \\
&\qquad\qquad \times P_n(m2^{-n}, x, dx').
\end{aligned}$$

Hence, we have now proved that

$$\int \mathbf{D}_{\mathbf{h}_n} \varphi\big(X_n(t,x,\boldsymbol{\omega})\big)\, \Gamma(d\boldsymbol{\omega})$$

tends to the left-hand side of (3.1.4).

To complete the proof, we apply Lemma 3.1.1 to see that

$$\int \mathbf{D}_{\mathbf{h}_n} \varphi\big(X_n(t,x,\boldsymbol{\omega})\big)\, \Gamma(d\boldsymbol{\omega})$$

$$= \int \varphi\big(X_n(t,x,\boldsymbol{\omega})\big) \sum_{m=1}^{2^n t} \big(h_m(\boldsymbol{\omega}), \omega_m\big)_{\mathbb{R}^N}\, \Gamma(d\boldsymbol{\omega}).$$

In addition, because $(b,\xi)_{\mathbb{R}^N} \equiv 0$ and $(x,\xi)_{\mathbb{R}^N} = 0$,

$$\sum_{m=1}^{2^n t} \big(h_m(\boldsymbol{\omega}), \omega_m\big)_{\mathbb{R}^N} = \big(\xi, X_n(t,x,\boldsymbol{\omega})\big)_{\mathbb{R}^N}.$$

Thus,

$$\int \mathbf{D}_{\mathbf{h}_n} \varphi\big(X_n(t,x,\boldsymbol{\omega})\big)\, \Gamma(d\boldsymbol{\omega}) = \int \varphi(y)\big(\xi, y\big)_{\mathbb{R}^N}\, P_n(t,x,dy),$$

and this tends to the right-hand side of (3.1.4). □

3.2 Application to the Backward Variable

By combining Corollary 2.2.8 and Theorem 3.1.3, in this section we will derive our first elliptic regularity result. Namely, we will show that when a and b are bounded and have bounded derivatives, and, in addition, a is *uniformly elliptic* (i.e., uniformly positive definite), derivatives of (cf. (2.0.2)) $u_\varphi(t, \cdot)$ for $t > 0$ can be estimated in terms of $\|\varphi\|_u$. Thus, throughout this section we will be assuming that a, b and all their derivatives are bounded and that

$$(3.2.1) \qquad \big(\xi, a(x)\xi\big)_{\mathbb{R}^N} \geq \epsilon|\xi|^2, \quad (x,\xi) \in \mathbb{R}^N \times \mathbb{R}^N,$$

for some $\epsilon > 0$. Also, we will take σ to be the non-negative definite, symmetric square root of a, and therefore $M = N$. By Lemma 2.3.1, we know that σ is also smooth and has bounded derivatives of all orders.

3.2.1. A Formula of Bismut Type: Before attempting to handle higher order derivatives, we will examine the gradient of $u_\varphi(t, \cdot)$, which turns out to admit a particularly pleasing expression, discovered originally, in a different setting, by J.-M. Bismut.

In order to state and prove the result, we need to make a few preparations. Recall (cf. § 2.2.2) that, because $M = N$ here, $E_\ell = \mathrm{Hom}(\mathbb{R}^N; \mathbb{R}^{N \otimes \ell})$ and $\mathbf{E}^{(\ell)} = \prod_{k=1}^{\ell} E_k$, and define $(x, \mathbf{J}) \in \mathbb{R}^N \times \mathbf{E}^{(\ell)} \longmapsto P^{(\ell)}\big(t, (x, \mathbf{J})\big) \in \mathbf{M}_1(\mathbb{R}^N \times \mathbf{E}^{(\ell)})$ accordingly, as in Theorem 2.2.7.

LEMMA 3.2.2. $P^{(1)}\big(t,(x,J)\big)$ is the distribution of $(x',J')\rightsquigarrow(x',J'J)$ under $P^{(1)}\big(t,(x,I)\big)$. Next, suppose that $\tau \in C^2\big(\mathbb{R}^N \times E_1; \mathrm{Hom}(\mathbb{R}^N;\mathbb{R}^N)\big)$ has bounded first order derivatives, set

$$\tilde{\sigma}(\tilde{x}) = \begin{pmatrix} \sigma(x) \\ \nabla\sigma(x)J \\ \tau(\tilde{x}) \end{pmatrix} \quad \text{and} \quad \tilde{b}(\tilde{x}) = \begin{pmatrix} b(x) \\ \nabla b(x)J \\ 0 \end{pmatrix}$$

for $\tilde{x} = (x,J,\chi) \in \mathbb{R}^N \times E_1 \times \mathbb{R}^N$, and define $(t,\tilde{x})\rightsquigarrow\tilde{P}(t,\tilde{x})$ and $(t,\tilde{x},\tilde{J})\rightsquigarrow \tilde{P}^{(1)}\big(t,(\tilde{x},\tilde{J})\big)$ accordingly. If

(3.2.3)
$$\tilde{J} = \begin{pmatrix} \tilde{J}_{(11)} & \tilde{J}_{(12)} & \tilde{J}_{(13)} \\ \tilde{J}_{(21)} & \tilde{J}_{(22)} & \tilde{J}_{(23)} \\ \tilde{J}_{(31)} & \tilde{J}_{(32)} & \tilde{J}_{(33)} \end{pmatrix},$$

with the blocks corresponding to $\mathbb{R}^N \times E_1 \times \mathbb{R}^N$, then

$$\tilde{P}^{(1)}\big(t,(x,J,\chi,\tilde{I}),\{\tilde{J}' : \tilde{J}'_{(ij)} = 0 \text{ for } 1 \le i < j \le 3 \ \& \ J' = \tilde{J}'_{(11)}J\}\big) = 1,$$

and the distribution of

$$(\tilde{x}',J',\chi',\tilde{J}')\rightsquigarrow \begin{pmatrix} x' \\ J' \\ \begin{pmatrix} \tilde{J}'_{(11)} & \tilde{J}'_{(12)} \\ \tilde{J}'_{(21)} & \tilde{J}'_{(22)} \end{pmatrix} \end{pmatrix} \quad \text{under } \tilde{P}^{(1)}\big(t,(x,I,\chi,\tilde{I})\big)$$

is the same as that of

$$(x',J'_1,J'_2)\rightsquigarrow \begin{pmatrix} x' \\ J' \\ \begin{pmatrix} J'_1 & 0 \\ J'_2 & J'_1 \otimes I \end{pmatrix} \end{pmatrix} \quad \text{under } P^{(2)}\big(t,(x,I)\big).$$

PROOF: To prove the first assertion, it suffices to first observe (cf. (2.2.1) and (2.2.4)) that $P_n^{(1)}\big(t,(x,J)\big)$ is the distribution of

$$\omega\rightsquigarrow \begin{pmatrix} X_n(t,x,\omega) \\ X_n^1(t,x,\omega)J \end{pmatrix} \quad \text{under } \Gamma,$$

and then to pass to the limit as $n \to \infty$.

To verify the second assertion, note that the distribution of

$$\tilde{J}'\rightsquigarrow\big(\tilde{J}'_{(11)},\tilde{J}'_{(12)},\tilde{J}'_{(13)},\tilde{J}'_{(23)}\big) \quad \text{under } \tilde{P}^{(1)}\big(t,(x,I,\chi,\tilde{I})\big)$$

is the limit as $n \to \infty$ of the distribution under Γ of

$$\omega\rightsquigarrow\big(\nabla_x X_n(t,x,\omega),\nabla_J X_n(t,x,\omega),\nabla_\chi X_n(t,x,\omega),\nabla_\chi X_n^1(t,x,\omega)\big)$$
$$= \big(X_n^1(t,x,\omega),0,0,0\big).$$

Turning to the final assertion, begin with the observation that, when $\tilde{X}_n\big(t, (x, J, \chi), \boldsymbol{\omega}\big)$ is defined from $\tilde{\sigma}$ and $\tilde{b}$ by the prescription in (2.2.1), its first and second components are, respectively, $X_n(t, x, \boldsymbol{\omega})$ and $X_n^1(t, x, \boldsymbol{\omega})J$. Thus, since

$$X_n^2(t, x, \boldsymbol{\omega}) = \nabla_x X_n^1(t, x, \boldsymbol{\omega}) \quad \text{and} \quad X_n^1(t, x, \boldsymbol{\omega}) \otimes I = \nabla_J\big(X_n^1(t, x, \boldsymbol{\omega})J\big),$$

the distribution of

$$(\tilde{x}', J', \chi', \tilde{J}') \rightsquigarrow \begin{pmatrix} x' \\ J' \\ \begin{pmatrix} \tilde{J}'_{(11)} & \tilde{J}'_{(12)} \\ \tilde{J}'_{(21)} & \tilde{J}'_{(22)} \end{pmatrix} \end{pmatrix} \quad \text{under } \tilde{P}^{(1)}\big(t, (x, I, \chi, \tilde{I})\big)$$

is the limit as $n \to \infty$ of the distribution of

$$\boldsymbol{\omega} \rightsquigarrow \begin{pmatrix} X_n(t, x, \boldsymbol{\omega}) \\ X_n^1(t, x, \boldsymbol{\omega}) \\ \begin{pmatrix} X_n^1(t, x, \boldsymbol{\omega}) & 0 \\ X_n^2(t, x, \boldsymbol{\omega}) & X_n^1(t, x, \boldsymbol{\omega}) \otimes I \end{pmatrix} \end{pmatrix} \quad \text{under } \Gamma,$$

and this latter limit is the distribution of

$$(x', J_1', J_2') \rightsquigarrow \begin{pmatrix} x' \\ J' \\ \begin{pmatrix} J_1' & 0 \\ J_2' & J_1' \otimes I \end{pmatrix} \end{pmatrix} \quad \text{under } P^{(2)}\big(t, (x, I, 0)\big). \quad \square$$

Now define $\tilde{\sigma} : \mathbb{R}^N \times E_1 \times \mathbb{R}^N \longrightarrow \mathrm{Hom}(\mathbb{R}^N; \mathbb{R}^N \times E_1 \times \mathbb{R}^N)$ and $\tilde{b} : \mathbb{R}^N \times E_1 \times \mathbb{R}^N \longrightarrow \mathbb{R}^N \times E_1 \times \mathbb{R}^N$ by

$$\tilde{\sigma}(x, J, \chi) = \begin{pmatrix} \sigma(x) \\ \nabla\sigma(x)J \\ J^\top \sigma(x)^{-1} \end{pmatrix} \quad \text{and} \quad \tilde{b}(x, J, \chi) = \begin{pmatrix} b(x) \\ \nabla b(x)J \\ 0 \end{pmatrix},$$

and let $\big(t, (x, J, \chi)\big) \rightsquigarrow \tilde{P}\big(t, (x, J, \chi)\big)$ be defined for $\tilde{\sigma}$ and $\tilde{b}$ as in §1.2. By Theorem 2.2.7, all the conditions in §3.1 are satisfied, and so Theorem 3.1.3 applies.

THEOREM 3.2.4. *Let $\varphi \in C_\mathrm{b}(\mathbb{R}^N; \mathbb{C})$ be given, and define u_φ as in (2.0.2). Then, for each $t > 0$, $u_\varphi(t, \cdot) \in C_\mathrm{b}^1(\mathbb{R}^N; \mathbb{C})$ and*

$$\nabla u_\varphi(t, x) = t^{-1} \int \varphi(x')\chi' \, \tilde{P}\big(t, (x, I, 0), dx' \times dJ' \times d\chi'\big).$$

In particular, there is a constant $C_1 < \infty$, depending only on the bounds on the first derivatives of a and b, such that (cf. (3.2.1))

$$|\nabla u_\varphi(t, x)| \leq \frac{C_1}{\epsilon^{\frac{1}{2}}(1 \wedge t^{\frac{1}{2}})} \|\varphi\|_\mathrm{u}.$$

PROOF: By an obvious approximation argument, it suffices for us to treat φ's which are smooth and have bounded derivatives of all orders. Thus, we will make this assumption about φ throughout.

The proof of the initial statement is mostly a matter of bookkeeping. Given $\xi \in \mathbb{R}^N$, use Corollary 2.2.8 with $\ell = 1$ to write

$$\big(\xi, \nabla u_\varphi(t,x)\big)_{\mathbb{R}^N} = \int \big(\nabla\varphi(x'), J'\xi\big)_{\mathbb{R}^N} P^{(1)}\big(t,(x,I),dx'\times dJ'\big),$$

where $\big(t,(x,J)\big) \in (0,\infty)\times(\mathbb{R}^N\times E_1) \longmapsto P^{(1)}\big(t,(x,J)\big) \in \mathbf{M}_1\big(\mathbb{R}^N\times E_1\big)$ is defined relative to σ and b. In order to see how Theorem 3.1.3 applies, take $\tilde\varphi(\tilde x) = \varphi(x)$ when $\tilde x = (x,J,\chi)$, and note that $\tilde\nabla\tilde\varphi(\tilde x) = (\nabla\varphi(x),0,0)$ and

$$(3.2.5)\quad \tilde a(\tilde x) \equiv \sum_{k=1}^{N}\big(\tilde\sigma(\tilde x)e_k\big)\otimes\big(\tilde\sigma(\tilde x)e_k\big) = \begin{pmatrix} \tilde a(\tilde x)_{(11)} & \tilde a(\tilde x)_{(12)} & \tilde a(\tilde x)_{(13)} \\ \tilde a(\tilde x)_{(21)} & \tilde a(\tilde x)_{(22)} & \tilde a(\tilde x)_{(23)} \\ \tilde a(\tilde x)_{(31)} & \tilde a(\tilde x)_{(32)} & \tilde a(\tilde x)_{(33)} \end{pmatrix},$$

where

$$\tilde a(\tilde x)_{(11)} = \sum_{k=1}^{N}\sigma(x)e_k\otimes\sigma(x)e_k = a(x),$$

$$\tilde a(\tilde x)_{(12)} = \sum_{k=1}^{N}\big(\sigma(x)e_k\big)\otimes\big(\nabla(\sigma(x)e_k)J\big),$$

$$\tilde a(\tilde x)_{(13)} = \sum_{k=1}^{N}\big(\sigma(x)e_k\big)\otimes\big(J^\top\sigma(x)^{-1}e_k\big) = J,$$

$$\tilde a(\tilde x)_{(22)} = \sum_{k=1}^{N}\big(\nabla(\sigma(x)e_k)J\big)^{\otimes 2},$$

$$\tilde a(\tilde x)_{(23)} = \sum_{k=1}^{N}\big(\nabla(\sigma(x)e_k)J\big)\otimes\big(J^\top\sigma(x)^{-1}e_k\big),$$

$$\tilde a(\tilde x)_{(33)} = \sum_{k=1}^{N}\big(J^\top\sigma(x)^{-1}\big)^{\otimes 2} = J^\top a(x)^{-1}J,$$

$\{e_k : 1\le k\le N\}$ is an orthonormal basis in $\mathbb{R}^N$, and the block structure corresponds to $\mathbb{R}^N\times E_1\times\mathbb{R}^N$. Hence, if $\tilde\xi = \begin{pmatrix} 0 \\ 0 \\ \xi \end{pmatrix}$, then, Lemma 3.2.2,

$$\int \big(\tilde\nabla\varphi(\tilde x''), \tilde J''\tilde a(x')\tilde\xi\big)_{\mathbb{R}^N\times E_1\times\mathbb{R}^N} \tilde P^{(1)}\big(t,(x',J',0,\tilde I),d\tilde x''\times d\tilde J''\big)$$

$$= \int \big(\nabla\varphi(x''), J''J'\xi\big)_{\mathbb{R}^N} P^{(1)}\big(t,(x',I),dx''\times dJ''\big)$$

$$= \int \big(\nabla\varphi(x''), J''\xi\big)_{\mathbb{R}^N} P^{(1)}\big(t,(x',J'),dx''\times dJ''\big).$$

Thus, by the Chapman-Kolmogorov equation for $P^{(1)}$ and Corollary 2.2.8 with $\ell = 1$,

$$\int_0^t \left(\int \left(\int \left(\tilde{\nabla}\varphi(\tilde{x}''), \tilde{J}''\tilde{a}(x')\tilde{\xi} \right)_{\mathbb{R}^N \times E_1 \times \mathbb{R}^N} \tilde{P}^{(1)}\left(t - \tau, (x', I, 0, \tilde{I}), d\tilde{x}'' \times d\tilde{J}'' \right) \right) \right.$$

$$\left. \times \tilde{P}^{(1)}\left(\tau, (x, I, 0, \tilde{I}), dx' \times dJ' \times d\chi \times d\tilde{J}' \right) \right) d\tau$$

$$= t \int \left(\nabla\varphi(x'), J'\xi \right)_{\mathbb{R}^N} P^{(1)}\left(t, (x, I), dx' \times dJ' \right) = t\left(\xi, \nabla u_\varphi(t, x) \right)_{\mathbb{R}^N},$$

and therefore the first assertion follows from Theorem 3.1.3.

To complete the proof, first note that it suffices to handle $t \in (0, 1]$. Indeed, if $t > 1$, then, by the Chapman–Kolmogorov equation,

$$u_\varphi(t, x) = \int u_\varphi(t - 1, x') P(1, x, dx'),$$

and $\|u_\varphi(t - 1, \cdot)\|_u \leq \|\varphi\|_u$. Thus, all that we have to show is that

$$\int |\chi'|^2 \tilde{P}\left(t, (x, I), dx' \times dJ' \times d\chi' \right) \leq \frac{C_1 t}{\epsilon}, \quad t \in (0, 1].$$

To this end, note that

$$\int |\chi'|^2 \tilde{P}\left(t, (x, I, 0), dx' \times dJ' \times d\chi' \right)$$

$$= \int_0^t \left(\int \|\sigma(x')^{-1} J'\|_{\text{H.S.}}^2 P^{(1)}\left(\tau, (x, I), dx' \times dJ' \right) \right) d\tau,$$

and apply the final estimate in Theorem 2.2.7 with $\ell = 1$, remembering that the first derivatives of σ are bounded by $\epsilon^{-\frac{1}{2}}$ times the first derivatives of a. $\square$

3.2.2. Higher Order Derivatives: Although expressions of the sort in Theorem 3.2.4 can be developed for higher derivatives of u_φ, they become increasingly unwieldy and less useful. For this reason, we will use an inductive procedure which gives us estimates but not formulas. Specifically, we are going to use induction on $\ell \geq 1$ to prove the following statement.

THEOREM 3.2.6. *For each $\ell \geq 1$ there exists a constant $K_\ell < \infty$, depending only on the $\epsilon > 0$ in (3.2.1) and the bounds on the derivatives of a and b up through order ℓ, such that*

$$\|\nabla^\ell u_\varphi(t, x)\|_{\text{H.S.}} \leq \frac{K_\ell}{1 \wedge t^{\frac{\ell}{2}}} \|\varphi\|_u \quad \text{for } \varphi \in C_b(\mathbb{R}^N; \mathbb{C}).$$

In particular, $u_\varphi \in C^\infty\big((0,\infty) \times \mathbb{R}^N; \mathbb{C}\big)$, for each $m \geq 1$, $\partial_t^m u_\varphi = L^m u_\varphi$, and so, for all $\alpha \in \mathbb{N}^N$, there exists a $K_{m,\alpha} < \infty$, depending only on ϵ and the derivatives of a and b up through order $2m + \|\alpha\|$, such that

$$\big|\partial_t^m \partial_x^\alpha u_\varphi(t,x)\big| \leq \frac{K_{m,\alpha}}{1 \wedge t^{\frac{2m+\|\alpha\|}{2}}} \|\varphi\|_{\mathrm{u}}.$$

PROOF: Using the Chapman–Kolmogorov equation as we did at the end of the proof of Theorem 3.2.4, we can reduce to the case when $t \in (0,1]$. Thus, $t \in (0,1]$ in what follows.

First note that, by Theorem 3.2.4, there is nothing to do when $\ell = 1$. Next, assume the estimate holds for derivatives of order $k \leq \ell - 1$, where $\ell \geq 2$. To prove that it holds for ℓ, we begin with the formula

$$\nabla^\ell u_\varphi(t,x) = \sum_{k=1}^\ell \sum_{\beta \in \mathfrak{B}(k,\ell)} c_\beta \int \nabla^k u_\varphi\big(\tfrac{t}{2}, x'\big) \mathbf{J}'^{\otimes\beta}\, \mathbf{P}^{(\ell)}\big(\tfrac{t}{2}, (x, \mathbf{J}_0), dx' \times d\mathbf{J}'\big),$$

which follows after one combines the result in Corollary 2.2.8 with the Chapman–Kolmogorov equation. By the induction hypothesis and our estimates in the last part of Theorem 2.2.7, the terms

$$\int \nabla^k u_\varphi\big(\tfrac{t}{2}, x'\big) \mathbf{J}'^{\otimes\beta}\, \mathbf{P}^{(\ell)}\big(\tfrac{t}{2}, (x, \mathbf{J}_0), dx' \times d\mathbf{J}'\big)$$

for $1 \leq k < \ell$ are dominated by a constant, having the required dependence, times $\big(\tfrac{t}{2}\big)^{-\frac{k}{2}} \|\varphi\|_{\mathrm{u}}$. Hence, all that remains is to handle the term corresponding to $k = \ell$, which we can write as

$$\int \nabla^\ell u_\varphi\big(\tfrac{t}{2}, x'\big) J'^{\otimes\ell}\, P^{(1)}\big(\tfrac{t}{2}, (x, I), dx' \times dJ'\big).$$

The entries in this matrix are finite sums of the form

$$\sum_{i_2,\dots,i_\ell=1}^N \int \Big(\nabla\big[\partial_{x_{i_2}} \cdots \partial_{x_\ell} u_\varphi\big(\tfrac{t}{2}, x'\big)\big] J'\Big)_{j_1}$$
$$\times J'_{i_2 j_2} \cdots J'_{i_\ell j_\ell}\, P^{(1)}\big(\tfrac{t}{2}, (x, I), dx' \times dJ'\big).$$

Thus, we will be done once we show that, for any $\psi \in C_{\mathrm{b}}^1(\mathbb{R}^N; \mathbb{C})$ and $\xi \in \mathbb{S}^{N-1}$,

$$\int \big(\nabla\psi(x'), J'\xi\big)_{\mathbb{R}^N} J'_{i_2 j_2} \cdots J'_{i_\ell j_\ell}\, \mathbf{P}^{(1)}\big(t, (x, I), dx' \times dJ'\big)$$

can be dominated by a constant with the required dependence times $\frac{\|\psi\|_{\mathrm{u}}}{t^{\frac{1}{2}}}$. For this purpose, set

$$\tilde{\psi}(\tilde{x}) = \psi(x) J_{i_2 j_2} \cdots J_{i_\ell j_\ell} \quad \text{for } \tilde{x} = (x, J, \chi) \in \mathbb{R}^N \times E_1 \times \mathbb{R}^N.$$

Referring to the notation introduced in the preceding section and applying Theorem 3.1.3, we see that

$$\int_0^t \left(\int \left(\int \left(\tilde{\nabla}\tilde{\psi}(\tilde{x}''), \tilde{J}''\tilde{a}(\tilde{x}')\tilde{\xi} \right)_{\mathbb{R}^N \times E_1 \times \mathbb{R}^N} \right. \right.$$
$$\left. \left. \times \tilde{P}^{(1)}\left(t - \tau, (\tilde{x}', \tilde{I}), d\tilde{x}'' \times d\tilde{J}'' \right) \right) \tilde{P}\left(\tau, (x, I, 0), d\tilde{x}' \right) \right) d\tau$$
$$= \int \tilde{\psi}(\tilde{x}')(\chi', \xi)_{\mathbb{R}^N} \, \tilde{P}\left(t, (x, I, 0), d\tilde{x}' \right),$$

where $\tilde{\xi} \in \mathbb{R}^N \times E_1 \times \mathbb{R}^N$ is chosen, as it was in the proof of Theorem 3.2.4, so that its first two components are 0 and its final component is ξ. Finally, splitting the components of $\tilde{\nabla}\tilde{\psi}$ into two groups, those corresponding to derivatives with respect to x and the other corresponding to derivatives with respect to the coordinates of J, we can use the same reasoning as we did in the proof of Theorem 3.2.4 to see that

$$\int \left(\int \left(\left(\tilde{\nabla}\tilde{\psi}(\tilde{x}''), \tilde{J}''\tilde{a}(\tilde{x}')\tilde{\xi} \right)_{\mathbb{R}^N \times E_1 \times \mathbb{R}^N} \right. \right.$$
$$\left. \left. \times \tilde{P}^{(1)}\left(t - \tau, (\tilde{x}', \tilde{I}), d\tilde{x}'' \times d\tilde{J}'' \right) \right) \tilde{P}\left(\tau, (x, I, 0), d\tilde{x}' \right) \right)$$

can be written as the finite sum of

$$\int \left(\nabla\psi(x'), J'\xi \right)_{\mathbb{R}^N} J'_{i_2 j_2} \cdots J'_{i_\ell j_\ell} \, P^{(1)}\left(t, (x, I), dx' \times dJ' \right),$$

which is the quantity we want to estimate, terms of type $\mathbf{T}_1(\tau)$ (cf. (3.2.5) and (3.2.3)) given by

$$\int \left(\int \psi(x'') J''_{i_3, j_3} \cdots J''_{i_\ell, j_\ell} (\tilde{J}''_{(21)} \nabla\sigma(x') J'\sigma(x')^\top \xi)_{j_\ell} \right.$$
$$\left. \times \tilde{P}^{(1)}\left(t - \tau, (x', I, 0, \tilde{I}), dx'' \times d\mathbf{J}'' \right) \right) \tilde{P}\left(\tau, (x, I, 0), dx' \times d\tilde{J}' \times d\chi' \right),$$

and those of type $\mathbf{T}_2(\tau)$ given by

$$\int \left(\int \psi(x'') J''_{i_3, j_3} \cdots J''_{i_\ell, j_\ell} (\tilde{J}''_{(22)} \nabla\sigma(x') J'(J')^\top \nabla\sigma(x')^\top \sigma(x')\xi)_{j_\ell} \right.$$
$$\left. \times \tilde{P}^{(1)}\left(t - \tau, (x', I, 0, \tilde{I}), dx'' \times d\tilde{J}'' \right) \right) \tilde{P}\left(\tau, (x, I, 0), dx' \times d\tilde{J}' \times d\chi' \right).$$

Hence,

$$\int \left(\nabla\psi(x'), J'\xi \right)_{\mathbb{R}^N} J'_{i_2 j_2} \cdots J'_{i_\ell j_\ell} \, P^{(1)}\left(t, (x, I), dx' \times dJ' \right)$$

equals

$$t^{-1} \int \psi(x') J'_{i_1 j_1} \cdots J'_{i_\ell j_\ell}(\chi', \xi)_{\mathbb{R}^N} \tilde{P}(t, (x, I, 0), dx' \times dJ' \times d\chi')$$

minus a finite sum of terms of the form

$$t^{-1} \int_0^t \mathbf{T}_1(\tau) \, d\tau \quad \text{or} \quad t^{-1} \int_0^t \mathbf{T}_2(\tau) \, d\tau.$$

By the estimates in Theorem 2.2.7 and the reasoning used at the end of the proof of Theorem 3.2.4, the first of these is dominated by a constant of the required sort times $t^{-\frac{1}{2}} \|\psi\|_{\mathrm{u}}$. To handle the terms containing $\mathbf{T}_1(\tau)$ and $\mathbf{T}_2(\tau)$, one can apply Lemma 3.2.2 to see that both these can be bounded by of $\|\psi\|_{\mathrm{u}}$ times $P^{(2)}\big(t - \tau, (x', I, 0)\big)$ moments of $\|\mathbf{J}''\|_{\mathrm{H.S.}}$, which, by the second estimate in Theorem 2.2.7, means that they are bounded, independent of $\tau \in [0, t]$, by constants with the required dependence times $\|\psi\|_{\mathrm{u}}$. Thus, these terms cause no problems.

To complete the proof, let $t > 0$ be given and set $\psi = u_\varphi\big(\frac{t}{2}, \cdot\big)$. Then, $\psi \in C_{\mathrm{b}}^\infty(\mathbb{R}^N; \mathbb{C})$ and so, using induction on m, we see that

$$\partial_t^m u_\varphi(t, x) = \partial_t^m \int u_\psi\big(\tfrac{t}{2}, y\big) P\big(\tfrac{t}{2}, x, dy\big) = 2^{-m} u_{L^m \psi}\big(\tfrac{t}{2}, x\big),$$

which shows that u_φ is smooth in (t, x). Finally, knowing that u_φ is smooth as a function of (t, x), we can use induction on m to check that $\partial_t^m u_\varphi = L^{m-1} \partial_t u_\varphi = L^m u_\varphi$. Hence

$$\big|\partial_t^m \partial_x^\alpha u_\varphi(t, x)\big| = \big|\partial_x^\alpha \partial_t^m u_\varphi(t, x)\big| = \big|\partial_x^\alpha L^m u_\varphi\big|$$

$$\leq C_m \sum_{k=1}^{2m + \|\alpha\|} \big\|\nabla^k u_\varphi(t, x)\big\|_{\mathrm{H.S.}},$$

where $C_m < \infty$ depends only on the bounds on the derivatives of a and b through order $2m + \|\alpha\|$. $\quad\square$

3.3 Application to the Forward Variable

We continue in the setting of § 3.2. In particular, a satisfies (3.2.1) and $\sigma = a^{\frac{1}{2}}$ is the non-negative definite, symmetric square root of a.

In this section, we will use the results in § 3.1 to study regularity properties of the transition probability function $P(t, x, dy)$ as a function of the forward variable y. Specifically, we will show that there is a continuous function $(t, x, y) \in (0, \infty) \times \mathbb{R}^N \times \mathbb{R}^N \longmapsto p(t, x, y) \in [0, \infty)$ which, for each (t, x), is the density of $P(t, x, dy)$ with respect Lebesgue measure. That is, $P(t, x, dy) = p(t, x, y) \, dy$. Moreover, we will show that, for each y, $(t, x) \rightsquigarrow p(t, x, y)$ is a smooth solution to the backward equation $\partial_t u = Lu$. Thus, since $P(t, x, \cdot) \longrightarrow \delta_x$ as $t \searrow 0$, this will show that $(t, x, y) \rightsquigarrow p(t, x, y)$ is a fundamental solution to the backward equation.

3.3.1. A Criterion for Absolute Continuity: In order to carry out our program, we will need to have a criterion for recognizing when a probability measure admits a density. The one which we will use is the following Sobolev type embedding theorem.

LEMMA 3.3.1. *Let $\mu \in \mathbf{M}_1(\mathbb{R}^N)$ and assume that, for some $r \in (N, \infty)$, there is a constant $C < \infty$ for which*

$$\left| \int \nabla\varphi(y)\, \mu(dy) \right| \leq C \|\varphi\|_{L^{r'}(\mu;\mathbb{R})}, \quad \varphi \in C_b^1(\mathbb{R}^N; \mathbb{R}),$$

where $r' = \frac{r}{r-1}$ is the Hölder conjugate of r. Then there exists a unique, uniformly continuous, bounded $f : \mathbb{R}^N \longrightarrow [0, \infty)$ such that $\mu(dy) = f(y)\, dy$. Moreover, both the bound on f and its modulus of continuity can be estimated in terms of r and C.

PROOF: First observe that, by elementary functional analysis, the condition given guarantees the existence of a $\Psi \in L^r(\mu; \mathbb{R}^N)$ with the properties that $\|\Psi\|_{L^r(\mu;\mathbb{R}^N)} \leq C$ and

$$\int \nabla\varphi(y)\, \mu(dy) = \int \varphi(y)\Psi(y)\, \mu(dy).$$

Hence, the hypothesis can be restated in the language of Schwartz distributions as the statement that $\nabla\mu = -\Psi\mu$. Thinking in terms of Schwartz tempered distributions (cf. § 7.1.1), one has that

$$(*) \quad \mu = R \star (\mu - \Delta\mu) = R \star \mu + \sum_{i=1}^N R_i \star \partial_{x_i}\mu = R \star \mu + \sum_{i=1}^N R_i \star (\psi_i\mu),$$

where

$$R(y) = \int_0^\infty e^{-t} g(2t, y)\, dt, \quad \text{with } g(t, y) \equiv (2\pi t)^{-\frac{N}{2}} e^{-\frac{|y|^2}{2t}}\, dt,$$

is the kernel for the resolvent operator $(I - \Delta)^{-1}$,

$$R_i(y) = \partial_{y_i} R(y) = -\int_0^\infty e^{-t} \frac{y_i}{2t} g(2t, y)\, dt,$$

and ψ_i is the ith coordinate of Ψ. By Minkowski's inequality,

$$A \equiv \|R\|_{L^{r'}(\mathbb{R})} \leq \int_0^\infty e^{-t} \left(\int g(2t, y)^{r'}\, dy \right)^{\frac{1}{r'}} dt$$

$$= \int e^{-t} t^{-\frac{N}{2r}}\, dt \left(\int g(2, y)^{r'}\, dy \right)^{\frac{1}{r'}} < \infty,$$

and similarly

$$B \equiv \|R_i\|_{L^{r'}(\mathbb{R})} \leq \frac{1}{2} \int_0^\infty e^{-t} t^{-\frac{N}{r}} \, dt \left(\int |y_1|^{r'} e^{-\frac{r'|y|^2}{4}} \, dy \right)^{\frac{1}{r'}} < \infty$$

since $r > N$. In particular, this means that the right-hand side of $(*)$ determines an $f \in L^{r'}(\mathbb{R}^N; \mathbb{R})$, and so $\mu(dy) = f(y) \, dy$. Clearly, f can be chosen to be non-negative, and $\int f(y) \, dy = 1$.

Next set $\varphi = f^{\frac{1}{r}}$. We want to check that, as a tempered distribution, $\nabla \varphi = -\frac{1}{r} \varphi \Psi$. For this purpose, take $\varphi_{\epsilon,\delta} = (\rho_\epsilon \star f + \delta)^{\frac{1}{r}}$, where $\rho \in C_c^\infty(B(0,1); [0,\infty))$ has integral 1 and $\rho_\epsilon(x) = e^{-N} \rho(\epsilon^{-1} x)$. Then, as $\epsilon \searrow 0$, $\varphi_{\epsilon,\delta} \longrightarrow (f + \delta)^{\frac{1}{r}}$ as tempered distributions, and so $\nabla \varphi_{\epsilon,\delta} \longrightarrow \nabla (f + \delta)^{\frac{1}{r}}$ as tempered distributions. But, as $\epsilon \searrow 0$,

$$\nabla \varphi_{\epsilon,\delta} = -\frac{1}{r} \varphi_{\epsilon,\delta}^{\frac{1}{r}-1} \rho_\epsilon \star (f\Psi) \longrightarrow -\frac{1}{r}(f+\delta)^{\frac{1}{r}-1} f\Psi,$$

and so $\nabla(f + \delta)^{\frac{1}{r}} = -\frac{1}{r}(f+\delta)^{\frac{1}{r}-1} f\Psi$. Finally, after $\delta \searrow 0$, this gives the result.

Knowing that $\nabla \varphi = -\frac{1}{r} \varphi \Psi$, we have that

$$\varphi = R \star \varphi + \frac{1}{r} \sum_{i=1}^N R_i \star (\psi_i \varphi).$$

Because $\|\varphi\|_{L^r(\mathbb{R};\mathbb{R})} = 1$ and $\|\psi_i \varphi\|_{L^r(\mathbb{R};\mathbb{R})} = \|\psi_i\|_{L^r(\mu;\mathbb{R})}$, this shows that φ is a bounded, uniformly continuous function. In fact,

$$\|\varphi\|_{\mathrm{u}} \leq A + \frac{B}{r} \sum_{i=1}^N \|\psi_i\|_{L^r(\mu;\mathbb{R})}$$

and

$$|\varphi(y + h) - \varphi(y)| \leq \|R(\cdot + h) - R\|_{L^{r'}(\mathbb{R})}$$

$$+ \frac{1}{r} \sum_{i=1}^N \|R_i(\cdot + h) - R_i\|_{L^{r'}(\mathbb{R})} \|\psi_i\|_{L^r(\mu;\mathbb{R})},$$

and so we know that both $\|\varphi\|_{\mathrm{u}}$ and the modulus of continuity of φ can be estimated in terms of r and $\|\Psi\|_{L^r(\mu;\mathbb{R}^N)}$. Obviously, since $f = \varphi^r$, these conclusions transfer to f. □

3.3.2. The Inverse of the Jacobian: In view of Lemma 3.3.1, what we need to prove is an estimate of the form $\left| \langle \nabla \varphi, P(t,x) \rangle \right| \leq C \|\varphi\|_{L^{r'}(P(t,x);\mathbb{R})}$, and the result in Theorem 3.2.4 seems to provide a starting place. Namely, by combining the formula there with the one in Corollary 2.2.8 (with $\ell = 1$), it says that

$$\int \nabla\varphi(y) J \, P^{(1)}\big(1, (x, I), dy \times dJ\big) = \int \varphi(y)\chi \, \tilde{P}\big(1, (x, 0), dy \times d\chi\big).$$

Since the right-hand side can be estimated by

$$\left(\int |\chi|^r \, \tilde{P}\big(1, (x, 0), dy \times d\chi\big) \right)^{\frac{1}{r}} \|\varphi\|_{L^{r'}(P(t,x);\mathbb{R})},$$

it should be clear that the only problem is that of removing the J from the left-hand side, and for this purpose we have to investigate the distribution of J^{-1}, the inverse of J, under $P^{(1)}\big(1, (x, I)\big)$.

With this goal in mind, remember that the distribution of J under $P^{(1)}\big(1, (x, I)\big)$ is the limit as $n \to \infty$ of the distribution of (cf. (2.2.4)) $\omega \in \Omega \longmapsto X_n^1(1, x, \omega)$ under Γ, where $X_n^1(t, x, \omega) = \nabla X_n(t, x, \omega)$. In the case when $\sigma = 0$, $X_n(t, x) = X_n(t, x, \omega)$ is independent of ω and $X_n(t, x)$ tends to the solution of $\dot{X}(t, x) = b\big(X(t, x)\big)$ with $X(0, x) = x$. Since one can easily check that $X^1(t, x)^{-1}$ is the solution to $\dot{Y}(t, x) = -Y(t, x)\nabla b\big(X(t, x)\big)$ with $Y(0, x) = I$, one might be tempted to guess that the distribution of the inverse of J should be the limit of the distributions of $Y_n(t, x, \omega)$, where $Y_n(0, x, \omega) = I$ and $Y_n(t, x, \omega) - Y_n(m2^{-n}, x, \omega)$ equals

$$-(t-m2^{-n})^{\frac{1}{2}} Y_n(m2^{-n}, x\omega)\nabla\sigma\big(X_n(m2^{-n}, x, \omega)\big)\omega_{m+1}$$
$$- (t - m2^{-n})Y_n(m2^{-n}, x, \omega)\nabla b\big(X_n(m2^{-n}, x, \omega)\big)$$

for $m2^{-n} \le t \le (m+1)2^{-n}$. However, if one makes this guess and tries to show that $\|Y_n(t, x, \omega)X_n(t, x, \omega) - I\|_{\text{H.S.}}$ tends to zero, one realizes that, due to the presence of the square root on $(t - m2^{-n})$, this guess is wrong. The reason is that there will be approximately 2^n terms of order 2^{-n}, and, although they contain random factors, these random factors will not have mean value 0. Thus, in order to handle these terms, one has to center their random factors by introducing what practitioners of stochastic integration would call their "Itô correction."

For the reason just given, the preceding guess has to be replaced by $Y_n(0, x, \omega) = I$ and

$$Y_n\big((m + 1)2^{-n}, x, \omega\big) - Y_n(m2^{-n}, x, \omega)$$
$$= -2^{-\frac{n}{2}} Y_n(m2^{-n}, x, \omega)\big(\nabla(\sigma\omega_{m+1})\big)\big(X_n(m2^{-n}, x, \omega)\big)$$
$$+ 2^{-n}Y_n(m2^{-n}, x, \omega)\big(-\nabla b + (\nabla\sigma)^2\big)\big(X_n(m2^{-n}, x, \omega)\big)\Big),$$

where

$$(\nabla\sigma)^2 \equiv \sum_{k=1}^{N}\big(\nabla(\sigma e_k)\big)^2.$$

Clearly,

$$\boldsymbol{\omega} \rightsquigarrow \begin{pmatrix} X_n(t, x, \boldsymbol{\omega}) \\ Y_n(t, x, \boldsymbol{\omega}) \end{pmatrix}$$

corresponds to the coefficients

$$(x, \check{J}) \rightsquigarrow \begin{pmatrix} \sigma(x) \\ -\check{J}\nabla\sigma(x) \end{pmatrix} \quad \text{and} \quad (x, \check{J}) \rightsquigarrow \begin{pmatrix} b(x) \\ \check{J}(-\nabla b(x) + (\nabla\sigma)^2(x)) \end{pmatrix},$$

with initial condition (x, I), and so, for each $r \in [1, \infty)$ there exists an $A_r < \infty$, depending only on the bounds on σ, b, and their first derivatives, such that

$$\sup_{n \geq 0} \int \|Y_n(1, x, \boldsymbol{\omega})\|_{\text{H.S.}}^{2r} \, \Gamma(d\boldsymbol{\omega}) \leq A_r.$$

Next, set $\Delta_n(m, \boldsymbol{\omega}) = Y_n(m2^{-n}, x, \boldsymbol{\omega})X_n^1(m2^{-n}, x, \boldsymbol{\omega}) - I$. Then

$$\Delta_n(m+1, \boldsymbol{\omega}) - \Delta_n(m, \boldsymbol{\omega})$$

$$= -2^{-n}Y_n(m2^{-n}, x, \boldsymbol{\omega})\Bigg(\sum_{k,k'=1}^{N} \Sigma_n(m, \boldsymbol{\omega})e_k\Sigma_n(m, \boldsymbol{\omega})e_{k'}$$

$$\times \left[(e_k, \boldsymbol{\omega}_{m+1})_{\mathbb{R}^N} (e_{k'}, \boldsymbol{\omega}_{m+1})_{\mathbb{R}^N} - \delta_{k,k'} \right] \Bigg) X_n^1(m2^{-n}, x, \boldsymbol{\omega})$$

$$- 2^{-\frac{3n}{2}} Y_n(m2^{-n}, x, \boldsymbol{\omega}) \Big(\left[\Sigma_n(m, \boldsymbol{\omega})\boldsymbol{\omega}_{m+1}, B_n(m, \boldsymbol{\omega}) \right]$$

$$+ \Sigma_n(m, \boldsymbol{\omega})\boldsymbol{\omega}_{m+1}(\nabla\sigma)^2\big(X_n(m2^{-n}, x, \boldsymbol{\omega})\big) \Big) X_n^1(m2^{-n}, x, \boldsymbol{\omega})$$

$$- 4^{-n}Y_n(m2^{-n}, x, \boldsymbol{\omega})B_n(m, \boldsymbol{\omega})\big[(B_n(m, \boldsymbol{\omega})$$

$$+ (\nabla\sigma)^2\big(X_n(m2^{-n}, x, \boldsymbol{\omega})\big) \big] X_n^1(m2^{-n}, x, \boldsymbol{\omega}),$$

where $[A_1, A_2] = A_1 A_2 - A_2 A_1$ is the commutator of A_1, $A_2 \in \text{Hom}(\mathbb{R}^N; \mathbb{R}^N)$,

$$\Sigma_n(m, \boldsymbol{\omega})\xi \equiv \big(\nabla(\sigma\xi)\big)\big(X_n(m2^{-n}, x, \boldsymbol{\omega})\big),$$
$$\text{and } B_n(m, \boldsymbol{\omega}) \equiv \nabla b\big(X_n(m2^{-n}, x, \boldsymbol{\omega})\big).$$

Because $(e_k, \boldsymbol{\omega}_{m+1})_{\mathbb{R}^N} (e_{k'}, \boldsymbol{\omega}_{m+1})_{\mathbb{R}^N} - \delta_{k,k'}$ has mean value 0 and $\boldsymbol{\omega}_{m+1}$ is independent of $X_n(m2^{-n}, x, \boldsymbol{\omega})$, $Y_n(m2^{-n}, x, \boldsymbol{\omega})$, $\check{X}_n^1(m2^{-n}, x, \boldsymbol{\omega})$, and $\Delta_n(m, \boldsymbol{\omega})$, we see that

$$\int \|\Delta_n(m+1, \boldsymbol{\omega})\|_{\text{H.S.}}^2 \, \Gamma(d\boldsymbol{\omega}) \leq \int \|\Delta_n(m, \boldsymbol{\omega})\|_{\text{H.S.}}^2 \, \Gamma(d\boldsymbol{\omega}) + C4^{-n}$$

for some $C < \infty$. Hence, if $\Delta_n(\boldsymbol{\omega}) = \Delta_n(2^n, \boldsymbol{\omega}) = Y_n(1, x, \boldsymbol{\omega})X_n(t, x, \boldsymbol{\omega}) - I$, then

$$\int \|\Delta_n(\boldsymbol{\omega})\|_{\text{H.S.}}^2 \, \Gamma(d\boldsymbol{\omega}) \leq C2^{-n}.$$

Now set $B_n = \{\boldsymbol{\omega} : \|\Delta_n(\boldsymbol{\omega})\|_{\text{H.S.}} \geq \frac{1}{2}\}$. Off B_n, $I + \Delta_n(\boldsymbol{\omega})$ is invertible and its inverse is given by the Newman series $\sum_{m=0}^{\infty} (-\Delta_n(\boldsymbol{\omega}))^m$, which has Hilbert–Schmidt norm less than or equal to 2. Moreover, from $Y_n(1, x, \boldsymbol{\omega}) X_n(1, x, \boldsymbol{\omega}) = I + \Delta_n(\boldsymbol{\omega})$, it is clear that $X_n(1, x, \boldsymbol{\omega})$ is invertible when $\boldsymbol{\omega} \notin B_n$ and that $X_n(1, x, \boldsymbol{\omega})^{-1} = (I + \Delta_n(\boldsymbol{\omega}))^{-1} Y_n(1, x, \boldsymbol{\omega})$ there. In particular, this means that

$$\left| \det(X_n(1, x, \boldsymbol{\omega})) \right|^{-1} \leq 2 \left| \det(Y_n(1, x, \boldsymbol{\omega})) \right| \leq 2N! \|Y_n(1, x, \boldsymbol{\omega})\|_{\text{H.S.}}^N$$

for $\boldsymbol{\omega} \notin B_n$. Finally, define $f_\epsilon : \text{Hom}(\mathbb{R}^N; \mathbb{R}^N) \longrightarrow (0, \infty)$ by $f_\epsilon(J) = (\epsilon + |\det(J)|)^{-1}$. Then, by the preceding, for any $q \in [1, \infty)$,

$$\int f_\epsilon(X_n(1, x, \boldsymbol{\omega}))^q \, \Gamma(d\boldsymbol{\omega})$$

$$= \int_{B_n\complement} f_\epsilon(X_n(1, x, \boldsymbol{\omega}))^q \, \Gamma(d\boldsymbol{\omega}) + \int_{B_n} f_\epsilon(X_n(1, x, \boldsymbol{\omega}))^q \, \Gamma(d\boldsymbol{\omega})$$

$$\leq (2N!)^q \int \|Y_n(1, x, \boldsymbol{\omega})\|_{\text{H.S.}}^{Nq} \, \Gamma(d\boldsymbol{\omega}) + 2\epsilon^{-q} \int \|\Delta_n(\boldsymbol{\omega})\|_{\text{H.S.}}^2 \, \Gamma(d\boldsymbol{\omega}),$$

and so

$$\int |\det(J)|^{-q} P^{(1)}(1, (x, I), dy \times dJ) = \lim_{\epsilon \searrow 0} \int f_\epsilon(J) \, P^{(1)}(1, (x, I), dy \times dJ)$$

$$\leq (2N!)^q \sup_{n \geq 0} \int \|Y_n(1, x, \boldsymbol{\omega})\|_{\text{H.S.}}^{Nq} \, \Gamma(d\boldsymbol{\omega}).$$

As a consequence, $P^{(1)}(1, (x, I))$-almost surely, J is invertible. Furthermore, since $\|J^{-1}\|_{\text{H.S.}} \leq N! |\det(J)|^{-1} \|J\|_{\text{H.S.}}^N$, we also know that, for each $r \in [1, \infty)$, there is an $A_r < \infty$ such that

$$(3.3.2) \qquad \int \|J^{-1}\|_{\text{H.S.}}^{2r} \, P^{(1)}(1, x, dy \times dJ) \leq A_r.$$

3.3.3. The Fundamental Solution: We now have all the machinery that we need to prove the following result.

LEMMA 3.3.3.   *For each $x \in \mathbb{R}^N$ there is a uniformly continuous $y \in \mathbb{R}^N \longmapsto p(1, x, y) \in [0, \infty)$ such that $P(1, x, dy) = p(1, x, y) \, dy$. Moreover, $\|p(1, x, \cdot)\|_u$ as well as the modulus of continuity of $y \rightsquigarrow p(1, x, y)$ can be estimated, independent of x, in terms of the ϵ in (3.2.1) and the bounds on a and b and their first derivatives.*

PROOF: In view of Lemma 3.3.1, it suffices for us to prove that, for some $r \in (N, \infty)$, there is a constant C, with the required dependence, such that

$$\left| \int \nabla \varphi(y) \, P(1, x, dy) \right| \leq C \|\varphi\|_{L^{r'}(P(1,x); \mathbb{R})}.$$

To this end, take $\tilde{\sigma}$ and $\tilde{b}$ as in Theorem 3.2.4, and apply Theorem 3.1.3 with

$$\tilde{\varphi}_{k,\ell}(\tilde{x}) = \varphi(x)\psi_{k,\ell}(J), \text{ where}^{[1]} \; \psi_{k,\ell}(J) \equiv \big(e_k, J^{-1}e_\ell\big)_{\mathbb{R}^N} \text{ and } \tilde{\xi}_k = \begin{pmatrix} 0 \\ 0 \\ e_k \end{pmatrix},$$

to obtain

$$\int \varphi(x')\psi_{k,\ell}(J')\big(\chi', e_k\big)_{\mathbb{R}^N} \tilde{P}\big(1, (x, I, 0), dx' \times dJ' \times d\chi'\big)$$

$$= \int_0^1 \left(\int \left(\int \big(\tilde{\nabla}\tilde{\varphi}(\tilde{x}''), \tilde{J}''\tilde{a}(\tilde{x}'')\tilde{\xi}_k\big)_{\mathbb{R}^N \times E_1 \times \mathbb{R}^N} \right. \right.$$

$$\left. \left. \times \tilde{P}^{(1)}\big(1 - \tau, (\tilde{x}', \tilde{I}), d\tilde{x}'' \times d\tilde{J}''\big) \right) \tilde{P}\big(\tau, (x, I, 0), d\tilde{x}'\big) \right) d\tau$$

$$= \int \big(\nabla\varphi(x'), J'e_k\big)_{\mathbb{R}^N} \big(e_k, (J')^{-1}e_\ell\big)_{\mathbb{R}^N} P^{(1)}\big(1, (x, I), dx' \times dJ'\big)$$

$$+ \int_0^1 \left(\int \left(\int \varphi(x'')\big(\tilde{\nabla}\psi_{k,\ell}(J''), \tilde{J}''\tilde{a}(\tilde{x}'')e_\ell\big)_{\mathbb{R}^N \times E_1 \times \mathbb{R}^N} \right. \right.$$

$$\left. \left. \times \tilde{P}^{(1)}\big(1 - \tau, (\tilde{x}', \tilde{I}), d\tilde{x}'' \times d\tilde{J}''\big) \right) \tilde{P}\big(\tau, (x, I, 0), d\tilde{x}'\big) \right) d\tau.$$

Thus, after summing over $1 \leq k \leq N$, we find that

$$\int \big(\nabla\varphi(y), e_\ell\big)_{\mathbb{R}^N} P(1, x, dy)$$

is the sum of N terms of the form

$$\int \varphi(x')\psi_{k,\ell}(J')\big(\chi', e_k\big)_{\mathbb{R}^N} \tilde{P}\big(1, (x, I, 0), dx' \times dJ' \times d\chi'\big)$$

minus the sum of N terms of the form

$$\int_0^1 \left(\int \left(\int \varphi(x'')\big(\tilde{\nabla}\psi_{k,\ell}(J''), \tilde{J}''\tilde{a}(\tilde{x}'')e_\ell\big)_{\mathbb{R}^N \times E_1 \times \mathbb{R}^N} \right. \right.$$

$$\left. \left. \times \tilde{P}^{(1)}\big(1 - \tau, (\tilde{x}', \tilde{I}), d\tilde{x}'' \times d\tilde{J}''\big) \right) \tilde{P}\big(\tau, (x, I, 0), d\tilde{x}'\big) \right) d\tau.$$

Finally, given any $r \in (1, \infty)$, apply Hölder's inequality to see that terms of the first sort are dominated by

$$\left(\int |\chi'|^r |\psi_{k,\ell}(J')|^r \tilde{P}\big(1, (x, I, 0), dx' \times dJ' \times d\chi'\big) \right)^{\frac{1}{r}} \|\varphi\|_{L^{r'}(P(t,x);\mathbb{R})}.$$

---

[1] Strictly speaking, we should use an approximation procedure here since we only know that J^{-1} exists off a set of measure 0. However, we have ignored this technicality and leave the concerned reader to fill in the details for himself.

As for terms of the second sort, Hölder's inequality allows one to dominate each of them by

$$
\int_0^1 \left(\int \left(\int \left| \left(\tilde{\nabla} \psi_{k,\ell}(J''), \tilde{J}'' \tilde{a}(\tilde{x}'') e_\ell \right)_{\mathbb{R}^N \times E_1 \times \mathbb{R}^N} \right|^r \right. \right.
$$
$$
\left. \left. \times \tilde{P}^{(1)}\left(1 - \tau, (\tilde{x}', \tilde{I}), d\tilde{x}'' \times d\tilde{J}''\right) \right)^{\frac{1}{r}} \tilde{P}\left(\tau, (x, I, 0), d\tilde{x}'\right) \right) d\tau
$$

times

$$
\int_0^1 \left(\int \left(\int |\varphi(x'')|^{r'} P(1 - \tau, x', dx'') \right)^{\frac{1}{r'}} P(\tau, x, dx') \right) d\tau
$$
$$
\leq \int_0^1 \left(\int \left(\int |\varphi(x'')|^{r'} P(1 - \tau, x', dx'') \right) P(\tau, x, dx') \right)^{\frac{1}{r'}} d\tau = \|\varphi\|_{L^{r'}(P(1,x))}.
$$

Because of (3.2.5) and the estimates in Theorem 2.2.7 and (3.3.2), it remains only to note that $\tilde{\nabla} \psi_{k,\ell}(J) = -\left((J^{-1})^\top e_k \right) \otimes \left((J^{-1})^\top e_\ell \right)$. $\square$

We now use Lemma 3.3.3 to produce the fundamental solution advertised in the introduction to this section.

It should be clear that the argument used to get $p(1, x, \cdot)$ could have been used to produce a bounded, uniformly continuous $p(t, x, \cdot)$ such that $P(t, x, dy) = p(t, x, y)\, dy$ for all $(t, x) \in (0, \infty) \times \mathbb{R}^N$. However, the estimate which we would have gotten on $p(t, x, \cdot)$ as a function of $t > 0$ would have been far from optimal as $t \searrow 0$. For this reason, we will take a different tack, one which is based on elementary *scaling* considerations.

Given $\lambda \in (0, \infty)$, set $a_\lambda(x) = a(\lambda^{\frac{1}{2}} x)$ and $b_\lambda(x) = \lambda^{\frac{1}{2}} b(\lambda^{\frac{1}{2}} x)$, and let $(t, x) \rightsquigarrow P_\lambda(t, x)$ be the associated transition probability function. We want to show that

$$
(*) \qquad \int \varphi(y)\, P(t, x, dy) = \int \varphi(\lambda^{\frac{1}{2}} y)\, P_\lambda\left(\lambda^{-1} t, \lambda^{-\frac{1}{2}} x, dy\right).
$$

For this purpose, fix $x \in \mathbb{R}^N$ and define $\mu_{t,\lambda} \in \mathbf{M}_1(\mathbb{R}^N)$ so that $\langle \varphi, \mu_{t,\lambda} \rangle$ is given by the right-hand side of (*). Then

$$
\frac{d}{dt} \langle \varphi, \mu_{t,\lambda} \rangle = \langle L\varphi, \mu_{t,\lambda} \rangle \quad \text{and} \quad \varphi(x) = \lim_{t \searrow 0} \langle \varphi, \mu_{t,\lambda} \rangle
$$

for all $\varphi \in C_c^\infty(\mathbb{R}^N; \mathbb{C})$. Hence, by Corollary 2.1.6, $\mu_{t,\lambda} = P(t, x)$, which is to say that (*) holds.

Now set

$$
p(t, x, y) = t^{-\frac{N}{2}} p_t\left(1, t^{-\frac{1}{2}} x, t^{-\frac{1}{2}} y\right),
$$

where $y \rightsquigarrow p_t(1, x, y)$ is the density for $P_\lambda(1, x)$ with $\lambda = t$. Then, from (*), it is easy to check that

$$
(3.3.4) \qquad\qquad P(t, x, dy) = p(t, x, y)\, dy.
$$

Moreover, because, for $t \in (0,1]$, a_t, b_t, and their derivatives are bounded by the corresponding quantities for a and b, the last part of Lemma 3.3.1 says that

$$\sup_{(t,x)\in(0,1]\times\mathbb{R}^N} t^{\frac{N}{2}} \|p(t,x,\cdot)\|_u < \infty$$

and

$$\lim_{|h|\searrow 0} \sup_{(t,x,y)\in(0,1]\times\mathbb{R}^N\times\mathbb{R}^N} t^{\frac{N}{2}} |p(t,x,y+h) - p(t,x,y)| = 0.$$

We next want to show that $(t,x,y) \rightsquigarrow p(t,x,y)$ is continuous and smooth as a function of (t,x). To this end, first note that, by the Chapman–Kolmogorov equation,

$$p(s+t,x,y) = \int p(s,x',y)\, P(t,x,dx').$$

Combined with the results in Theorem 3.2.6, this proves that $(t,x) \rightsquigarrow p(t,x,y)$ is smooth and that, for each $m \in \mathbb{N}$ and $\alpha \in \mathbb{N}^N$, there is a $K_{m,\alpha} < \infty$, with the dependence described in that theorem, such that

$$(3.3.5) \qquad |\partial_t^m \partial_x^\alpha p(t,x,y)| \le \frac{K_{m,\alpha}}{1 \wedge t^{\frac{2m+\|\alpha\|+N}{2}}}.$$

In particular, by putting this together with the final part of Lemma 3.3.3, one sees that $(t,x,y) \rightsquigarrow p(t,x,y)$ is continuous on $(0,\infty) \times \mathbb{R}^N \times \mathbb{R}^N$. At the same time, one sees that $(t,x) \rightsquigarrow p(t,x,y)$ solves the backward equation. Hence, for each $y \in \mathbb{R}^N$,

$$(3.3.6) \qquad \begin{aligned} \partial_t p(t,x,y) &= \big[Lp(t,\cdot,y)\big](x) \quad \text{on } (0,\infty)\times\mathbb{R}^N \\ &\text{and } \lim_{t\searrow 0} p(t,\cdot,y) = \delta_y, \end{aligned}$$

which is to say that $p(t,x,y)$ is the *fundamental solution* to Kolmogorov's backward equation (2.0.1). Finally, observe that the Chapman–Kolmogorov relation above can be rewritten as

$$(3.3.7) \qquad p(s+t,x,y) = \int p(s,x,x')p(t,x',y)\, dx',$$

which is the *Chapman–Kolmogorov equation* for the density of the transition probability function $(t,x) \rightsquigarrow P(t,x)$.

3.3.4. Smoothness in the Forward Variable: It is possible to prove the smoothness of $y \rightsquigarrow p(t,x,y)$ by reasoning analogous to that in §3.3.3 applied to higher derivatives. However, the expressions which one gets are very cumbersome, and so we will use a different approach. Namely, by combining (3.3.4) with Theorem 2.5.8, we see that

$$(3.3.8) \qquad P^\top(t,y,dx) = p(t,y,x)\, dx.$$

Thus, we can prove the smoothness of $p(t, x, y)$ as a function of its forward variable y by treating y as the backward variable for the adjoint transition function $P^\top(t, y)$.

To be more precise, define $\tilde{\sigma} : \mathbb{R}^N \times \mathbb{R} \times E_1 \times \mathbb{R}^N \longrightarrow \mathrm{Hom}\big(\mathbb{R}^N; \mathbb{R}^N \times \mathbb{R} \times E_1 \times \mathbb{R}^N\big)$ and $\tilde{b} : \mathbb{R}^N \times \mathbb{R} \times E_1 \times \mathbb{R}^N \longrightarrow \mathbb{R}^N \times \mathbb{R} \times E_1 \times \mathbb{R}^N$ by (cf. (2.5.2)),

$$\tilde{\sigma}(\tilde{y}) = \begin{pmatrix} \sigma(y) \\ 0 \\ \nabla\sigma(y)J \\ J^\top \sigma(y)^{-1} \end{pmatrix} \quad \text{and} \quad \tilde{b}(\tilde{y}) = \begin{pmatrix} b^\top(y) \\ V^\top(y) \\ \nabla b^\top(y)J \\ 0 \end{pmatrix}$$

for $\tilde{y} = (y, \eta, \tilde{J}, \chi)$. Then, taking $u_\psi^\top(t, x) = \langle \psi, P^\top(t, x) \rangle$ and proceeding in exactly the same way as we did in the derivation of Lemma 3.3.3, only now taking account the presence of $e^{\eta'}$ in equation (2.5.5) for $P^\top(t, y)$ in terms of $\tilde{P}\big(t, (x, 0)\big)$ by defining $\tilde{\psi}(\tilde{y}) = \psi(y)e^\eta$ for $\tilde{y} = (y, \eta, J, \chi)$, we find that

$$t\nabla u_\psi^\top(t, y) = \int e^{\eta'} \psi(y')\chi' \, \tilde{P}\big(t, (y, 0, I, 0), dy' \times d\eta' \times dJ' \times d\chi'\big)$$
$$- \int_0^t \left(\int \left(\int e^{\eta''} \psi(y'') \tilde{J}_{(21)}'' J' \tilde{P}^{(1)}\big(t - \tau, (\tilde{y}', \tilde{J}'), d\tilde{y}'' \times dJ''\big) \right) \right.$$
$$\left. \times \tilde{P}\big(\tau, (y, 0, I, 0), d\tilde{y}'\big) \right) d\tau.$$

From this expression, we see that there is a $K_1^\top$, depending only on ϵ and the bounds on the first derivatives of a, $b^\top$, and $V^\top$, such that $|\nabla u_\psi^\top(t, y)| \leq K_1^\top t^{-\frac{1}{2}} \|\psi\|_\mathrm{u}$ for $(t, y) \in (0, 1] \times \mathbb{R}^N$. Hence, because

$$u_\psi^\top(s + t, y) = \int u_\psi^\top(t, y') \, P^\top(s, y, dy'),$$

we can use the same reasoning as we used in §3.2.2, especially the proof of Theorem 3.2.6, to arrive at

$$(*) \qquad \|\nabla^\ell u_\psi^\top(t, y)\|_\mathrm{H.S.} \leq \frac{K_\ell^\top e^{t\|V^\top\|_\mathrm{u}}}{1 \wedge t^{\frac{\ell}{2}}} \|\psi\|_\mathrm{u},$$

for a $K_\ell^\top < \infty$ which depends only on ϵ and the bounds on the a, $b^\top$, and $V^\top$ and their derivatives through order $\ell + 1$.

Our interest in the preceding is that it allows us to prove that, for each $n \geq 1$, there exists a $A_n < \infty$, depending only on the ϵ in (3.2.1) and the bounds on the derivatives of a, b, $b^\top$, and $V^\top$ of orders through $n + 1$, such that

$$(3.3.9) \qquad |\partial_t^m \partial_x^\alpha \partial_y^\beta p(t, x, y)| \leq \frac{A_n e^{t\|V^\top\|_\mathrm{u}}}{1 \wedge t^{\frac{n+N}{2}}} \quad \text{when } 2m + \|\alpha\| + \|\beta\| = n.$$

To see this, note that

$$\partial_t^m \partial_x^\alpha p(2t, x, y) = \int \partial_t^m \partial_x^\alpha p(t, x, y') p(t, y', y) \, dy' = u_\psi^\top(t, y),$$

where $\psi = \partial_t^m \partial_x^\alpha p(t, x, \cdot)$. Hence, (3.3.9) follows when one combines (*) with (3.3.5).

Knowing that $p(t, x, \cdot)$ is smooth and using

$$\frac{d}{dt} \int \varphi(y) \, P(t, x, dy) = \int L\varphi(y) \, P(t, x, dy), \quad \varphi \in C_c^2(\mathbb{R}^N; \mathbb{R}),$$

to see that $(t, y) \rightsquigarrow p(t, x, y)$ is a solution of Kolmogorov's forward equation

$$(3.3.10) \qquad \partial_t p(t, x, y) = [L^\top p(t, x, \cdot)](y) \quad \text{with } \lim_{t \searrow 0} p(t, x, \cdot) = \delta_x.$$

3.3.5. Gaussian Estimates: We will complete this preliminary discussion of the fundamental solution by proving the following statement.

THEOREM 3.3.11.   *There exists an $A_0 < \infty$, depending only on the ϵ in (3.2.1), the bounds on a, b, and (cf. (2.5.2)) $b^\top$ and $V^\top$, and their first derivatives, such that*[2]

$$p(t, x, y) \le \frac{A_0}{1 \wedge t^{\frac{N}{2}}} \exp\left(-\left(A_0 t - \frac{|y - x|^2}{A_0 t} \right)^- \right).$$

Moreover, for each $n \ge 1$, there exists an $A_n < \infty$, depending only on ϵ and the bounds on a, b, $b^\top$, $V^\top$, and their derivatives through order $(n + 1)$, such that

$$\left| \partial_t^m \partial_x^\alpha \partial_y^\beta p(t, x, y) \right| \le \frac{A_n}{1 \wedge t^{\frac{n+N}{2}}} \exp\left(-\left(A_n t - \frac{|y - x|^2}{A_n t} \right)^- \right)$$

when $2m + \|\alpha\| + \|\beta\| = n$

The proof of the first estimate is an application of Theorem 2.2.7, the Chapman–Kolmogorov, the estimates we already have on $p(t, x, y)$, and the following lemma. In the computations which follow, it is helpful to remember (1.3.1).

LEMMA 3.3.12.   *If $A = \|\mathrm{Trace}(a)\|_u$ and $B = \|b\|_u$, then*

$$\int e^{\frac{|y-x|^2}{4(A+B)t}} \, P(t, x, dy) \le K e^{Bt} \quad \text{where } K \equiv \sum_{n=0}^\infty \binom{2n}{n} 2^{-n} < \infty,$$

---

[2] We use $\xi^- = -(\xi \wedge 0)$ to denote the negative part of $\xi \in \mathbb{R}$.

and so

$$P\big(t, x, B(x, R)\complement\big) \le K \exp\left(\left(Bt - \frac{R^2}{4(A+B)t}\right)^-\right).$$

PROOF: Given $x \in \mathbb{R}^N$, set

$$u_n(t) = \int |y - x|^{2n}\, P(t, x, dy) \quad \text{and} \quad v_n(t) = e^{-Bt} u_n(t).$$

Then, applying (1.1.3) and using (1.3.1), one sees that

$$\dot{u}_n(t) = \int \Big(2n(n-1)|y-x|^{2(n-2)}\big((y-x), a(y)(y-x)\big)_{\mathbb{R}^N}$$
$$+ n|y-x|^{2(n-1)}\big[\mathrm{Trace}\big(a(y)\big) + 2\big(y-x, b(y)\big)_{\mathbb{R}^N}\big]\Big) P(t, x, dy)$$
$$\le An(2n-1)u_{n-1}(t) + 2Bn\int |y-x|^{2n-1}\, P(t, x, dy),$$

and so

$$\dot{v}_n(t) \le e^{-Bt}\bigg(An(2n-1)u_{n-1}(t)$$
$$+ B\int \big(2n|y-x| - |y-x|^2\big)|y-x|^{2(n-1)}\bigg) P(t, x, dy)$$
$$\le \big(An(2n-1) + Bn^2\big)v_{n-1}(t) \le (A+B)n(2n-1)v_{n-1}(t)$$

for $n \ge 1$. Hence, since $v_0(t) \le 1$, we can use induction on n to get

$$v_n(t) \le \big((A+B)t\big)^n \prod_{m=1}^{n}(2m-1) = \binom{2n}{n}\left(\frac{(A+B)t}{2}\right)^n,$$

from which the desired conclusion is an easy step since, by Stirling's formula, $\binom{2n}{n} \sim \frac{4^n}{\sqrt{\pi n}}$. $\square$

Now set $R = \frac{|y-x|}{2}$ and use the Chapman–Kolmogorov equation to write

$$p(2t, x, y) = \int p(t, x, x')p(t, x', y)\, dx'$$
$$\le \int_{B(x,R)\complement} p(t, x, x')p(t, x', y)\, dx' + \int_{B(y,R)\complement} p(t, x, x')p(t, x', y)\, dx'.$$

By (3.3.5) with $m = 0$ and $\alpha = 0$, there exists a $C < \infty$, with the required dependence, such that the right-hand side is dominated by

$$\frac{C}{1 \wedge t^{\frac{N}{2}}}\Big(P\big(t, x, B(x, R)\complement\big) + P^\top\big(t, y, B(y, R)\complement\big)\Big).$$

Finally, by the estimate in Lemma 3.3.12, the $P\big(t, x, B(x, R)\complement\big)$ is dominated by $C \exp\left(Ct - \frac{R^2}{Ct}\right)$ for another $C < \infty$ depending only on bounds on a and b. At the same time, (cf. the notation in §2.5),

$$
\begin{aligned}
P^\top\big(t, y, B(y, R)\complement\big) &= \int_{B(y,R)\complement} e^{\eta'}\, \bar{P}^\top\big(t, (y, 0), dy' \times d\eta'\big) \\
&\leq e^{t\|V^\top\|_u}\bar{P}^\top\big(t, (y, 0), B(y, R)\complement \times \mathbb{R}\big),
\end{aligned}
$$

and another application of Lemma 3.3.12, this time to $\bar{\sigma}\bar{\sigma}^\top$ and $\bar{b}^\top$, completes the proof of the first estimate in Theorem 3.3.11.

To prove the estimates for derivatives of $p(t, x, y)$, we will use a simple, but rather crude, *interpolation* argument.

LEMMA 3.3.13. *Let $n \geq 1$, and suppose that $F : \mathbb{R}^N \longrightarrow \mathbb{R}$ is an $(n+1)$-times continuously differentiable function. Given a closed cube Q in $\mathbb{R}^N$ of side length 2 and an α with $\|\alpha\| = n$,*

$$
\|\partial^\alpha F\|_{u,Q} \leq 9\|F\|_{u,Q}^{2^{-n}} \max_{\|\beta\|\leq n+1} \|\partial^\beta F\|_{u,Q}^{1-2^{-n}},
$$

where $\|\cdot\|_{u,Q}$ denotes the uniform norm on Q.

PROOF: We begin with the case when $n = N = 1$. Given $x \in [-1, 0]$, Taylor's Theorem says that, for $h \in (0, 1]$,

$$
F(x+h) = F(x)+hF'(x)+\frac{h^2}{2}F''\big(x+\theta(x, h)\big) \quad \text{for some } \theta(x, h) \in [x, x+h].
$$

Hence,

$$
|F'(x)| \leq \frac{2}{h}\|F\|_{u,Q} + \frac{h}{2}\|F''\|_{u,Q} \quad \text{for all } h \in (0, 1].
$$

In particular, by taking $h^2 = \frac{\|F\|_{u,Q}}{\|F\|_{u,Q}\vee\|F''\|_{u,Q}}$, we see that

$$
|F'(x)| \leq 3\|F\|_{u,Q}^{\frac{1}{2}}\big(\|F\|_{u,Q} \vee \|F''\|_{u,Q}\big)^{\frac{1}{2}}.
$$

The same argument, with $h \in [-1, 0)$, applies to $x \in [0, 1]$, and, by working with one coordinate at a time, one sees that, for any N,

$$
(*) \qquad \max_{\|\alpha\|=1} \|\partial^\alpha F\|_{u,Q} \leq 3\|F\|_{u,Q}^{\frac{1}{2}} \max_{\|\beta\|\leq 2} \|\partial^\beta F\|_{u,Q}^{\frac{1}{2}}.
$$

To complete the proof, we work by induction on $n \geq 1$. Thus, assume the result holds for n, let $\|\beta\| = n$, and suppose α is obtained from β by adding 1 to precisely one of its coordinates. Then, by $(*)$ applied to $\partial^\beta F$ and the induction hypothesis, we know that

$$\|\partial^\alpha f\|_{u,Q} \le 3\|\partial^\beta F\|_{u,Q}^{\frac{1}{2}} \max_{\|\beta'\|\le n+1} \|\partial^{\beta'} F\|_{u,Q}^{\frac{1}{2}}$$

$$\le 9\|F\|_{u,Q}^{2-2^{-(n+2)}} \max_{\|\beta'\|\le n+2} \|\partial^{\beta'} F\|_{u,Q}^{1-2^{-(n+1)}}. \quad \square$$

Returning to the proof of Theorem 3.3.11, let $(t,x,y) \in [1,\infty) \times \mathbb{R}^N \times \mathbb{R}^N$ be given, and set $R = |y-x|$. Then we can choose cubes Q_1 and Q_2, each of side length 2, so that $x \in Q_1$, $y \in Q_2$, and $|x'-y'| \ge R$ for all $x' \in Q_1$ and $y' \in Q_2$. Hence, by Lemma 3.3.13, we know that

$$\left|\partial_x^\alpha \partial_y^\beta p(t,x,y)\right| \le 9\|\partial_y^\beta p(t,\cdot,y)\|_{u,Q}^{2-\|\alpha\|} \max_{\|\alpha'\|\le\|\alpha\|+1} \|\partial_x^{\alpha'}\partial_y^\beta p(t,\cdot,y)\|_{u,Q}^{1-2^{-\|\alpha\|}}$$

$$\le 81\|p(t,\cdot,*)\|_{u,Q_1\times Q_2}^{2-\|\alpha\|-\|\beta\|} \max_{\substack{\|\alpha'\|\le\|\alpha\|+1 \\ \|\beta'\|\le\|\beta\|+1}} \|\partial_x^{\alpha'}\partial_y^{\beta'} p(t,\cdot,*)\|_{u,Q_1\times Q_2}^{1-2^{-\|\alpha\|-\|\beta\|}},$$

which, together with (3.3.9) and the estimate already proved for $p(t,x,y)$, gives the required result when $t \ge 1$ and $m=0$. Next let $t \in (0,1]$, and remember (cf. the scaling argument in § 3.3.3) that

$$p(t,x,y) = t^{-\frac{N}{2}} p_t(1, t^{-\frac{1}{2}}x, t^{-\frac{1}{2}}y).$$

Hence, because $t \le 1$ and therefore the coefficients determining p_t as well as their derivatives are bounded by the corresponding quantities for p, the result is proved for all (t,x,y) when $m=0$. Finally, to handle $m \ne 0$, recall that $\partial_t^m p(t,x,y) = [L^m p(t,\cdot,y)](x)$.

3.4 Hypoellipticity

In this subsection we will show how our results thus far, and particularly Theorem 3.3.11, lead to a proof that L is *hypoelliptic*. That is, we are going to show that if u is a distribution (in the sense of L. Schwartz) and Lu is smooth on an open set W, then u is smooth on W. In fact, after proving that L itself is hypoelliptic, we will show that the associated heat operator $L + \partial_t$ is also hypoelliptic.

3.4.1. Hypoellipticity of L: After an elementary translation and rescaling, one sees that it suffices to prove that if u is a distribution and $Lu \restriction B(0,5) = f \in C_b^\infty(B(0,5);\mathbb{R})$, then $u \restriction B(0,1) \in C^\infty(B(0,1);\mathbb{R})$. With this in mind, choose $\eta \in C_c^\infty(B(0,4);[0,1])$ so that $\eta = 1$ on $\overline{B(0,3)}$, and set $\tilde{u} = \eta u$. Obviously, it suffices for us to show that $\tilde{u} \restriction B(0,1)$ is smooth. To this end, define $\tilde{u}_\theta(x) = \langle p(\theta,x,\cdot),\tilde{u}\rangle$ for $(\theta,x) \in (0,1] \times \mathbb{R}^N$. Because $\tilde{u}$ has compact support and $p(\theta,x,\cdot) \in C^\infty(\mathbb{R}^N;\mathbb{R})$, there is no question that $\tilde{u}_\theta$ is well defined. In fact, because $(\theta, x, y) \leadsto p(\theta, x, y)$ is smooth, one can easily check that $(\theta, x) \leadsto \tilde{u}_\theta(x)$ is smooth and that (cf. (3.3.10))

$$\frac{d}{d\theta}\tilde{u}_\theta(x) = \langle \partial_\theta p(\theta, x, \cdot\,), \tilde{u}\rangle$$

$$= \langle L^T\big(p(\theta, x, \cdot\,)\big), \tilde{u}\rangle = \langle p(\theta, x, \cdot\,), L\tilde{u}\rangle.$$

Now, write $L\tilde{u} = \tilde{f} + \tilde{v}$, where $\tilde{f} = \eta f$. Clearly, $\tilde{v}$ is a distribution with compact support in the open annular region $A = B(0, 4) \setminus \overline{B(0, 2)}$, and

$$(*) \qquad\qquad \tilde{u}_\theta(x) = \tilde{u}_1 - \int_\theta^1 u_{\tilde{f}}(\xi, x)\,d\xi - \int_\theta^1 \tilde{v}_\xi(x)\,d\xi,$$

where $\tilde{v}_\xi = \langle p(\xi, x, \cdot\,), \tilde{v}\rangle$. The first term on the right-hand side of $(*)$ is smooth, and, because $\tilde{f} \in C_c^\infty(\mathbb{R}^N; \mathbb{R})$, we know (cf. Corollary 2.2.8) that the second term is also smooth and has derivatives which can be bounded independent of $\theta \in (0, 1]$. To handle the third term on the right of $(*)$, notice that

$$\partial_x^\alpha \tilde{v}_\theta(x) = \langle \partial_x^\alpha \bar{p}(\theta, x, \cdot\,), \tilde{v}\rangle,$$

where $\bar{p}(\theta, x, y) = \bar{\eta}(y)p(\theta, x, y)$ with an $\bar{\eta} \in C_c^\infty(A; [0, 1])$ which is 1 on an open neighborhood of $\mathrm{supp}(\tilde{v})$. Since, by Theorem 3.3.11, all derivatives of $(x, y) \in B(0, 1) \times \mathbb{R}^N \longmapsto \bar{p}(\theta, x, y)$ are bounded independent of $\theta \in (0, 1]$, it follows that all derivatives of the third term are bounded independent of $\theta \in (0, 1]$. Thus, we have now shown that, for every $\alpha \in \mathbb{N}^N$,

$$\sup_{\theta \in (0,1]} \sup_{x \in B(0,1)} |\partial_x^\alpha \tilde{u}_\theta(x)| < \infty.$$

With the preceding result at hand, all that remains is to check that $\tilde{u}_\theta$ tends, in the sense of distributions, to $\tilde{u}$ as $\theta \searrow 0$. For this purpose, let $\varphi \in C_c^\infty(\mathbb{R}^N; \mathbb{R})$ be given, and set $u_\varphi^T(t, y) = \langle \varphi, P^T(t, y)\rangle$. Then $\langle \varphi, \tilde{u}_\theta\rangle = \langle u_\varphi^T(\theta, \cdot\,), \tilde{u}\rangle$, and so (remember that $\tilde{u}$ has compact support) we will be done once we show that $u_\varphi^T(\theta, \cdot\,) \longrightarrow \varphi$ in $C^\infty(\mathbb{R}^N; \mathbb{R})$. But, using (2.5.5) and applying Corollary 2.2.8 to the transition probability function $(t, \bar{y}) \leadsto \bar{P}^T(t, \bar{y})$, this is easily verified.

3.4.2. Hypoellipticity of $L + \partial_t$: In this section we will prove that the heat operator $L + \partial_t$ is also hypoelliptic. The basic idea is essentially the same as in § 3.5.1, although the details are slightly more intricate.

THEOREM 3.4.1. *Let W be a non-empty, open subset of $\mathbb{R}^N$; assume that $a : W \longrightarrow \mathrm{Hom}(\mathbb{R}^N; \mathbb{R}^N)$ and $b : W \longrightarrow \mathbb{R}^N$ are smooth functions with the property that, for each $x \in W$, $a(x)$ is symmetric and strictly positive definite; and define the operator L accordingly, as in (1.1.8), on $C^2(W; \mathbb{R})$. If, for some open interval $J \neq \emptyset$ and distribution u on $J \times W$, $(L + \partial_t)u = f \in C^\infty(J \times W; \mathbb{R})$, then $u \in C^\infty(J \times W; \mathbb{R})$.*

PROOF: Without loss of generality, we will assume that a and b are defined and smooth on all of $\mathbb{R}^N$; that a, b, and all their derivatives are bounded; and that, for each $x \in \mathbb{R}^N$, $a(x)$ is symmetric and, for some $\epsilon > 0$, $a(x) \geq \epsilon I$. Further, after translation and scaling, we may and will assume that $J = (-5, 5)$ and $W = B(0, 5)$. What we have to show is that u is smooth on $(-1, 1) \times B(0, 1)$.

We begin by choosing $\eta \in C_c^\infty\big((-4, 4) \times B(0, 4); [0, 1]\big)$ so that $\eta = 1$ on $[-3, 3] \times \overline{B(0, 3)}$ and setting $\tilde{u} = \eta u$, $\tilde{f} = \eta f$, and $\tilde{v} = \tilde{f} - (L + \partial_t)\tilde{u}$. Clearly, $\tilde{v}$ has compact support in $A = [(-4, 4) \times B(0, 4)] \setminus [[-2, 2] \times \overline{B(0, 2)}]$. Next, choose $\rho \in C_c^\infty\big((1, 2); [0, \infty)\big)$ with total integral 1, and set $\rho_\theta(t) = \theta^{-1}\rho(\theta^{-1}t)$ for $\theta \in (0, 1)$. Finally, set

$$\tilde{u}_\theta(t, x) = \big\langle \rho_\theta(\,\cdot\, - t)p(\,\cdot\, - t, x, *), \tilde{u} \big\rangle,$$

where the notation is meant to indicate that, for a fixed (t, x), the action of $\tilde{u}$ is on the function $(\tau, y) \rightsquigarrow \rho_\theta(\tau - t)p(\tau - t, x, y)$. In order to compute $\frac{d}{d\theta}\tilde{u}_\theta(t, x)$, observe that

$$\frac{d}{d\theta}\rho_\theta(t) = -\dot{\psi}_\theta(t), \quad \text{where } \psi(t) = t\rho(t) \text{ and } \psi_\theta(t) = \theta^{-1}\psi(\theta^{-1}t).$$

Hence, $\frac{d}{d\theta}\tilde{u}_\theta(t, x)$ equals

$$-\big\langle \partial.\big(\psi_\theta(\,\cdot\, - t)p(\,\cdot\, - t, x, *)\big), \tilde{u} \big\rangle + \big\langle \psi_\theta(\,\cdot\, - t)L_*^\top p(\,\cdot\, - t, x, *), \tilde{u} \big\rangle$$
$$= \big\langle \psi_\theta(\,\cdot\, - t)p(\,\cdot\, - t, x, *), \tilde{f} \big\rangle + \big\langle \psi_\theta(\,\cdot\, - t)p(\,\cdot\, - t, x, *), \tilde{v} \big\rangle,$$

and so

$$\tilde{u}_\theta(t, x) = \tilde{u}_1(t, x) - \int_\theta^1 \big\langle \psi_\xi(\,\cdot\, - t)p(\,\cdot\, - t, x, *), \tilde{f} \big\rangle \, d\xi - \int_\theta^1 \tilde{v}_\xi(t, x) \, d\xi,$$

$$\text{where } \tilde{v}_\theta(t, x) \equiv \big\langle \psi_\theta(\,\cdot\, - t)p(\,\cdot\, - t, x, *), \tilde{v} \big\rangle.$$

Just as before, the first and second terms on the right have derivatives of all orders which are bounded independent of $\theta \in (0, 1)$. As for the third term, $\tilde{v}_\theta$ poses no problems as long as $\theta \in [\frac{1}{2}, 1)$. Thus, suppose that $\theta \in (0, \frac{1}{2})$, and consider $(t, x) \in (-1, 1) \times B(0, 1)$. Next, choose $\bar{\eta} \in C_c^\infty\big(A; [0, 1]\big)$ so that $\bar{\eta} = 1$ on an open neighborhood of supp$(\tilde{v})$, set $\bar{p}_\theta(t, x; \tau, y) = \bar{\eta}(\tau, y)\rho_\theta(\tau - t)p(\tau - t, x, y)$, and note that $\tilde{v}_\theta(t, x)$ is equal to $\big\langle \bar{p}_\theta(t, x; \,\cdot\,, *), \tilde{u} \big\rangle$. The key to controlling derivatives of this quantity is the observation that $\bar{p}_\theta(t, x; \tau, y) = 0$ unless $|y - x| \geq 1$. Indeed, because $|x| < 1$, $|y - x| < 1 \implies |y| < 2$. Thus, if $|y - x| < 1$ and $|\tau| \leq 2$, then $\bar{\eta}(\tau, y) = 0$. On the other hand, if $|\tau| > 2$, then, because $|t| < 1$ and $\theta \leq \frac{1}{2}$, $\rho_\theta(\tau - t) = 0$. Knowing that $\bar{p}_\theta(t, x'\tau, y) = 0$ unless $|y - x| \geq 1$, we can now use Theorem 3.3.11 to check that all derivatives of

$$(t, x; \tau, y) \in \big((-1, 1) \times B(0, 1)\big) \times \big(\mathbb{R} \times \mathbb{R}^N\big) \longmapsto \bar{p}_\theta(t, x; \tau, y) \in \mathbb{R}$$

are bounded independent of $\theta \in (0, \frac{1}{2}]$, the point being that although derivatives of $\bar{p}_\theta(t, x; \tau, y)$ produce factors which are reciprocal powers of θ and $\tau - t \in (\theta, 2\theta)$, such factors are canceled by the exponential factor in the estimates from Theorem 3.3.11.

Given the preceding, it remains to check that $\tilde{u}_\theta \longrightarrow \tilde{u}$ in the sense of Schwartz distributions, and this is done by the same line of reasoning for the analogous result earlier. □

REMARK 3.4.2. There are two remarks worth making about the preceding. The first is that the same reasoning allows one to conclude that if L is as in Theorem 3.4.1 and $c \in C^\infty(W; \mathbb{R})$, then the same conclusion holds when $L + \partial_t$ is replaced by $L + c + \partial_t$. The idea is that, by the technique with which we handled the 0th order term $V^\top$ of $L^\top$ in § 3.3.4, one can produce a kernel which plays the same role for $L + c$ as $p(t, x, y)$ plays for L. The second remark is of a less practical nature. Namely, it is interesting to note that the estimates in Theorem 3.3.11 do not, by themselves, seem sufficient to conclude hypoellipticity. Indeed, at least our proof requires that one also knows the "do no harm" sort of results contained in Chapter 2.

3.5 Historical Notes and Commentary

As far as I know, elliptic regularity theory began with H. Weyl's paper [57], in which he proved that weak (in the sense of L^2) solutions to Laplace's equations are classical solutions, thereby laying the groundwork for a rigorous treatment of Hodge theory. Since Weyl, a small industry devoted to extensions and refinements of his result has grown and flourished. Indeed, these extensions have become essential ingredients in the work of geometers as well as traditional analysts. For example, the Index Theorem of Atiyah and Singer and its descendants make heavy use of them.

In this chapter, the approach I have adopted makes minimal use of the powerful techniques (some of which are explained in Chapter 7) which analysts have developed. Instead, in keeping with the contents of Chapters 1 and 2, I have based the treatment on Itô's representation of solutions, to which I have applied a technique which, a long time ago, I dubbed "Malliavin's calculus." In fact, just as Chapter 1 is a poor man's version of Itô's theory, the present chapter is an equally poor man's version of Malliavin's. Malliavin's seminal idea in [37] was that Itô's representation provides a vehicle for transferring smoothness properties of Gaussian measures to the distribution of a diffusion process. If one is doing true Malliavin calculus, implementation of his idea requires one to come to terms with all sorts of technical difficulties[3] which arise when doing analysis in infinite dimensions. In this chapter I have avoided those difficulties by doing all the analysis on

---

[3] All these come down to the uncomfortable fact that, in infinite dimensions, Sobolev's embedding theorems are false. Specifically, although the solution to an Itô stochastic integral equation is infinitely smooth in the sense of Sobolev, it need not be even continuous in a classical sense.

the approximating quantities, which are finite dimensional. A similar, but somewhat different, approach was taken by D. Bell in [7].

The treatment of hypoellipticity in § 3.4 is taken from [32]. In that it is potential theoretic, it bears greater resemblance to Weyl's [57] than most modern, analytic approaches, like the one described in Chapter 7.

CHAPTER 4

Nash Theory

In this chapter we will develop a theory which, at least from the probabilistic perspective underlying our thinking up to now, gives results which are surprising. On the other hand, since, from the outset, the reasoning here has minimal probabilistic input, maybe the surprise should not be too great.

Although, as explained in §1.1, (1.1.8) was the probabilistically natural way for Kolmogorov to write his operators, there are good physical as well as mathematical reasons (cf. Remark 4.3.10 below) for thinking that it is preferable to write our operator in *divergence form*, that is, to write

$$(4.0.1) \qquad L\varphi = \tfrac{1}{2}\nabla \cdot \left(a\nabla\varphi\right) = \frac{1}{2}\sum_{i,j=1}^{N} \partial_{x_i}\left(a_{ij}\partial_{x_j}\varphi\right).$$

This is the form in which the operator arises when one is thinking about conservation laws in physics or about Euler–Legrange equations when doing variational calculus, and, from a purely mathematical standpoint, it has the obvious advantage of being *formally self-adjoint*. In fact, under mild decay conditions on φ and ψ,

$$(4.0.2) \qquad \left(\varphi, L\psi\right)_{L^2(\mathbb{R}^N;\mathbb{R})} = -\tfrac{1}{2}\left(\nabla\varphi, a\nabla\psi\right)_{L^2(\mathbb{R}^N;\mathbb{R}^N)},$$

which is the crucial fact on which everything else in this chapter turns.

Under the assumption that a is symmetric and satisfies (3.2.1), we will show how to systematically exploit the advantages of writing L in divergence form. The most important conclusion which we will draw is that our upper bound in Theorem 3.3.11 on the fundamental solution can be complemented with a commensurate lower bound. Of secondary, but considerable, interest will be the freedom of our conclusions from regularity properties of a. Namely, all our estimates will depend only on N and the upper and lower bounds on a.

Once we have dealt with operators of the form in (4.0.1), we will show how to deal with operators containing lower order terms.

4.1 The Upper Bound

Let L be the operator in (4.0.1). Besides (3.2.1), we will be assuming that a is smooth and has bounded derivatives of all orders. Thus, by the results in § 3.3, we know that there is a smooth $(t, x, y) \in (0, \infty) \times \mathbb{R}^N \times \mathbb{R}^N \longmapsto$ $p(t, x, y) \in [0, \infty)$ which is a fundamental solution to the heat equation $\partial_t u = Lu$.

Our first application of (4.0.2) is to check that

$$(4.1.1) \qquad p(t, x, y) = p(t, y, x) \quad \text{for all } (t, x, y) \in (0, \infty) \times \mathbb{R}^N \times \mathbb{R}^N.$$

One way to check (4.1.1) is to note that $L^\top = L$ and therefore, by (3.3.8), § 3.3.4) that $P^\top(t, x) = P(t, x)$. Alternatively, for $\varphi, \psi \in C_c^2(\mathbb{R}^N; \mathbb{R})$, one can use the estimates in Theorem 3.3.11 and (4.0.2) to justify (cf. (2.0.2))

$$\frac{d}{ds} \big(u_\varphi(t-s), u_\psi(s) \big)_{L^2(\mathbb{R}^N; \mathbb{R})}$$
$$= - \big(L u_\varphi(t-s), u_\psi(s) \big)_{L^2(\mathbb{R}^N; \mathbb{R})} + \big(u_\varphi(t-s), L u_\psi(s) \big)_{L^2(\mathbb{R}^N; \mathbb{R})}$$
$$= \big(\nabla u_\varphi(t-s), a \nabla u_\psi(s) \big)_{L^2(\mathbb{R}^N; \mathbb{R}^N)} - \big(\nabla u_\varphi(t-s), a \nabla u_\psi(s) \big)_{L^2(\mathbb{R}^N; \mathbb{R}^N)} = 0,$$

which means that

$$\iint \varphi(x) p(t, x, y) \psi(y) \, dx dy = \big(\varphi, u_\psi(t) \big)_{L^2(\mathbb{R}^N; \mathbb{R})}$$
$$= \big(\psi, u_\phi(t) \big)_{L^2(\mathbb{R}^N; \mathbb{R})} = \iint \psi(y) p(t, y, x) \varphi(x) \, dx dy.$$

As in § 2.5, it is best to employ operator notation here. Thus, we will write

$$\mathbf{P}_t \varphi(x) = u_\varphi(t, x) = \int \varphi(y) \, P(t, x, dy) = \int \varphi(y) p(t, x, y) \, dy.$$

Because $L = L^\top$, and therefore $V^\top = 0$, Theorem 2.5.8 tells us that, for each $r \in [1, \infty]$, $\mathbf{P}_t$ determines a contraction on $L^r(\mathbb{R}^N; \mathbb{R})$, which, by (4.1.1), is self-adjoint on $L^2(\mathbb{R}^N; \mathbb{R})$. In particular, these facts have the consequences that

$$(4.1.2) \quad \|\mathbf{P}_t \varphi\|_{L^r(\mathbb{R}^N)} \leq \|\varphi\|_{L^r(\mathbb{R}^N)} \text{ and } \big(\psi, \mathbf{P}_t \varphi \big)_{L^2(\mathbb{R}^N; \mathbb{R})} = \big(\varphi, \mathbf{P}_t \psi \big)_{L^2(\mathbb{R}^N; \mathbb{R})}$$

extend to all $\varphi \in L^r(\mathbb{R}^N; \mathbb{R})$ and $\psi \in L^{r'}(\mathbb{R}^N; \mathbb{R})$ when r' is the Hölder conjugate of $r \in [1, \infty)$.

4.1.1. Nash's Inequality: As we have just seen, $\mathbf{P}_t$ is a contraction on $L^r(\mathbb{R}^N; \mathbb{R})$ for each $r \in [1, \infty]$. However, an examination of our proof reveals that this property relies only on $L = L^\top$ and has nothing to do with ellipticity. To bring ellipticity into play, we exploit (4.0.2), and the following Sobolev type result is the one which will enable us to do so.

Lemma 4.1.3.   If $\varphi \in C^1(\mathbb{R}^N; \mathbb{R}) \cap L^1(\mathbb{R}^N; \mathbb{R})$ and $\nabla\varphi \in L^2(\mathbb{R}^N; \mathbb{R}^N)$, then

$$\|\varphi\|_{L^2(\mathbb{R}^N;\mathbb{R})}^{2+\frac{4}{N}} \le C_N \|\nabla\varphi\|_{L^2(\mathbb{R}^N;\mathbb{R}^N)}^2 \|\varphi\|_{L^1(\mathbb{R}^N)}^{\frac{4}{N}},$$

where $C_N < \infty$ is universal.

Proof: Without loss of generality, we may and will assume that $\varphi \in C_c^\infty(\mathbb{R}^N; \mathbb{R})$.

There are several ways to prove the asserted inequality. One of them is to use Parseval's identity and $|\widehat{\nabla\varphi}(\xi)| = |\xi| |\hat{\varphi}(\xi)|$ to write, for each $R > 0$,

$$
\begin{aligned}
(2\pi)^N \|\varphi\|_{L^2(\mathbb{R}^N;\mathbb{R})}^2 &= \int_{B(0,R)} |\hat{\varphi}(\xi)|^2 \, d\xi + \int_{B(0,R)\complement} |\hat{\varphi}(\xi)|^2 \, d\xi \\
&\le \Omega_N R^N \|\varphi\|_{L^1(\mathbb{R}^N)}^2 + (2\pi)^N R^{-2} \|\nabla\varphi\|_{L^2(\mathbb{R}^N;\mathbb{R}^N)}^2,
\end{aligned}
$$

and to then minimize the right-hand side with respect to R. A second and, in view of the context, perhaps more revealing proof is to set $\varphi(t) = g(t) \star \varphi$, where $g(t,x) = (2\pi t)^{-\frac{N}{2}} \exp\left(-\frac{|x|^2}{2t}\right)$ is the standard heat kernel, and write

$$\big(\varphi, \varphi(t)\big)_{L^2(\mathbb{R}^N;\mathbb{R})} = \|\varphi\|_{L^2(\mathbb{R}^N;\mathbb{R})}^2 + \frac{1}{2}\int_0^t \big(\varphi, \Delta\varphi(\tau)\big)_{L^2(\mathbb{R}^N;\mathbb{R})} \, d\tau.$$

Note that, because all the operations are translation invariant and therefore commute with one another,

$$
\begin{aligned}
\big|\big(\varphi, \Delta\varphi(\tau)\big)_{L^2(\mathbb{R}^N;\mathbb{R})}\big| &= \big|\big(\nabla\varphi, \nabla\varphi(\tau)\big)_{L^2(\mathbb{R}^N;\mathbb{R}^N)}\big| \\
&\le \|\nabla\varphi\|_{L^2(\mathbb{R}^N;\mathbb{R}^N)} \|g(\tau)\star\nabla\varphi\|_{L^2(\mathbb{R}^N;\mathbb{R}^N)} \le \|\nabla\varphi\|_{L^2(\mathbb{R}^N;\mathbb{R}^N)}^2,
\end{aligned}
$$

where, in the final step, we have used the fact that, since $\|g(\tau)\|_{L^1(\mathbb{R}^N)} = 1$, convolution with $g(\tau)$ is a contraction in $L^2(\mathbb{R}^N; \mathbb{R}^N)$. At the same time,

$$\big(\varphi, \varphi(t)\big)_{L^2(\mathbb{R}^N;\mathbb{R})} \le \|\varphi\|_{L^1(\mathbb{R}^N)} \|\varphi(t)\|_{L^\infty(\mathbb{R}^N)} \le (2\pi t)^{-\frac{N}{2}} \|\varphi\|_{L^1(\mathbb{R}^N)}^2.$$

Hence,

$$\|\varphi\|_{L^2(\mathbb{R}^N;\mathbb{R})}^2 \le (2\pi t)^{-\frac{N}{2}} \|\varphi\|_{L^1(\mathbb{R}^N)}^2 + \frac{t}{2} \|\nabla\varphi\|_{L^2(\mathbb{R}^N;\mathbb{R}^N)}^2$$

for all $t > 0$. Now minimize with respect to t. $\square$

To give an immediate demonstration of the power of Lemma 4.1.3, let $\varphi \in C_c^\infty(\mathbb{R}^N; \mathbb{R})$ with $\|\varphi\|_{L^1(\mathbb{R}^N)} \le 1$ be given, and set $u_\varphi(t) = \mathbf{P}_t\varphi$. Since $\|u_\varphi(t)\|_{L^1(\mathbb{R}^N)} \le 1$,

$$
\begin{aligned}
\frac{d}{dt}\|u_\varphi(t)\|_{L^2(\mathbb{R}^N;\mathbb{R})}^2 &= 2\big(u_\varphi(t), Lu_\varphi(t)\big)_{L^2(\mathbb{R}^N;\mathbb{R})} = -\big(\nabla u_\varphi, a\nabla u_\varphi(t)\big)_{L^2(\mathbb{R}^N;\mathbb{R}^N)} \\
&\le -\epsilon\|\nabla u_\varphi\|_{L^2(\mathbb{R}^N;\mathbb{R}^N)}^2 \le -\frac{\epsilon}{C_N}\|u_\varphi(t)\|_{L^2(\mathbb{R}^N;\mathbb{R})}^{2+\frac{4}{N}}.
\end{aligned}
$$

Hence, $\|u_\varphi(t)\|_{L^2(\mathbb{R}^N;\mathbb{R})} \le \left(\frac{NC_N}{2\epsilon}\right)^{\frac{N}{4}} t^{-\frac{N}{4}}$, and so we have shown that the operator norm $\|\mathbf{P}_t\|_{1\to 2}$ of $\mathbf{P}_t$ as a mapping from $L^1(\mathbb{R}^N;\mathbb{R})$ to $L^2(\mathbb{R}^N;\mathbb{R})$ satisfies

$$\|\mathbf{P}_t\|_{1\to 2} \le \left(\frac{NC_N}{2\epsilon}\right)^{\frac{N}{4}} t^{-\frac{N}{4}}.$$

Working by duality, this also shows that, as a mapping from $L^2(\mathbb{R}^N;\mathbb{R})$ to $L^\infty(\mathbb{R}^N;\mathbb{R})$, the norm $\|\mathbf{P}_t\|_{2\to\infty}$ of $\mathbf{P}_t$ satisfies the same estimate. That is, by (4.1.2) and Schwarz's inequality,

$$\left|(\psi,\mathbf{P}_t\varphi)_{L^2(\mathbb{R}^N;\mathbb{R})}\right| = \left|(\mathbf{P}_t\psi,\varphi)_{L^2(\mathbb{R}^N;\mathbb{R})}\right|$$
$$\le \|\mathbf{P}_t\psi\|_{L^2(\mathbb{R}^N;\mathbb{R})}\|\varphi\|_{L^2(\mathbb{R}^N;\mathbb{R})} \le \|\mathbf{P}_t\|_{1\to 2}\|\psi\|_{L^1(\mathbb{R}^N)}\|\varphi\|_{L^2(\mathbb{R}^N;\mathbb{R})},$$

and so $\|\mathbf{P}_t\varphi\|_{L^\infty(\mathbb{R}^N;\mathbb{R})} \le \|\mathbf{P}_t\|_{1\to 2}\|\varphi\|_{L^2(\mathbb{R}^N;\mathbb{R})}$. Finally, because $\mathbf{P}_t = \mathbf{P}_{\frac{t}{2}} \circ \mathbf{P}_{\frac{t}{2}}$, this means that the norm $\|\mathbf{P}_t\|_{1\to\infty}$ of $\mathbf{P}_t$ as a mapping from $L^1(\mathbb{R}^N;\mathbb{R})$ to $L^\infty(\mathbb{R}^N;\mathbb{R})$ satisfies $\|\mathbf{P}_t\|_{1\to\infty} \le A_N(\epsilon t)^{-\frac{N}{2}}$ for some universal $A_N < \infty$. Equivalently,

$$\left|\iint \varphi(x)p(t,x,y)\psi(y)\,dxdy\right| \le A_N(\epsilon t)^{-\frac{N}{2}}\|\varphi\|_{L^1(\mathbb{R}^N;\mathbb{R})}\|\psi\|_{L^1(\mathbb{R}^N;\mathbb{R})},$$

from which, because $p(t,x,y)$ is continuous, it is an easy step to the conclusion that

$$(4.1.4) \qquad p(t,x,y) \le \frac{A_N}{(\epsilon t)^{\frac{N}{2}}}.$$

REMARK 4.1.5. Although (4.1.4) looks like a deficient version of the estimate in Theorem 3.3.11, there is a significant difference. Namely, the estimate in (4.1.4) is independent of the smoothness of a. To understand why this is important, remember the scaling argument which we used in §3.3.3. Applying it to an operator in divergence form, we see that

$$(4.1.6) \qquad p(t,x,y) = \lambda^{-\frac{N}{2}}p_\lambda\big(\lambda^{-1}t, \lambda^{-\frac{1}{2}}x, \lambda^{-\frac{1}{2}}y\big),$$

where p_λ is the density corresponding to $a_\lambda(x) = a(\lambda^{\frac{1}{2}}x)$. When our estimates depended on smoothness properties, we could not afford to take λ large, and that is why our estimates in Chapter 3 were actually estimates for $t \in (0,1]$. However, when smoothness does not enter we can take any $\lambda \in (0,\infty)$. Thus, for example, scaling allows us to derive (4.1.4) for all $t \in (0,\infty)$ from (4.1.4) for $t = 1$.

A second remark is that (4.1.4) is an estimate for $p(t,x,y)$ on the diagonal $y = x$. To see this, use the Chapman–Kolmogorov equation and (4.1.1) to write

$$p(t,x,y) = \int p\big(\tfrac{t}{2},x,\xi\big)p\big(\tfrac{t}{2},y,\xi\big)\,d\xi$$
$$\le \left(\int p\big(\tfrac{t}{2},x,\xi\big)^2\,d\xi\right)^{\frac{1}{2}}\left(\int p\big(\tfrac{t}{2},y,\xi\big)^2\,d\xi\right)^{\frac{1}{2}} = p(t,x,x)^{\frac{1}{2}}p(t,y,y)^{\frac{1}{2}}.$$

Thus, when L is in divergence form, $\sup_{x,y} p(t,x,y) = \sup_x p(t,x,x)$.

4.1.2. Off-Diagonal Upper Bound: We will next show how to replace
(4.1.4) with an off-diagonal estimate like the one in Theorem 3.3.11, only
now the estimate will be sharper and not depend on the smoothness of a.

Given a $\psi \in C^\infty(\mathbb{R}^N; \mathbb{R})$ having bounded, continuous first derivatives,
set

$$p^\psi(t, x, y) = e^{-\psi(x)} p(t, x, y) e^{\psi(y)},$$

and define the operator $\mathbf{P}_t^\psi$ by

$$\mathbf{P}_t^\psi \varphi(x) = \int \varphi(y) p^\psi(t, x, y) \, dy$$

for $\varphi \in C_b(\mathbb{R}^N; \mathbb{R})$. Because of the estimate in Theorem 3.3.11, there are no
integrability problems in the definition of $\mathbf{P}_t^\psi$. In addition, one can easily
check that $\mathbf{P}_{s+t}^\psi = \mathbf{P}_s^\psi \circ \mathbf{P}_t^\psi$ and

$$\frac{d}{dt} \mathbf{P}_t^\psi \varphi = L^\psi \mathbf{P}_t^\psi \varphi, \quad \text{where } L^\psi \varphi = e^{-\psi} L(e^\psi \varphi).$$

What we want to do is apply the technique in §4.1.1 to get an estimate
on $\|\mathbf{P}_t^\psi\|_{2\to\infty}$. However, there is a problem here which we did not have
there. Namely, we must find out how to avoid the use which we made there
of $\mathbf{P}_t \mathbf{1} = \mathbf{1}$.

LEMMA 4.1.7. *Set*

(4.1.8) $$D(\psi) = \sup_{x \in \mathbb{R}^N} \sqrt{\left(\nabla\psi(x), a(x)\nabla\psi(x)\right)_{\mathbb{R}^N}}.$$

There is a $K_N < \infty$, depending only on N, such that

$$\|\mathbf{P}_t^\psi\|_{2\to\infty} \leq \frac{K_N}{(\epsilon t)^{\frac{N}{4}}} e^{\frac{(1+\delta)tD(\psi)^2}{2}}$$

for every $\delta \in (0, 1]$.

PROOF: Let $\varphi \in C_c^2\big(\mathbb{R}^N; [0, \infty)\big) \setminus \{0\}$ be given, and set $u(t) = \mathbf{P}_t^\psi \varphi$ and
$v_r(t) = \|u(t)\|_{L^r(\mathbb{R}^N)}$ for $r \in [1, \infty)$. Because

$$\|\mathbf{P}_t^\psi \varphi\|_{L^1(\mathbb{R}^N)} = \big(e^{-\psi}, \mathbf{P}_t(e^\psi \varphi)\big)_{L^2(\mathbb{R}^N; \mathbb{R})} = \big(\mathbf{P}_t e^{-\psi}, e^\psi \varphi\big)_{L^2(\mathbb{R}^N; \mathbb{R})} > 0,$$

it is clear that $v_r(t) > 0$ for all r and t. Next,

$$\dot{v}_{2r}(t) = v_{2r}(t)^{1-2r} \big(u(t)^{2r-1}, L^\psi u(t)\big)_{L^2(\mathbb{R}^N; \mathbb{R})},$$

and[1]

$$\left(u(t)^{2r-1}, L^\psi u(t)\right)_{L^2(\mathbb{R}^N;\mathbb{R})} = -\frac{1}{2}\left(\nabla\left(e^{-\psi}u^{2r-1}(t)\right), a\nabla\left(e^\psi u(t)\right)\right)_{L^2(\mathbb{R}^N;\mathbb{R}^N)}$$

$$= -\frac{2r-1}{2}\left(u(t)^{2(r-1)}\nabla u(t), a\nabla u(t)\right)_{L^2(\mathbb{R}^N;\mathbb{R})}$$

$$\quad - (r-1)\left(u(t)^{2r-1}\nabla u(t), a\nabla\psi\right)_{L^2(\mathbb{R}^N;\mathbb{R}^N)} + \frac{1}{2}\left(u(t)^{2r}\nabla\psi, a\nabla\psi\right)_{L^2(\mathbb{R}^N;\mathbb{R}^N)}$$

$$\leq -\frac{r}{2}\left(u(t)^{2(r-1)}\nabla u(t), a\nabla u(t)\right)_{L^2(\mathbb{R}^N;\mathbb{R}^N)} + \frac{r}{2}\left(u(t)^{2r}\nabla\psi, a\nabla\psi\right)_{L^2(\mathbb{R}^N;\mathbb{R}^N)}$$

$$\leq -\frac{\epsilon}{2r}\|\nabla u(t)^r\|_{L^2(\mathbb{R}^N;\mathbb{R}^N)}^2 + \frac{rD(\psi)^2}{2}v_{2r}(t)^{2r}.$$

At the same time, by Lemma 4.1.3 applied to $u(t)^r$,

$$C_N\|\nabla u(t)^r\|_{L^2(\mathbb{R}^N;\mathbb{R}^N)}^2 \geq \frac{\|u(t)^r\|_{L^2(\mathbb{R}^N;\mathbb{R})}^{2+\frac{4}{N}}}{\|u(t)^r\|_{L^1(\mathbb{R}^N;\mathbb{R})}^{\frac{4}{N}}} = \frac{v_{2r}(t)^{2r+\frac{4r}{N}}}{v_r(t)^{\frac{4r}{N}}},$$

and so we can now say that

$$(*) \qquad \dot{v}_{2r}(t) \leq -\frac{\epsilon}{2C_N r}\frac{v_{2r}(t)^{1+\frac{4r}{N}}}{v_r(t)^{\frac{4r}{N}}} + \frac{rD(\psi)^2}{2}v_{2r}(t).$$

Applying (*) when $r = 1$ and ignoring the first term on the right, one sees that $v_2(t) \leq e^{\frac{tD(\psi)^2}{2}}\|\varphi\|_{L^2(\mathbb{R}^N;\mathbb{R})}$. Before applying (*) when $r \geq 2$, set

$$w_r(t) = \sup_{\tau\in[0,t]} \tau^{\frac{N}{4r}(r-2)}v_r(\tau).$$

Then, $w_2(t) \leq e^{\frac{tD(\psi)^2}{2}}\|\varphi\|_{L^2(\mathbb{R}^N;\mathbb{R})}$ and (*) can be replaced by

$$\dot{v}_{2r}(t) \leq -\frac{\epsilon}{2C_N r}\frac{t^{r-2}v_{2r}(t)^{1+\frac{4r}{N}}}{w_r(t)^{\frac{4r}{N}}} + \frac{rD(\psi)^2}{2}v_{2r}(t).$$

After elementary manipulation, the preceding can be rewritten as

$$\frac{d}{dt}\left(e^{-\frac{trD(\psi)^2}{2}}v_{2r}(t)\right)^{-\frac{4r}{N}} \geq \frac{2\epsilon}{NC_N w_r(t)^{\frac{4r}{N}}}t^{r-2}e^{\frac{t2r^2D(\psi)^2}{N}},$$

---

[1] Here and elsewhere, we make frequent, and occasionally creative, use of the quadratic inequality $|(\xi, a\eta)_{\mathbb{R}^N}| \leq \frac{\delta}{2}(\xi, a\xi) + \frac{1}{2\delta}(\eta, a\eta)_{\mathbb{R}^N}$ for any symmetric, non-negative definite $a \in \mathrm{Hom}(\mathbb{R}^N;\mathbb{R}^N)$, $\xi, \eta \in \mathbb{R}^N$, and $\delta > 0$.

and therefore, for any $\delta \in (0, 1]$,

$$e^{\frac{t2r^2D(\psi)^2}{N}} v_{2r}(t)^{-\frac{4r}{N}} \geq \frac{2\epsilon}{NC_N w_r(t)^{\frac{4r}{N}}} \int_{(1-\frac{\delta}{r^2})t}^{t} \tau^{r-2} e^{\frac{\tau 2r^2 D(\psi)^2}{N}} d\tau$$

$$\geq \frac{2\epsilon}{NC_N w_r(t)^{\frac{4r}{N}}} t^{r-1} e^{\frac{t2r^2D(\psi)^2}{N}} e^{-\frac{t2\delta D(\psi)^2}{N}} \int_{1-\frac{\delta}{r^2}}^{1} \tau^{r-2} d\tau.$$

Hence, since

$$\int_{1-\frac{\delta}{r^2}}^{1} \tau^{r-2} d\tau = \frac{1}{r-1}\left(1 - \left(1 - \frac{\delta}{r^2}\right)^{r-1}\right) \geq \frac{\delta}{r^2},$$

for $r \geq 2$, we can say first that

$$v_{2r}(t) \leq \left(\frac{NC_N r^2}{\epsilon\delta}\right)^{\frac{N}{4r}} t^{\frac{N}{4r}(1-r)} e^{\frac{t\delta D(\psi)^2}{2r}} w_r(t),$$

and then that

$$w_{2r}(t) \leq \left(\frac{NC_N r^2}{\epsilon\delta}\right)^{\frac{N}{4r}} e^{\frac{t\delta D(\psi)^2}{2r}} w_r(t), \quad r \geq 2.$$

Working by induction on $n \geq 2$, we arrive at

$$w_{2^n}(t) \leq 4^{\frac{N}{4}\sum_{m=1}^{n-1} m2^{-m}} \left(\frac{NC_N}{\epsilon\delta}\right)^{\frac{N}{4}(1-2^{1-n})} e^{\frac{(1-2^{1-n})t\delta D(\psi)^2}{2}} w_2(t),$$

and so, after letting $n \to \infty$ and using our estimate on $w_2(t)$, we get

$$\|\mathbf{P}_t^{\psi}\varphi\|_u \leq 4^N \left(\frac{NC_N}{\epsilon\delta t}\right)^{\frac{N}{4}} e^{\frac{t(1+\delta)D(\psi)^2}{2}} \|\varphi\|_{L^2(\mathbb{R}^N;\mathbb{R})}. \quad \square$$

Because the adjoint of $\mathbf{P}_t^{\psi}$ is $\mathbf{P}_t^{-\psi}$, we can use duality and the result in Lemma 4.1.7 to get the same estimate for $\|\mathbf{P}^{\psi}\|_{1\to2}$, and therefore, since $\mathbf{P}_t^{\psi} = (\mathbf{P}_{\frac{t}{2}}^{\psi})^2$,

$$\|\mathbf{P}_t^{\psi}\|_{1\to\infty} \leq \frac{2^{\frac{N}{2}} K_N^2}{(\epsilon\delta t)^{\frac{N}{2}}} e^{\frac{t(1+\delta)D(\psi)^2}{2}}.$$

Equivalently,

$$p(t, x, y) \leq \frac{2^{\frac{N}{2}} K_N^2}{(\epsilon\delta t)^{\frac{N}{2}}} \exp\left(\psi(x) - \psi(y) + \frac{t(1+\delta)D(\psi)^2}{2}\right).$$

Obviously, what we want to do now is minimize the right-hand side of the preceding with respect to ψ and to interpret what that minimization yields. For this purpose, recall that the *Riemannian distance* from x to y relative to the Riemann metric a^{-1} is

(4.1.9)
$$d(x,y) = \inf\left\{ \sqrt{\int_0^1 \left(\dot{\pi}(\tau), a(\pi(\tau))^{-1}\dot{\pi}(\tau)\right)_{\mathbb{R}^N} d\tau} \right.$$
$$\left. : \pi(0) = x \ \& \ \pi(1) = y \right\},$$

where $\pi \in C^1([0,1];\mathbb{R}^N)$.

LEMMA 4.1.10.   *For each* $\alpha > 0$,

$$\sup\left\{ \psi(y) - \psi(x) - \frac{\alpha D(\psi)^2}{2} : \psi \in C^\infty(\mathbb{R}^N;\mathbb{R}) \right\} = \frac{d(x,y)^2}{2\alpha}.$$

PROOF: Set

$$D(x,y) = \sup\{|\psi(y) - \psi(x)| : \psi \in C^\infty(\mathbb{R}^N;\mathbb{R}) \ \& \ D(\psi) \le 1\}.$$

Then, after replacing ψ by $\lambda\psi$ and minimizing over λ, one sees that

$$\sup\left\{ \psi(y) - \psi(x) - \frac{\alpha D(\psi)^2}{2} : \psi \in C^\infty(\mathbb{R}^N;\mathbb{R}) \right\}$$
$$= \sup\left\{ \frac{(\psi(y) - \psi(x))^2}{2\alpha D(\psi)^2} : \psi \in C^1(\mathbb{R}^N;\mathbb{R}) \right\} = \frac{D(x,y)^2}{2\alpha}.$$

Hence, we will be done once we show that $d(x,y) = D(x,y)$. To this end, first suppose that $\psi \in C^\infty(\mathbb{R}^N;\mathbb{R})$ with $D(\psi) \le 1$, and note that for any $\pi \in C^1([0,1];\mathbb{R}^N)$ with $\pi(0) = x$ and $\pi(1) = y$,

$$|\psi(y) - \psi(x)| = \left| \int_0^1 \left(\nabla\psi(\pi(\tau)), \dot{\pi}(\tau)\right)_{\mathbb{R}^N} d\tau \right|$$
$$\le \int_0^1 |(a^{\frac{1}{2}}\nabla\psi)(\pi(\tau))| |a(\pi(\tau))^{-\frac{1}{2}}\dot{\pi}(\tau)| d\tau$$
$$\le \sqrt{\int_0^1 \left(\dot{\pi}(\tau), a(\pi(\tau))^{-1}\dot{\pi}(\tau)\right)_{\mathbb{R}^N} d\tau}.$$

Hence, after minimizing with respect to π, we see that $D(x,y) \le d(x,y)$.

To get the opposite inequality, we must check first that $D(x,y) \ge |\psi(y) - \psi(x)|$ for any ψ satisfying the Lipschitz condition $|\psi(\eta) - \psi(\xi)| \le d(\xi,\eta)$ for all $\xi, \eta \in \mathbb{R}^N$. For this purpose, we begin by showing that if $\psi \in C^\infty(\mathbb{R}^N;\mathbb{R})$

and $|\psi(\eta) - \psi(\xi)| \leq C d(\xi, \eta)$, then $D(\psi) \leq C$. Indeed, given $\theta \in \mathbb{S}^{N-1}$, set $\pi_t(\tau) = x + t\tau a(x)^{\frac{1}{2}}\theta$. Then

$$
\begin{aligned}
\left(\psi\left(x + ta(x)^{\frac{1}{2}}\theta\right) - \psi(x)\right)^2 &\leq C^2 d\left(x, x + ta(x)^{\frac{1}{2}}\theta\right)^2 \\
&\leq C^2 \int_0^1 \left(\dot{\pi}_t(\tau), a\left(\pi_t(\tau)\right)^{-1}\dot{\pi}_t(\tau)\right)_{\mathbb{R}^N} d\tau \\
&= C^2 t^2 \int_0^1 \left(a(x)^{\frac{1}{2}}\theta, a\left(\pi_t(\tau)\right)^{-1}a(x)^{\frac{1}{2}}\theta\right)_{\mathbb{R}^N} d\tau,
\end{aligned}
$$

and so

$$
\left(a(x)^{\frac{1}{2}}\nabla\psi(x), \theta\right)_{\mathbb{R}^N}^2 = \lim_{t\searrow 0} t^{-2}\left(\psi\left(x + ta(x)^{\frac{1}{2}}\theta\right) - \psi(x)\right)^2 \leq C^2,
$$

from which it is clear that $D(\psi) \leq C$. Next, let ψ be any function satisfying $|\psi(\eta) - \psi(\xi)| \leq d(\xi, \eta)$ for all $\xi, \eta \in \mathbb{R}^N$ be given. We want to show that $\psi = \lim_{s\searrow 0}\psi_s$, where $\{\psi_s : s \in (0,1]\} \subseteq C^\infty(\mathbb{R}^N; \mathbb{R})$ satisfies $\varlimsup_{s\searrow 0} D(\psi_s) \leq 1$. For this purpose, choose $\rho \in C_c^\infty\left(B(0,1); [0,\infty)\right)$ with integral 1, and set $\psi_s = \rho_s \star \psi$, where $\rho_s(\xi) = s^{-N}\rho(s^{-1}\xi)$. Then each ψ_s is smooth, $\psi_s \longrightarrow \psi$ uniformly on compacts as $s \searrow 0$, and

$$
|\psi_s(\eta) - \psi_s(\xi)| \leq \int \rho_s(\zeta) d(\xi - \zeta, \eta - \zeta)\, d\zeta.
$$

Hence, since

$$
\int \rho_s(\zeta) d(\xi - \zeta, \eta - \zeta)\, d\zeta \leq C_s d(\xi, \eta),
$$

where

$$
C_s \equiv \sup\left\{\left\|a(\xi' - \zeta)^{-1}a(\xi')\right\|_{\mathrm{op}} : \xi' \in \mathbb{R}^N \ \& \ |\zeta| \leq s\right\} \searrow 1,
$$

we have that $\varlimsup_{s\searrow 0} D(\psi_s) \leq 1$. Finally, this means that

$$
D(x, y) \geq \varlimsup_{s\searrow 0} \frac{|\psi_s(y) - \psi_s(x)|}{D(\psi_s)} \geq |\psi(y) - \psi(x)|.
$$

Now take $\psi(\xi) = d(x, \xi)$. By the triangle inequality, $|\psi(\eta) - \psi(\xi)| \leq d(\xi, \eta)$, and so, by the preceding, $D(x, y) \geq |\psi(y) - \psi(x)| = d(x, y)$. $\square$

By combining Lemma 4.1.10 with the estimate which we already have on $p(t, x, y)$, we get the following upper bounds.

THEOREM 4.1.11. *Assume that L is given by (4.0.1), where $a : \mathbb{R}^N \longrightarrow \mathrm{Hom}(\mathbb{R}^N; \mathbb{R}^N)$ is a smooth function which, along with each of its derivatives, is bounded. Further, assume that, for all $x \in \mathbb{R}^N$, $a(x)$ is symmetric and satisfies $a(x) \geq \epsilon I$ for some $\epsilon > 0$. Then, for each $\delta \in (0, 1]$,*

$$
p(t, x, y) \leq \frac{K_N}{(\epsilon\delta t)^{\frac{N}{2}}} \exp\left(-\frac{d(x, y)^2}{2(1+\delta)t}\right),
$$

where $K_N < \infty$ is a universal constant depending only on N and $d(x, y)$ is the Riemannian distance described in (4.1.9). In particular,

$$\overline{\lim_{t \searrow 0}} \, t \log p(t, x, y) \leq -\frac{d(x, y)^2}{2},$$

and

$$p(t, x, y) \leq \frac{K_N}{(\epsilon t)^{\frac{N}{2}}} \exp\left(-\frac{|y - x|^2}{4At}\right),$$

where $A \equiv \sup_{x \in \mathbb{R}^N} \|a(x)\|_{\mathrm{op}}$.

PROOF: The first assertion is exactly what we just finished proving, and the second assertion is an immediate consequence of the first. Finally, to prove the last assertion, all that one needs to do is take $\delta = 1$ and note that $d(x, y)^2 \geq A^{-1}|y - x|^2$. This last fact is perhaps most easily checked by using the equality $d(x, y) = D(x, y)$, which was established in the proof of Lemma 4.1.10, and taking test functions of the form $\psi(x) = (\theta, x)_{\mathbb{R}^N}$ to estimate $D(x, y)$. $\square$

4.2 The Lower Bound

In this section we will do our best to develop a lower bound for $p(t, x, y)$ which complements the upper bound in Theorem 4.1.11. Before getting down to business, it may help to recognize that the upper bound immediately gives a lower bound at the diagonal. Namely, from the last estimate in Theorem 4.1.11, we know that

$$\int |y - x|^2 p(t, x, y) \, dy \leq 2K_N \left(\frac{4\pi A}{\epsilon}\right)^{\frac{N}{2}} t.$$

Hence, because $p(t, x, \cdot)$ has integral 1,

(4.2.1)
$$\int_{B(x, t^{\frac{1}{2}} R)} p(t, x, y) \, dy \geq \frac{1}{2}$$

$$\text{where } R = R\left(\tfrac{A}{\epsilon}\right) \equiv (4K_N)^{\frac{1}{2}} \left(\frac{4\pi A}{\epsilon}\right)^{\frac{N}{4}}.$$

In particular, if we use $|\Gamma|$ to denote the Lebesgue measure of $\Gamma \in \mathcal{B}_{\mathbb{R}^N}$, then

$$p(2t, x, x) = \int p(t, x, y)^2 \, dy \geq \int_{B(x, t^{\frac{1}{2}} R)} p(t, x, y)^2 \, dy$$

$$\geq \frac{1}{|B(x, t^{\frac{1}{2}} R)|} \left(\int_{B(x, t^{\frac{1}{2}} R)} p(t, x, y) \, dy\right)^2 \geq \frac{1}{4\Omega_N R^N} t^{-\frac{N}{2}}.$$

That is, there is an $\alpha_N \in (0, 1]$, depending only on N, such that

$$p(t, x, x) \geq \alpha_N \left(\frac{\epsilon}{A}\right)^{\frac{N^2}{4}} t^{-\frac{N}{2}}.$$

The simplicity with which one gets the preceding estimate is deceptive because it avoids the real issue. The real problem is to get away from the diagonal, and the solution to this problem requires us to prove a local "ergodic" result. That is, we must show that, by time t, the diffusion with transition density $p(t, x, y)$ has had time to spread more or less evenly over a ball centered at x of radius proportional to $t^{\frac{1}{2}}$.

4.2.1. A Poincaré Inequality: The ergodic result on which our proof relies is the Poincaré inequality for the Gauss distribution γ. More precisely, we will need to know that

$$(4.2.2) \qquad \|\varphi - \langle \varphi, \gamma \rangle\|_{L^2(\gamma;\mathbb{R})}^2 = \langle \varphi^2, \gamma \rangle - \langle \varphi, \gamma \rangle^2 \leq \|\nabla \varphi\|_{L^2(\gamma;\mathbb{R}^N)}^2$$

for $\varphi \in C^1(\mathbb{R}^N; \mathbb{R})$ with slowly increasing derivatives.

There are many ways to prove (4.2.2). Among the most elementary is to use the Mehler kernel,

$$m(t, x, y) = (2\pi)^{\frac{N}{2}} g\left(1 - e^{-2t}, y - e^{-t}x\right) e^{\frac{|y|^2}{2}}$$
$$= \left(1 - e^{-2t}\right)^{-\frac{N}{2}} \exp\left(-\frac{e^{-2t}|x|^2 - 2e^{-t}(x, y)_{\mathbb{R}^N} + e^{-2t}|y|^2}{2(1 - e^{-2t})}\right),$$

where, as usual, $g(t, x) = (2\pi t)^{-\frac{N}{2}} e^{-\frac{|x|^2}{2t}}$ is the heat kernel for $\frac{1}{2}\Delta$. Next define

$$\mathbf{M}_t \varphi(x) = \int m(t, x, y) \varphi(y) \, \gamma(dy) = \int g\left(1 - e^{-2t}, y - e^{-t}x\right) \varphi(y) \, dy.$$

Notice that, because $m(t, x, y)$ is symmetric and, for each (t, x), is a probability density with respect to γ, we can repeat the argument given at the beginning of § 4.1 to see that $\mathbf{M}_t$ is a self-adjoint contraction on $L^2(\gamma; \mathbb{R})$. Next, assuming that $\varphi \in C(\mathbb{R}^N; \mathbb{R})$ is slowly increasing, one can easily check that $(t, x) \rightsquigarrow \mathbf{M}_t \varphi(x)$ is a smooth function on $(0, \infty) \times \mathbb{R}^N$ with, for each $t > 0$, slowly increasing derivatives, and that

$$\frac{d}{dt} \mathbf{M}_t \varphi(x) = (\Delta - x \cdot \nabla) \mathbf{M}_t \varphi(x) = e^{\frac{|x|^2}{2}} \nabla \cdot \left(e^{-\frac{|x|^2}{2}} \nabla \mathbf{M}_t \varphi\right)(x).$$

Hence, if $\varphi \in C^1(\mathbb{R}^N; \mathbb{R})$ has slowly increasing derivatives, then

$$\frac{d}{dt} \left(\varphi, \mathbf{M}_t \varphi\right)_{L^2(\gamma)} = -\left(\nabla \varphi, \nabla \mathbf{M}_t \varphi\right)_{L^2(\gamma;\mathbb{R}^N)}.$$

At the same time, $\nabla \mathbf{M}_t \varphi = e^{-t} \mathbf{M}_t \nabla \varphi$. Thus, because $\mathbf{M}_t$ is a contraction on $L^2(\gamma; \mathbb{R}^N)$,

$$-\frac{d}{dt}\big(\varphi, \mathbf{M}_t \varphi\big)_{L^2(\gamma)} = e^{-t}\big(\nabla\varphi, \mathbf{M}_t \nabla\varphi\big)_{L^2(\gamma;\mathbb{R}^N)} \leq e^{-t}\|\nabla\varphi\|^2_{L^2(\gamma;\mathbb{R}^N)}.$$

Finally, it is easy to check that

$$\lim_{t\searrow 0} \mathbf{M}_t \varphi = \varphi \quad \text{and} \quad \lim_{t\to\infty} \mathbf{M}_t \varphi = \langle\varphi,\gamma\rangle$$

in $L^2(\gamma; \mathbb{R})$, and so (4.2.2) follows when we integrate the preceding over $t \in (0,\infty)$.

4.2.2. Nash's Other Inequality: In this section we will prove the estimates on which our lower bounds will be based. Again, the first of these was found by Nash.

LEMMA 4.2.3.   For $\theta \in \left[0, \frac{1}{2}\right]$, set $p_\theta(t, x, y) = (1-\theta)p(t,x,y) + \theta$. There is a non-decreasing $\kappa : [1,\infty) \longrightarrow (0,\infty)$ with the property that

$$\int \log p_\theta(1, y+\xi, y)\,\gamma(d\xi) \geq -\kappa(A) \quad \text{for all } y \in \mathbb{R}^N \text{ and } \theta \in \big(0, \tfrac{1}{2}\big]$$

whenever $I \leq a(x) \leq AI$ for all $x \in \mathbb{R}^N$.

PROOF: By an obvious translation argument, we may and will assume that $y = 0$.

Given $\theta \in \big(0, \frac{1}{2}\big]$, set $u_\theta(t, \xi) = (1-\theta)p(t, \xi, 0) + \theta$ and $w_\theta(t) = \langle \log u_\theta(t), \gamma\rangle$, and note that, by Jensen's inequality and (4.1.1),

$$w_\theta(t) \leq \log\big(\langle u_\theta(t), \gamma\rangle\big) \leq \log\big((1-\theta)(2\pi)^{-\frac{N}{2}} + \theta\big) < 0.$$

Next,

$$\begin{aligned}
2\dot{w}_\theta(t) &= \int g(1,\xi)\frac{2Lu_\theta(t,\xi)}{u_\theta(t,\xi)}\,d\xi \\
&= \int \big(\xi, a(\xi)\nabla\log u_\theta(t,\xi)\big)_{\mathbb{R}^N} \gamma(d\xi) \\
&\qquad + \int \big(\nabla\log u_\theta(t,\xi), a(\xi)\nabla\log u_\theta(t,\xi)\big)_{\mathbb{R}^N} \gamma(d\xi) \\
&\geq -\frac{1}{2}\int \big(\xi, a(\xi)\xi\big)_{\mathbb{R}^N} \gamma(d\xi) \\
&\qquad + \frac{1}{2}\int \big(\nabla\log u_\theta(t,\xi), a(\xi)\nabla\log u_\theta(t,\xi)\big)_{\mathbb{R}^N} \gamma(d\xi) \\
&\geq -\frac{AN}{2} + \frac{1}{2}\big\|\nabla\log u_\theta(t)\big\|^2_{L^2(\gamma;\mathbb{R}^N)} \\
&\geq -\frac{AN}{2} + \frac{1}{2}\int \big(\log u_\theta(t,\xi) - w_\theta(t)\big)^2 \gamma(d\xi),
\end{aligned}$$

where, at the final step, we have used (4.2.2).

Remembering that $w_\theta < 0$, we see that

$$\left(\log u_\theta(t,\xi) - w_\theta(t)\right)^2 \geq \frac{w_\theta(t)^2}{2} - \left(\log^- u_\theta(t,\xi)\right)^2.$$

Hence, for any $\lambda > 0$ and $\theta \in \left(0, \frac{1}{2}\right]$,

$$\int \left(\log u_\theta(t,\xi) - w_\theta(t)\right)^2 \gamma(d\xi) \geq -\lambda^2 + \frac{1}{2}\gamma\left(\{\xi : p(t,\xi,0) \geq 2e^{-\lambda}\}\right)w_\theta(t)^2.$$

Now take R as in (4.2.1) with $\epsilon = 1$. Then, for $t \in \left[\frac{1}{2}, 1\right]$,

$$\frac{1}{2} \leq \int_{B(0,t^{\frac{1}{2}}R)} p(t,\xi,0)\, d\xi \leq \int_{B(0,R)} p(t,\xi,0)\, d\xi$$

$$\leq 2\Omega_N R^N e^{-\lambda} + 2^{\frac{N}{2}} K_N \left|\{\xi \in B(0,R) : p(t,\xi,0) \geq 2e^{-\lambda}\}\right|,$$

where we have used the last inequality in Theorem 4.1.11 with $\epsilon = 1$. Hence, if we take $\lambda = \log\left(8\Omega_N R^N\right)$, we get

$$\gamma\left(\{\xi : p(t,\xi,0) \geq 2e^{-\lambda}\}\right) \geq \gamma\left(\{\xi \in B(0,R) : p(t,\xi,0) \geq 2e^{-\lambda}\}\right)$$

$$\geq (2\pi)^{-\frac{N}{2}} e^{-\frac{R^2}{2}} \left|\{\xi \in B(0,R) : p(t,\xi,0) \geq 2e^{-\lambda}\}\right| \geq \frac{e^{-\frac{R^2}{2}}}{2^{N+2}\pi^{\frac{N}{2}} K_N}.$$

Thus, we have now shown that

$$\dot{w}_\theta(t) \geq -\frac{AN + \lambda^2}{2} + \frac{e^{-\frac{R^2}{2}}}{2^{N+3}\pi^{\frac{N}{2}} K_N} w_\theta(t)^2 \quad \text{for all } (\theta,t) \in \left(0, \frac{1}{2}\right] \times \left[\frac{1}{2}, 1\right].$$

In particular, if $w_\theta(t)^2 \leq 2^{N+3}\pi^{\frac{N}{2}} K_N e^{\frac{R^2}{2}}(AN + 2\lambda^2)$ for some $t \in \left[\frac{1}{2}, 1\right]$, then

$$w_\theta(1) \geq -\frac{AN + \lambda^2}{4} - \sqrt{2^{N+3}\pi^{\frac{N}{2}} K_N e^{\frac{R^2}{2}}(AN + \lambda^2)}.$$

On the other hand, if the opposite inequality holds for all $t \in \left[\frac{1}{2}, 1\right]$, then

$$\dot{w}_\theta(t) \geq \frac{e^{-\frac{R^2}{2}}}{2^{N+4}\pi^{\frac{N}{2}} K_N} w_\theta(t)^2, \quad t \in \left[\frac{1}{2}, 1\right],$$

and so $w_\theta(1) \geq -2^{N+5}\pi^{\frac{N}{2}} K_N e^{\frac{R^2}{2}}$. Therefore, in either case, we have that there is a $\kappa(A) \in [1, \infty)$ such that $w_\theta(1) \geq -\kappa(A)$ for all $\theta \in \left(0, \frac{1}{2}\right]$. $\square$

In addition to the preceding, we will need the following variation on the same line of reasoning. In the statement below, $\rho \in C_b^2(\mathbb{R}^N; [0, \infty))$ has integral 1 and

$$H(\rho) \equiv \int \frac{|\nabla\rho(\xi)|^2}{\rho(\xi)}\, d\xi < \infty.$$

For example, one can take $\rho = f^2$, where $f \in C^1(\mathbb{R}^N; \mathbb{R})$ with $\|f\|_{L^2(\mathbb{R}^N;\mathbb{R})} = 1$ and $\|\nabla f\|_{L^2(\mathbb{R}^N;\mathbb{R}^N)} < \infty$, in which case $H(\rho) = 4\|\nabla f\|^2_{L^2(\mathbb{R}^N;\mathbb{R}^N)}$. Given such a ρ and a continuously differentiable path $\pi : [0, t] \longrightarrow \mathbb{R}^N$, set

$$(4.2.4) \qquad E_\rho(t, \pi) = \int_0^t \left(\dot{\pi}(\tau), \rho \star a^{-1}\big(\pi(\tau)\big)\dot{\pi}(\tau) \right)_{\mathbb{R}^N} d\tau.$$

LEMMA 4.2.5. *Let $p_\theta(t, x, y)$ be as in Lemma 4.2.3. Then, for all $(t, y) \in (0, \infty) \times \mathbb{R}^N$, $\tau \in (0, \infty)$, $\theta \in \left(0, \frac{1}{2}\right]$, $\delta \in (0, 1]$, and $\pi \in C^1\big([0, t]; \mathbb{R}^N\big)$ with $\pi(0) = y$,*

$$\int \rho(\xi) \log \frac{p_\theta\big(t + \tau, \pi(t) + \xi, y\big)}{p_\theta(\tau, y + \xi, y)} \, d\xi \geq -\frac{AH(\rho)t}{2\delta} - \frac{(1 + \delta)E_\rho(t, \pi)}{2}.$$

PROOF: We begin by observing that, by (2.3.2), $|\nabla\rho| \leq C\rho^{\frac{1}{2}}$, where $C < \infty$ depends only on the bounds on the second derivatives of ρ. Hence, it is an easy matter to check that we need only prove the result when $\rho > 0$ everywhere, and so we will make this assumption here.

Now, without loss of generality, assume that $y = 0$, and set $u_\theta(s, \xi) = p_\theta\big(s + \tau, \xi + \pi(s), 0\big)$. Clearly, it suffices for us to show that

$$\frac{d}{ds} \int \rho(\xi) \log u_\theta(s, \xi) \, d\xi \geq -\frac{AH(\rho)}{2\delta} - \frac{1 + \delta}{2}\left(\dot{\pi}(s), \rho \star a\big(\pi(s)\big)^{-1}\dot{\pi}(s) \right)_{\mathbb{R}^N}.$$

But if $a(s, \xi) = a\big(\xi + \pi(s)\big)$, then the left-hand side of the preceding is equal to

$$-\frac{1}{2}\left(\nabla\frac{\rho}{u_\theta(s)}, a(s)\nabla u_\theta(s) \right)_{L^2(\mathbb{R}^N;\mathbb{R})} + \big(\dot{\pi}(s), \nabla\log u_\theta(s) \big)_{L^2(\mathbb{R}^N;\mathbb{R})}$$

$$= \frac{1}{2}\Bigg[\left(\rho^{\frac{1}{2}}\nabla\log u_\theta(s), a(s)\rho^{\frac{1}{2}}\nabla\log u_\theta(s) \right)_{L^2(\mathbb{R}^N;\mathbb{R})}$$

$$+ 2\left(\rho^{\frac{1}{2}}a(s)^{-1}\dot{\pi}(s) - \frac{\nabla\rho}{2\rho^{\frac{1}{2}}}, a(s)\rho^{\frac{1}{2}}\nabla\log u_\theta(s) \right)_{L^2(\mathbb{R}^N;\mathbb{R})}\Bigg]$$

$$\geq -\frac{1}{2}\left(\left[\rho^{\frac{1}{2}}a(s)^{-1}\dot{\pi}(s) - \frac{\nabla\rho}{2\rho^{\frac{1}{2}}}\right], a(s)\left[\rho^{\frac{1}{2}}a(s)^{-1}\dot{\pi}(s) - \frac{\nabla\rho}{2\rho^{\frac{1}{2}}}\right] \right)_{L^2(\mathbb{R}^N;\mathbb{R})}$$

$$= -\frac{1}{2}\big(\dot{\pi}(s), \rho a(s)^{-1}\dot{\pi}(s) \big)_{L^2(\mathbb{R}^N;\mathbb{R}^N)} + \frac{1}{2}\big(\dot{\pi}(s), \nabla\rho \big)_{L^2(\mathbb{R}^N;\mathbb{R}^N)}$$

$$- \frac{1}{8}\big(\nabla\rho, \rho^{-1}a(s)\nabla\rho \big)_{L^2(\mathbb{R}^N;\mathbb{R}^N)}$$

$$\geq -\frac{1 + \delta}{2}\big(\dot{\pi}(s), \rho \star a^{-1}\big(\pi(s)\big)\dot{\pi}(s) \big)_{\mathbb{R}^N} - \frac{AH(\rho)}{2\delta}. \qquad \square$$

4.2.3. Proof of the Lower Bound: We at last have all the ingredients needed to prove our lower bound on $p(t, x, y)$, and we begin with a preliminary version of our final result. For now, and until further notice, we will be assuming that $I \leq a(x) \leq AI$.

Let x, $y \in \mathbb{R}^N$ be given, and set $z = \frac{x+y}{2}$. Then, by Lemma 4.2.5, with $\rho = g(1, \cdot)$, $\tau = \delta = 1$, $t = \frac{1}{2}$, and $\pi(s) = y + s(x - y)$,

$$\int \log p_\theta\left(\tfrac{3}{2}, \xi + z, y\right) \gamma(d\xi) \geq \int \log p_\theta(1, \xi + y, y) \gamma(d\xi) - \frac{AN}{4} - \frac{|y - x|^2}{2}$$

since $H(g(1)) = N$. Similarly, by symmetry,

$$\int \log p_\theta\left(\tfrac{3}{2}, x, z + \xi\right) \gamma(d\xi) \geq \int \log p_\theta(1, x, \xi + x) \gamma(d\xi) - \frac{AN}{4} - \frac{|y - x|^2}{2}.$$

Hence, by the Chapman–Kolmogorov equation, Fatou's Lemma, Jensen's inequality, and Lemma 4.2.3,

$$\begin{aligned}
\log p(3, x, y) &= \log \left(\int p\left(\tfrac{3}{2}, x, \xi\right) p\left(\tfrac{3}{2}, y, \xi\right) d\xi \right) \\
&= \log \left(\int p\left(\tfrac{3}{2}, x, z + \xi\right) p\left(\tfrac{3}{2}, z + \xi, y\right) d\xi \right) \\
&\geq \varlimsup_{\theta \searrow 0} \log \left(\int p_\theta\left(\tfrac{3}{2}, x, z + \xi\right) p_\theta\left(\tfrac{3}{2}, z + \xi, y\right) \gamma(d\xi) \right) \\
&\geq \varlimsup_{\theta \searrow 0} \left[\int \log p_\theta\left(\tfrac{3}{2}, x, z + \xi\right) \gamma(d\xi) + \int \log p_\theta\left(\tfrac{3}{2}, z + \xi, y\right) \gamma(d\xi) \right] \\
&\geq -2\kappa(A) - \frac{AN}{2} - |y - x|^2,
\end{aligned}$$

and so

$$p(3, x, y) \geq e^{-2\kappa(A) - \frac{AN}{2}} e^{-|y - x|^2}.$$

Because this estimate depends only on A, we can use scaling (cf. (4.1.6)), as we have before, to obtain the somewhat crude lower bound

$$(4.2.6) \qquad p(t, x, y) \geq \frac{\alpha(A)}{t^{\frac{N}{2}}} e^{-\frac{3|y - x|^2}{t}}, \qquad (t, x, y) \in (0, \infty) \times \mathbb{R}^N \times \mathbb{R}^N,$$

where $\alpha(A) \equiv 3^{\frac{N}{2}} e^{-2\kappa(A) - \frac{AN}{2}}$.

To improve on (4.2.6), choose ρ as in Lemma 4.2.5, and note that, because of (4.2.6), we can now use that lemma to say that, for any $(t, x) \in (0, \infty) \times \mathbb{R}^N$, $\pi \in C^1([0, t]; \mathbb{R}^N)$ with $\pi(0) = x$, and $\delta \in (0, 1]$,

$$(*) \quad \begin{aligned}
&\int \rho(\xi) \log p\left((1 + \delta)t, x, \xi + \pi(t)\right) d\xi \\
&\qquad \geq \log \alpha(A) - \frac{N}{2} \log \delta t - \frac{3AS(\rho)}{\delta t} - \frac{AH(\rho)t}{2\delta} - \frac{(1 + \delta)E_\rho(t, \pi)}{2},
\end{aligned}$$

where $S(\rho) \equiv \int |\xi|^2 \rho(\xi) \, d\xi$. Next, let $x, y \in \mathbb{R}^N$ be given, and choose $\pi \in C^1\big([0, t]; \mathbb{R}^N\big)$ so that $\pi(0) = x$ and $\pi(t) = y$. Then, by (*) with t replaced by $\frac{t}{2}$,

$$\int \rho(\xi) \log p\big((1 + \delta)\tfrac{t}{2}, x, \xi + z)\big) \, d\xi$$

$$\geq \log \alpha_N(A) - \frac{N}{2} \log \frac{\delta t}{2} - \frac{6AS(\rho)}{\delta t} - \frac{AH(\rho)t}{4\delta} - \frac{(1 + \delta)E_\rho\big(\tfrac{t}{2}, \pi\big)}{2},$$

and

$$\int \rho(\xi) \log p\big((1 + \delta)\tfrac{t}{2}, \xi + z, y)\big) \, d\xi$$

$$\geq \log \alpha_N(A) - \frac{N}{2} \log \frac{\delta t}{2} - \frac{6AS(\rho)}{\delta t} - \frac{AH(\rho)t}{4\delta} - \frac{(1 + \delta)E_\rho\big(\tfrac{t}{2}, \check{\pi}\big)}{2},$$

where $z = \pi\big(\tfrac{t}{2}\big)$ and $\check{\pi}(s) = \pi(t - s)$. Proceeding as before via the Chapman–Kolmogorov equation, and noting that $E_\rho\big(\tfrac{t}{2}, \pi\big) + E_\rho\big(\tfrac{t}{2}, \check{\pi}\big) = E_\rho(t, \pi)$, one can pass from these to

$$\log p\big((1 + \delta)t, x, y\big)$$

$$\geq \log \left(\frac{1}{\|\rho\|_u} \int \rho(\xi) p\big((1 + \delta)\tfrac{t}{2}, x, \xi + z\big) p\big((1 + \delta)\tfrac{t}{2}, y, z + \xi\big) \, d\xi \right)$$

$$\geq -\log \|\rho\|_u + 2 \log \alpha_N(A) - N \log \frac{\delta t}{2} - \frac{12AS(\rho)}{\delta t}$$

$$- \frac{AH(\rho)t}{2\delta} - \frac{(1 + \delta)E_\rho(t, \pi)}{2}.$$

Now choose and fix a ρ with compact support in $B(0, 1)$, and set $\rho_t(\xi) = t^{-\frac{N}{2}} \rho(t^{-\frac{1}{2}}\xi)$. Then

$$(4.2.7) \qquad \|\rho_t\|_u = t^{-\frac{N}{2}} \|\rho\|_u, \quad S(\rho_t) = tS(\rho), \quad \text{and} \quad H(\rho_t) = \frac{H(\rho)}{t}.$$

Hence, after replacing ρ by ρ_t in the preceding, we find that

$$(**) \quad p\big((1 + \delta)t, x, y\big) \geq \frac{2^N \alpha_N(A)^2}{\delta^N \|\rho\|_u t^{\frac{N}{2}}} e^{-\frac{A(24S(\rho) + H(\rho))}{2\delta}} \exp\left(-\frac{(1 + \delta)E_{\rho_t}(t, \pi)}{2} \right)$$

for every $\pi \in C^1\big([0, t]; \mathbb{R}^N\big)$ with $\pi(0) = x$ and $\pi(t) = y$.

Before stating our final result, set

$$(4.2.8) \qquad d_{t,\rho}(x, y) = \inf \left\{ \sqrt{E_{\rho_t}(1, \pi)} : \pi(0) = x \ \& \ \pi(1) = y \right\},$$

and note that, because

$$\inf\left\{E_\rho(t,\pi):\ \pi(0) = x\ \&\ \pi(t) = y\right\}$$
$$= t^{-1}\inf\left\{E_\rho(1,\pi):\ \pi(0) = x\ \&\ \pi(1) = y\right\},$$

one can replace (**) by

$$p(t,x,y) \geq \frac{2^{\frac{3N}{2}}\alpha_N(A)^2}{\delta^N\|\rho\|_u t^{\frac{N}{2}}}e^{-\frac{\kappa(\rho)A}{\delta}}\exp\left(-\frac{(1+\delta)^2 d_{t,\rho}(x,y)^2}{2t}\right),$$

where $\kappa(\rho) \equiv 12S(\rho) + \frac{1}{2}H(\rho)$.

THEOREM 4.2.9.   *Assume that $\epsilon I \leq a(x) \leq (\epsilon\Lambda)I$, $x \in \mathbb{R}^N$, where $\epsilon > 0$ and $\Lambda \in [1,\infty)$. There is a universal $\beta_N : [1,\infty) \longrightarrow (0,1)$ such that*

$$p(t,x,y) \geq \frac{\beta_N(\Lambda)e^{-\frac{2\kappa(\rho)\Lambda}{\delta}}}{\|\rho\|_u(\epsilon\delta^2 t)^{\frac{N}{2}}}\exp\left(-\frac{(1+\delta)d_{\epsilon t,\rho}(x,y)^2}{2t}\right)$$

for all $\delta \in (0,1]$. In particular,

(4.2.10) $$\lim_{t\searrow 0} t\log p(t,x,y) = -\frac{d(x,y)^2}{2}$$

and, when $\kappa \equiv \kappa\big(g(1)\big) = \frac{25N}{8}$,

$$p(t,x,y) \geq \frac{\beta_N(\Lambda)e^{-\kappa\Lambda}}{(\epsilon t)^{\frac{N}{2}}}\exp\left(-\frac{2|y - x|^2}{\epsilon t}\right).$$

PROOF: In view of the preceding, the first assertion follows from (**) and the observation that if $p_\epsilon(t,x,y)$ is the transition probability density for $\epsilon^{-1}a$, then $p(t,x,y) = p_\epsilon(\epsilon t,x,y)$. As for the second assertion, we already know (cf. Theorem 4.1.11) that $\overline{\lim}_{t\searrow 0} t\log p(t,x,y) \leq -\frac{d(x,y)^2}{2}$, and clearly the estimate just proved gives $\underline{\lim}_{t\searrow 0} t\log p(t,x,y) \geq -\frac{d(x,y)^2}{2}$. Finally, the last estimate comes from taking $\rho = g(1)$, $\delta = 1$, and using $d_{t,\rho}(x,y)^2 \leq \epsilon^{-1}|y - x|^2$. □

4.3 Conservative Perturbations

As yet, the results in this chapter apply only to uniformly elliptic operators which are in divergence form (cf. (4.0.1)). Our goal now is to see what can be said about operators which are perturbations of operators in divergence form, and we begin with perturbations by vector fields which preserve the basic properties of the unperturbed operator.

Let $U \in C_b^\infty(\mathbb{R}^N;\mathbb{R})$ be given, and consider the operator L whose action is

(4.3.1) $$L\varphi = \tfrac{1}{2}\nabla\cdot\big(a\nabla\varphi\big) + \big(\nabla U, a\nabla\varphi\big)_{\mathbb{R}^N},$$

where a is as in the preceding sections. Our goal here is to obtain upper and lower bounds on the associated $p(t, x, y)$ and to show that these can be made to depend only on

$$(4.3.2) \qquad \delta(U) \equiv \sup_{x \in \mathbb{R}^N} U(x) - \inf_{x \in \mathbb{R}^N} U(x)$$

in addition to the upper and lower bounds on a.

The observation on which our argument rests is that the operator L is formally self-adjoint in $L^2(\lambda^U; \mathbb{R})$, where $\lambda^U(dx) = e^{2U} \, dx$. To see this, notice that

$$L\varphi = \tfrac{1}{2} e^{-2U} \nabla \cdot \left(e^{2U} a \nabla \varphi\right),$$

and conclude that

$$(4.3.3) \qquad \left(\varphi, L\psi\right)_{L^2(\lambda^U; \mathbb{R})} = -\tfrac{1}{2} \left(\nabla \varphi, a \nabla \psi\right)_{L^2(\lambda^U; \mathbb{R})}$$

for all $\varphi, \psi \in C^2(\mathbb{R}^N; \mathbb{R}) \cap L^2(\lambda^U; \mathbb{R})$ with $L\psi, |\nabla \varphi|, |\nabla \psi| \in L^2(\mathbb{R}^N; \mathbb{R})$. Starting from (4.3.3), one can proceed, as in the derivation of (4.1.2), to see that, for each $t > 0$, the operator $\mathbf{P}_t$ given by $\mathbf{P}_t \varphi(x) = \int \varphi(y) p(t, x, y) \, dy$ is symmetric in $L^2(\lambda^U; \mathbb{R})$ and is therefore a contraction in $L^r(\lambda^U; \mathbb{R})$ for each $r \in [1, \infty]$. Notice that this symmetry statement is equivalent to

$$(4.3.4) \qquad e^{2U(x)} p(t, x, y) = e^{2U(y)} p(t, y, x).$$

Knowing the preceding, we base our proof of an upper bound on precisely the same line of reasoning as we used in § 4.1. Namely, begin with the observation that, just as in the proof of Lemma 4.1.7, (4.3.3) leads to

$$\frac{d}{dt} \|\mathbf{P}_t^\psi \varphi\|_{L^{2r}(\lambda^U; \mathbb{R})}$$

$$\leq -\frac{\epsilon}{2r} \|\mathbf{P}_t^\psi \varphi\|_{L^{2r}(\lambda^U; \mathbb{R})}^{1-2r} \left\|\nabla (\mathbf{P}_t^\psi \varphi)^r\right\|_{L^2(\lambda^U; \mathbb{R}^N)} + \frac{rD(\psi)^2}{2} \|\mathbf{P}_t^\psi \varphi\|_{L^{2r}(\lambda^U; \mathbb{R})}.$$

Next, observe that Lemma 4.1.3 implies

$$\|\varphi\|_{L^2(\lambda^U; \mathbb{R})}^{2+\frac{4}{N}} \leq C_N(U) \|\nabla \varphi\|_{L^2(\lambda^U; \mathbb{R}^N)}^2 \|\varphi\|_{L^1(\lambda^U; \mathbb{R})}^{\frac{4}{N}},$$

where $C_N(U) \leq C_N e^{\frac{2(N+4)\delta(U)}{N}}$. Hence,

$$\frac{d}{dt} \|\mathbf{P}_t^\psi \varphi\|_{L^{2r}(\lambda^U)} \leq -\frac{\epsilon}{2C_N(U)r} \frac{\|\mathbf{P}_t^\psi \varphi\|_{L^{2r}(\lambda^U; \mathbb{R})}^{1+\frac{4r}{N}}}{\|\mathbf{P}^\psi \varphi\|_{L^r(\lambda^U; \mathbb{R})}^{\frac{4r}{N}}} + \frac{rD(\psi)^2}{2} \|\mathbf{P}_t^\psi \varphi\|_{L^{2r}(\lambda^U; \mathbb{R})}.$$

At this point, one can repeat the argument used after (*) in the proof of Lemma 4.1.7 and thereby conclude that, for every $\delta \in (0, 1]$,

$$\|\mathbf{P}_t^\psi\|_{2\to\infty} \leq \frac{K_N(U)}{(\epsilon\delta t)^{\frac{N}{4}}} e^{\frac{t(1+\delta)D(\psi)^2}{2}},$$

where we are now thinking of $\mathbf{P}_t^\psi$ as a mapping from $L^2(\lambda^U; \mathbb{R})$ to $L^\infty(\lambda^U; \mathbb{R})$ and $K_N(U) \equiv 4^N \big(NC_N(U)\big)^{\frac{N}{4}}$. Because $\mathbf{P}_t^{-\psi}$ is the adjoint of $\mathbf{P}_t^\psi$ in $L^2(\lambda^U; \mathbb{R})$, we also have that $\|\mathbf{P}_t^\psi\|_{1 \to 2}$ satisfies the same estimate and therefore that

$$\|\mathbf{P}_t^\psi \varphi\|_{1 \to \infty} \leq \frac{2^{\frac{N}{2}} K_N(U)^2}{(\epsilon \delta t)^{\frac{N}{2}}} e^{\frac{t(1+\delta)D(\psi)^2}{2}}$$

as an operator from $L^1(\lambda^U; \mathbb{R})$ to $L^\infty(\lambda^U; \mathbb{R})$. Equivalently,

$$\left| \iint e^{2U(x)-\psi(x)} \varphi_1(x) p(t, x, y) \varphi_2(y) e^{\psi(y)} \, dx dy \right|$$
$$\leq \frac{2^{\frac{N}{2}} N K_N(U)^2}{(\epsilon \delta t)^{\frac{N}{2}}} e^{\frac{t(1+\delta)D(\psi)^2}{2}} \|\varphi_1\|_{L^1(\lambda^U; \mathbb{R})} \|\varphi_1\|_{L^1(\lambda^U; \mathbb{R})},$$

and so

$$p(t, x, y) \leq \frac{e^{2\delta(U)} 2^{\frac{N}{2}} N K_N(U)^2}{(\epsilon \delta t)^{\frac{N}{2}}} \exp\left(\psi(x) - \psi(y) + \frac{t(1+\delta)D(\psi)^2}{2} \right).$$

Finally, just as in the analogous step at the end of §4.1.2, we arrive at the existence of a new $K_N(U)$, depending only on N and $\delta(U)$, such that

$$(4.3.5) \qquad p(t, x, y) \leq \frac{K_N(U)}{(\epsilon \delta t)^{\frac{N}{2}}} \exp\left(-\frac{d(x,y)^2}{2(1+\delta)t} \right),$$

where $d(x, y)$ is the Riemannian distance in (4.1.9). In particular,

$$(4.3.6) \qquad p(t, x, y) \leq \frac{K_N(U)}{(\epsilon t)^{\frac{N}{2}}} \exp\left(-\frac{|y-x|^2}{4At} \right),$$

where A is the upper bound on $\|a\|_{\mathrm{op}}$.

Our proof of the lower bound is a little more challenging and leads to a result which complements (4.3.6) but not (4.3.5). As in §4.2.3, we will begin with the case when $I \leq a(x) \leq AI$ and, at the end, will rescale time to reduce the general case to this one. In addition, because nothing is changed when we add a constant to U, we will assume that $\sup_{x \in \mathbb{R}^N} U(x) = 0$ and therefore that $U \leq 0$ and $\delta(U) = \|U\|_{\mathrm{u}}$.

Given $z \in \mathbb{R}^N$, set $Z(z) = \int e^{2U(z+\xi)} \gamma(d\xi)$. Clearly, $1 \geq Z(z) \geq e^{-2\delta(U)}$. Next, set $\gamma(z, d\xi) = Z(z)^{-1} e^{2U(z+\xi)} \gamma(d\xi)$. Our replacement for Lemma 4.2.3 will be

$$(4.3.7) \qquad \begin{aligned} &\int \log p_\theta(1, z+\xi, y) \gamma(z, d\xi) \geq -\kappa(A, U) \\ &\qquad\qquad\qquad\qquad\qquad\qquad \text{for } |x-z| \vee |y-z| \leq 1, \\ &\int \log p_\theta(1, x, z+\xi) \gamma(z, d\xi) \geq -\kappa(A, U) \end{aligned}$$

where $\kappa(A, U) < \infty$ depends only on A and $\delta(U)$. Without loss in generality, we will assume that $z = 0$ and $|x| \vee |y| \leq 1$. Finally, observe that we need only prove the first line in (4.3.7), since, by (4.3.4),

$$
\int \log p_\theta(1, x, \xi)\, \gamma(0, d\xi)
$$

$$
= \int \log \Big[e^{2U(\xi) - 2U(x)}(1 - \theta)p(1, \xi, x) + \theta \Big] \gamma(0, d\xi)
$$

$$
\geq \int \log \Big[e^{-2\delta(U)}p_\theta(1, x, \xi) \Big] \gamma(0, d\xi)
$$

$$
= \int \log p_\theta(1, \xi, x)\, \gamma(0, d\xi) - 2\delta(U).
$$

To prove the first line of (4.3.7), we want to use the same reasoning as we did in the proof of Lemma 4.2.3. Thus, set $u_\theta(t, \xi) = p_\theta(t, \xi, y)$, $w_\theta(t) = \int \log u_\theta(t, \xi)\, \gamma(0, d\xi)$, and proceed as we did there to get

$$
\dot{w}_\theta(t) \geq -\frac{1}{2} \int \big(y, a(y)y \big)_{\mathbb{R}^N} \gamma(0, dy) + \frac{1}{2} \big\| \nabla \log u_\theta(t) \big\|^2_{L^2(\gamma(0);\mathbb{R})}.
$$

Clearly,

$$
\frac{1}{2} \int \big(y, a(y)y \big)_{\mathbb{R}^N} \gamma(0, dy) \leq \frac{AN}{2Z(0)} \leq ANe^{2\delta(U)}.
$$

In order to handle the other term, we need the Poincaré inequality

$$
\big\| \varphi - \langle \varphi, \gamma(0) \rangle \big\|^2_{L^2(\gamma(0);\mathbb{R})} \leq e^{2\delta(U)} \big\| \nabla \varphi \big\|^2_{L^2(\gamma(0),\mathbb{R}^N)},
$$

which can be derived from Lemma (4.2.2) as follows:

$$
\big\| \varphi - \langle \varphi, \gamma(0) \rangle \big\|^2_{L^2(\gamma(0);\mathbb{R})} \leq \big\| \varphi - \langle \varphi, \gamma \rangle \big\|^2_{L^2(\gamma(0);\mathbb{R})}
$$

$$
\leq Z(0)^{-1} \big\| \varphi - \langle \varphi, \gamma \rangle \big\|^2_{L^2(\gamma;\mathbb{R})} \leq Z(0)^{-1} \big\| \nabla \varphi \big\|^2_{L^2(\gamma;\mathbb{R}^N)}
$$

$$
\leq e^{2\delta(U)} \big\| \nabla \varphi \big\|^2_{L^2(\gamma(0);\mathbb{R}^N)}.
$$

With this and the preceding, we now have

$$
\dot{w}_\theta(t) \geq -ANe^{2\delta(U)} + \frac{e^{-2\delta(U)}}{2} \big\| \log u_\theta(t) - w_\theta \big\|^2_{L^2(\gamma(0);\mathbb{R})}.
$$

From here, the rest of the argument is essentially the same as the one given at the end of the proof of Lemma 4.2.3. However, there are two points which require our attention. First, we cannot guarantee that $w_\theta(t) \leq 0$, but this causes no real problem since, if $w_\theta(t) \geq 0$ for some $t \in (0, 1]$, then the preceding says that $w_\theta(1) \geq -ANe^{2\delta(U)}$. Thus, we may assume that $w_\theta(t) \leq 0$ for all $t \in (0,1]$, in which case

$$\| \log u_\theta(t) - w_\theta(t) \|^2_{L^2(\gamma(0))} \geq \frac{w_\theta(t)^2}{2e^{2\delta(U)}} \gamma(\{\xi : p(t,\xi,y)(t,\xi) \geq 2e^{-\lambda}\}) - \lambda^2$$

for any $\lambda > 0$ and $\theta \in \left(0, \frac{1}{2}\right]$. If we can find $\lambda > 0$ and $\alpha \in (0,1]$, depending only on A and $\delta(U)$ so that

$$\gamma(\{\xi : p(t,\xi,(t,\xi,y) \geq 2e^{-\lambda}\}) \geq \alpha, \quad t \in \left[\tfrac{1}{2},1\right],$$

then we can repeat the argument used at end of the proof of Lemma 4.2.3 to conclude that $w_\theta(1)$ is bounded below by a finite, negative constant depending only on A and $\delta(U)$. To this end, first note that, by (4.3.4),

$$\int p(t,\xi,y)\,d\xi = e^{2U(y)} \int e^{-2U(\xi)} p(t,y,\xi)\,d\xi \geq e^{-2\delta(U)}.$$

Thus, for any $R > 0$,

$$e^{-2\delta(U)} \leq \int_{B(0,R)} p(t,\xi,y)\,d\xi + \int_{B(0,R)\complement} p(t,\xi,y)\,d\xi$$

$$\leq \Omega_N R^N e^{-\lambda} + \|p(t,\,\cdot\,,y)\|_u \frac{e^{\frac{R^2}{2}}}{(2\pi)^{\frac{N}{2}}} \gamma(\{\xi : p(t,\xi,y) \geq 2e^{-\lambda}\})$$

$$+ \int_{B(0,R)\complement} p(t,\xi,y)\,d\xi.$$

Remembering that $|y| \leq 1$ and using the estimate in (4.3.6), we can first choose R so that the last term is less than or equal $\frac{1}{4}e^{-2\delta(U)}$ for all $t \in (0,1]$ and then choose $\lambda > 0$ so that $\Omega_N R^N e^{-\lambda} \leq \frac{1}{8}e^{-\delta(U)}$, where both R and λ depend only A and $\delta(U)$. Finally, another application of (4.3.6) shows that $\sup_{t \in [\frac{1}{2},1]} \|p(t,\,\cdot\,,y)\|_u$ is bounded by a constant depending only on A and $\delta(U)$, and this gives us our α.

At this point, one should be looking for the analog of Lemma 4.2.5. However, that does not seem to be available unless one is willing to allow ones estimates to depend on $\|\nabla U\|_u$, in which case one is in the situation dealt with in the next section. Thus, we must take another, and less delicate, tack.

Let $x, y \in \mathbb{R}^N$ with $|y - x| \leq 2$ be given, and set $z = \frac{x+y}{2}$. By the Chapman–Kolmogorov equation, Jensen's inequality, and (4.3.7), we have

$$\log p(2,x,y) \geq \varlimsup_{\theta \searrow 0} \log \left(Z(z) \int p_\theta(1,x,z+\xi) p_\theta(1,z+\xi,y)\,\gamma(z,d\xi) \right)$$

$$\geq \varlimsup_{\theta \searrow 0} \log \left(\int p_\theta(1,x,z+\xi) p_\theta(1,z+\xi,y)\,\gamma(z,d\xi) \right) - 2\delta(U)$$

$$\geq -2\kappa(A,U) - 2\delta(U),$$

and therefore that there exists an $\alpha \in (0,1)$, depending only on A and $\delta(U)$, such that $p(2,x,y) \geq \alpha$ whenever $|y - x| \leq 2$. Next, note that (4.1.6) holds when $p_\lambda(t,x,y)$ corresponds to the operator with coefficients $a(\lambda \cdot)$ and $U(\lambda \cdot)$. Hence, from the preceding, we conclude that

$$p(t,x,y) \geq \frac{\beta}{t^{\frac{N}{2}}} \quad \text{when } |y - x|^2 \leq t,$$

where $\beta = 2^{\frac{N}{2}}\alpha$.

To get further away from the diagonal, we will use a *chaining argument*, *which, for future reference, we give as a lemma.*

LEMMA 4.3.8. *Assume that $q : (0,\infty) \times \mathbb{R}^N \times \mathbb{R}^N \longrightarrow [0,\infty)$ is a continuous function which satisfies the Chapman–Kolmogorov equation*

$$q(s + t, x, y) = \int q(s,x,\xi)q(t,\xi,y)\,d\xi.$$

If, for some $\beta > 0$, $q(t,x,y) \geq \beta t^{-\frac{N}{2}}$ whenever $|y - x|^2 \leq t$, then

$$q(t,x,y) \geq \frac{1}{Mt^{\frac{N}{2}}}e^{-\frac{M|y-x|^2}{t}},$$

where $M \in [1,\infty)$ depends only on N and β. If, on the other hand, $q(t,x,y) \geq \beta t^{-\frac{N}{2}}$ only for $|y - x|^2 \leq t \leq 1$, then

$$q(t,x,y) \geq \frac{1}{Mt^{\frac{N}{2}}}e^{-Mt - \frac{M|y-x|^2}{t}}$$

for an $M \in [1,\infty)$ with the same dependence as before.

PROOF: In the first case, note that there is nothing to do unless $\frac{|y-x|^2}{t} > 1$. Thus assume that $\frac{|y-x|^2}{t} > 1$, and take n to be the smallest integer dominating $\frac{4|x-y|^2}{t}$. Next, set $\tau = \frac{t}{n}$, $r = \frac{|y-x|}{2n}$, $z_m = x + \frac{m(y-x)}{n}$, and $B_m = B(z_m, r)$ for $0 \leq m \leq n$. If $\xi \in B_m$ and $\xi' \in B_{m+1}$, then $|\xi' - \xi|^2 \leq 16r^2 = \frac{4|y-x|^2}{n^2} \leq \frac{t}{n} = \tau$, and so, by hypothesis, $q(\tau,\xi,\xi') \geq \beta\tau^{-\frac{N}{2}}$. Thus, by the Chapman–Kolmogorov equation,

$$q(t,x,y) = \int \cdots \int q(\tau,x,\xi_1)q(\tau,\xi_1,\xi_2)\cdots q(\tau,\xi_{n-1},y)\,d\xi_1\cdots d\xi_{n-1}$$

$$\geq \int_{B_1}\cdots\int_{B_{n-1}} q(\tau,x,\xi_1)q(\tau,\xi_1,\xi_2)\cdots q(\tau,\xi_{n-1},y)\,d\xi_1\cdots d\xi_{n-1}$$

$$\geq \frac{\beta^n(\Omega_N r^N)^{n-1}}{\tau^{\frac{nN}{2}}} = \frac{\beta}{\tau^{\frac{N}{2}}}\left(\beta\Omega_N\left(\frac{r}{\tau^{\frac{1}{2}}}\right)^N\right)^{n-1}.$$

Since $|y - x| \geq \frac{1}{2}\sqrt{(n-1)t}$ and $n \geq 4$, $\frac{r}{\tau^{\frac{1}{2}}} \geq \frac{1}{4}$, and so

$$q(t, x, y) \geq \frac{\beta n^{\frac{N}{2}}}{t^{\frac{N}{2}}} e^{-\lambda(n-1)} \geq \frac{\beta}{t^{\frac{N}{2}}} \exp\left(-\frac{4\lambda|y-x|^2}{t}\right),$$

where $\lambda = -\log\left(4^{-N}\beta\Omega_N\right)$. Clearly, this completes the proof of the first case.

Turning to the second case, note that the argument just given works when either $t \in (0, 1]$ or $|x - y| \geq t \geq 1$. That is, we know that

$$q(t, x, y) \geq \frac{1}{Mt^{\frac{N}{2}}} e^{-\frac{M|y-x|^2}{t}} \quad \text{when either } t \leq 1 \text{ or } |y - x| \geq t.$$

Now assume that $t > |x - y| \vee 1$, and take n to be the smallest integer dominating t, $\tau = \frac{t}{n}$, $r = \tau^{\frac{1}{2}}$, $z_m = x + \frac{m}{n}(y - x)$, and $B_m = B(z_m, r)$. Then $\tau \leq 1$, $|\xi' - \xi| \leq 3\tau^{\frac{1}{2}}$ if $\xi \in B_m$ and $\xi' \in B_{m+1}$, and so, proceeding as we did before, we find that

$$q(t, x, y) \geq \left(\frac{e^{-9M}}{M\tau^{\frac{N}{2}}}\right)^n \left(\Omega_N r^N\right)^{n-1} = \frac{n^{\frac{N}{2}} e^{-9M}}{Mt^{\frac{N}{2}}} \left(\frac{\Omega_N e^{-9M}}{M}\right)^{n-1},$$

and therefore, after making a minor adjustment in M, one arrives at the desired result. $\square$

THEOREM 4.3.9. *Let* $a : \mathbb{R}^N \longrightarrow \text{Hom}(\mathbb{R}^N; \mathbb{R}^N)$ *be a smooth, symmetric matrix valued function satisfying* $\epsilon I \leq a(x) \leq (\epsilon\Lambda)I$ *for some* $\epsilon > 0$ *and* $\Lambda \in [1, \infty)$. *Further, assume that* $U : \mathbb{R}^N \longrightarrow \mathbb{R}$ *is a bounded, smooth function, and define* $\delta(U)$ *by* (4.3.2). *Finally, let* $(t, x, y) \in (0, \infty) \times \mathbb{R}^N \times \mathbb{R}^N \longmapsto p(t, x, y) \in [0, \infty)$ *be the transition probability density associated with the operator* L *in* (4.3.1). *Then there exists a constant* $M \in [1, \infty)$, *depending only on* Λ *and* $\delta(U)$, *such that*

$$\frac{1}{M(\epsilon t)^{\frac{N}{2}}} e^{-\frac{M|y-x|^2}{\epsilon t}} \leq p(t, x, y) \leq \frac{M}{(\epsilon t)^{\frac{N}{2}}} e^{-\frac{|y-x|^2}{M\epsilon t}}.$$

PROOF: In view of (4.3.6) and the first part of Lemma 4.3.8 plus the estimate which precedes it, there is nothing to do when $\epsilon = 1$. To handle general ϵ's, consider $(t, x, y) \rightsquigarrow p(\epsilon^{-1}t, x, y)$, and argue as we did in proof of Theorem 4.2.9. $\square$

REMARK 4.3.10. There are two good reasons why the results in this section may be confusing to people whose primary training is in probability theory. One of these reasons is terminology. Namely, when L is presented in the form given in (1.1.8), probabilists call a the *diffusion coefficient* and b the *drift coefficient*. This terminology undoubtedly derives from the fact it makes perfectly good sense from the probabilistic "path property" perspective: a governs the martingale part of the path and b the absolutely continuous part. However, this terminology is misleading if one wants the presence or absence of a drift to be reflected in the absence or presence of diffusive behavior. To be more precise, say that L has diffusive behavior if the associated diffusion spreads like $t^{\frac{1}{2}}$ in the sense that, for each $\theta \in (0, 1)$, there exists an $R \in (0, \infty)$ such that

$$\int_{B(x, t^{\frac{1}{2}} R)} p(t, x, y) \, dy \geq \theta \quad \text{for all } (t, x) \in (0, \infty) \times \mathbb{R}^N.$$

By Lemma 3.3.12, when L is given by (1.1.8), this property holds if "there is no drift" in the sense that $b = 0$. On the other hand, we now know that it also holds when L is given by (4.3.1) with $a \geq \epsilon I$ and U bounded. Thus, at least when dealing with operators L given by $L = \frac{1}{2} \nabla \cdot (a \nabla) + b \cdot \nabla$ with $a \geq \epsilon I$, diffusivity seems to have more to do with whether $b = a \nabla U$ for some bounded U. Hence, in this setting, there is reason for thinking that the "drift" component of L is the difference between b and its "projection" onto the space $\{a \nabla U : U \in C_{\mathrm{b}}^1(\mathbb{R}^N; \mathbb{R})\}$. Of course, before this can be made precise, one has to decide exactly how one is going to take this projection. Nonetheless, without ambiguity, one can say that there is no drift present when $b = a \nabla U$ for some bounded U. Moreover, this definition has the enormous advantage that it is invariant under coordinate changes by diffeomorphisms whose Jacobians are uniformly bounded and non-degenerate. Namely, if one adopts it, then the set of operators which will be "driftless" are precisely those of the form given by (4.3.1), and, as distinguished from those in (1.1.8) with $b = 0$, this set of operators is the same in all coordinate systems. To verify their coordinate invariance, use (4.3.3) to describe L and note that a change of coordinates results in a getting replaced by $J^\top a J$ and U by $U - \frac{1}{2} \log |\det(J)|$, where J is the Jacobian matrix of the coordinate transformation.

The second probabilistically confusing aspect will bother only aficionados of Brownian motion. Namely, because probabilists like to think of local time at 0 as the indefinite integral of δ_0 along a (one-dimensional) Brownian path, Tanaka's formula (cf. (6.1.7) in [55]) for reflected Brownian motion indicates that the generator of reflected Brownian motion is the operator $\frac{1}{2} \partial_x^2 + \delta_0 \partial_x$. Since one seems to get this same operator when one takes $U = \mathbf{1}_{[0, \infty)}$ in the theory here, and, because our estimates depend only on the size of U, as opposed to U', the fact that reflected Brownian motion never leaves the right half-line would seem to contradict the lower bound given in the preceding theorem. Of course, the contradiction is a consequence of being too casual about ones interpretation of the operator $\frac{1}{2} \partial_x^2 + \delta_0 \partial_x$, for which there are at least two rigorous interpretations. The interpretation which leads to reflected Brownian motion is that the associated heat flow $\{\mathbf{P}_t : t \geq 0\}$ satisfies

$$\frac{d}{dt} \int_{[0, \infty)} \psi(x) \mathbf{P}_t \varphi(x) \, dx \bigg|_{t=0} = -\frac{1}{2} \int_{[0, \infty)} \psi'(x) \varphi'(x) \, dx$$

for all $\varphi, \psi \in C^\infty([0,\infty);\mathbb{R})$ with compact support. The interpretation which corresponds to the heat flow in Theorem 4.3.9 is that

$$\frac{d}{dt}\int_\mathbb{R} \psi(x)\mathbf{P}_t\varphi(x)e^{2\mathbf{1}_{[0,\infty)}(x)}\,dx\bigg|_{t=0} = -\frac{1}{2}\int_\mathbb{R} \psi'(x)\varphi'(x)e^{2\mathbf{1}_{[0,\infty)}(x)}\,dx$$

for all $\varphi, \psi \in C_c^\infty(\mathbb{R};\mathbb{R})$.

4.4 General Perturbations of Divergence Form Operators

In this section we want to extend the results in §§ 4.1 and 4.2 to operators L given by

$$(4.4.1) \qquad L\varphi = \tfrac{1}{2}\nabla \cdot \big(a\nabla\varphi\big) + \big(b, a\nabla\varphi\big)_{\mathbb{R}^N},$$

where $a \in C_b^\infty\big(\mathbb{R}^N; \mathrm{Hom}(\mathbb{R}^N;\mathbb{R}^N)\big)$, $b \in C_b^\infty(\mathbb{R}^N;\mathbb{R}^N)$, and a is symmetric and uniformly positive definite. However, because our technique relies on duality, we are forced to consider a more general class of operators, a class which is closed under formal adjoints. Thus, in order to get our estimates, we will have to consider operators given by

$$(4.4.2) \qquad L\varphi = \tfrac{1}{2}\nabla \cdot (a\nabla\varphi) + \big(b, a\nabla\varphi\big)_{\mathbb{R}^N} - \nabla \cdot \big(\varphi a\check{b}\big) + c\varphi,$$

where $\check{b} \in C_b^\infty(\mathbb{R}^N;\mathbb{R}^N)$ and $c \in C_b^\infty(\mathbb{R}^N;\mathbb{R})$. Notice that the formal adjoint $L^\top$ of such an L is the operator obtained by interchanging b and $\check{b}$. Also, observe that an L of this sort can be rewritten in the form

$$L = \frac{1}{2}\sum_{i,j=1}^N a_{ij}\partial_{x_i}\partial_{x_j} + \sum_{i=1}^N \tilde{b}_i\partial_{x_i} + V,$$

where $\tilde{b} \in C_b^\infty(\mathbb{R}^N;\mathbb{R}^N)$ and $V \in C_b^\infty(\mathbb{R}^N;\mathbb{R}^N)$. Thus (cf. § 2.5 and the reasoning used in § 3.3.4), there exists a smooth $(t,x,y) \in (0,\infty)\times\mathbb{R}^N \longrightarrow q(t,x,y) \in [0,\infty)$ with the properties that

$$e^{-t\|V^-\|_u} \le \int q(t,x,y)\,dy \le e^{t\|V^+\|_u}\ and$$

$$q(s+t,x,y) = \int q(t,\xi,y)q(s,x,\xi)\,d\xi$$

and, for all $\varphi \in C_b(\mathbb{R}^N;\mathbb{R})$,

$$(t,x,y) \in (0,\infty)\times\mathbb{R}^N \longmapsto \mathbf{Q}_t\varphi(x) \equiv \int \varphi(y)q(t,x,y)\,dy$$

is a smooth solution to $\partial_t u = Lu$ with $\lim_{t\searrow 0} u(t,\cdot) = \varphi$. Furthermore, by the same sort of argument which we used to prove Theorem 2.5.8, we can *identify the adjoint $\mathbf{Q}_t^\top$ of $\mathbf{Q}_t$ in $L^2(\mathbb{R}^N; \mathbb{R})$ as the operator associated with $L^\top$, and, of course this means that*

$$\mathbf{Q}_t^\top \psi(y) = \int q^\top(t, x, y)\psi(y)\, dx \quad \text{where } q^\top(t, x, y) = q(t, y, x).$$

Finally, until further notice, we will be assuming that

$$I \leq a(x) \leq AI \text{ and that } \|b\|_u + \|\check{b}\|_u \leq B.$$

4.4.1. The Upper Bound: *With these preliminaries, we are ready to derive the upper bound for $q(t, x, y)$. Thus, let $\varphi \in C_c^\infty(\mathbb{R}^N; [0, \infty)) \setminus \{0\}$ be given, and set $u(t, \cdot) = \mathbf{Q}_t\varphi$ and $v_r(t) = \|u(t)\|_{L^r(\mathbb{R}^N)}$ for $r \in [0, \infty)$. By the remarks above,*

$$\int \mathbf{Q}_t\varphi(x)\, dx = \int \varphi(y)\left(\int q(t, x, y)\, dx\right) dy \geq e^{-t\|V^\top\|_u} \int \varphi(y)\, dy > 0,$$

where $V^\top$ is 0th order term in the operator $L^\top$ when it is expressed in the form in (1.1.8). Hence, $v_r(t) > 0$. Proceeding as we did in the proof of Lemma 4.1.7, we find that

$$\begin{aligned} v_{2r}^{2r-1}\dot{v}_{2r} &= -\frac{2r-1}{2r^2}\left(\nabla u^r, a\nabla u^r\right)_{L^2(\mathbb{R}^N;\mathbb{R}^N)} + \left(u^r b_r, a\nabla u^r\right)_{L^2(\mathbb{R}^N;\mathbb{R}^N)} \\ &\quad + \left(c, u^{2r}\right)_{L^2(\mathbb{R}^N;\mathbb{R})} \\ &\leq -\frac{r-1}{2r^2}\left(\nabla u^r, a\nabla u^r\right)_{L^2(\mathbb{R}^N;\mathbb{R}^N)} + r\left(c_r, u^{2r}\right)_{L^2(\mathbb{R}^N;\mathbb{R})}, \end{aligned}$$

where

$$b_r = \frac{b}{r} + \left(2 - \tfrac{1}{r}\right)\check{b} \text{ and } c_r = \frac{c}{r} + \frac{1}{2}\left(b_r, ab_r\right)_{\mathbb{R}^N}.$$

In particular, by taking $r = 1$, we get $\dot{v}_2 \leq \|c_1\|_u v_2$ and therefore that

$$(4.4.3) \qquad v_2(t) \leq e^{t\|c_1\|_u}\|\varphi\|_{L^2(\mathbb{R}^N;\mathbb{R})}.$$

At the same time, when $r \geq 2$, Lemma 4.1.3 allows us to say that

$$\dot{v}_{2r} \leq -\frac{r-1}{2C_N r^2}\frac{v_{2r}^{1+\frac{4r}{N}}}{v_r^{\frac{4r}{N}}} + r\|c_r\|_u v_{2r},$$

and so, by the same reasoning that we used in the proof of Lemma 4.1.7, for each $\delta \in (0, 1]$,

$$(4.4.4) \qquad \begin{aligned} w_{2r}(t) &\leq \left(\frac{2NC_N r^2}{\delta}\right)^{\frac{N}{4r}} e^{\frac{t\delta\|c_r\|_u}{r}} w_r(t) \quad \text{when} \\ w_r(t) &\equiv \sup_{\tau \in [0,t]} \tau^{\frac{N}{4}\left(1-\frac{2}{r}\right)} v_r(\tau). \end{aligned}$$

Rather than applying (4.4.3) and (4.4.4) to get a "diagonal" estimate for $q(t, x, y)$, we develop the corresponding estimate for

$$q^\psi(t, x, y) \equiv e^{-\psi(x)} q(t, x, y) e^{\psi(y)},$$

where ψ is a smooth function with bounded derivatives. But $q^\psi(t, x, y)$ is the heat kernel corresponding to L^ψ, where $L^\psi \varphi = e^{-\psi} L(e^\psi \varphi)$, and, after an elementary calculation, one sees that L^ψ is the operator in (4.4.2) when b, $\check{b}$, and c are replaced, respectively, by

$$b^\psi = b + \tfrac{1}{2}\nabla\psi, \quad \check{b}^\psi = \check{b} - \tfrac{1}{2}\nabla\psi, \quad and \quad c^\psi = c + \big(b - \check{b}, a\nabla\psi\big)_{\mathbb{R}^N} + \tfrac{1}{2}\big(\nabla\psi, a\nabla\psi\big)_{\mathbb{R}^N}.$$

Hence, if $\mathbf{Q}_\varphi^\psi = e^{-\psi} \mathbf{Q}_t(e^\psi \varphi)$ and

$$v_r^\psi(t) = \|\mathbf{Q}_t^\psi \varphi\|_{L^r(\mathbb{R}^N)} \quad and \quad w_r^\psi(t) \equiv \sup_{\tau \in [0,t]} \tau^{\frac{N}{4}(1 - \frac{2}{r})} v_r^\psi(\tau),$$

then (4.4.3) and (4.4.4) say that, for every $\delta \in (0, 1]$,

$$w_2^\psi(t) \le e^{t\|c_1^\psi\|_u} \quad and \quad w_{2^{n+1}}^\psi(t) \le \left(\frac{4^{n+1} N C_N}{\delta}\right)^{\frac{2^{-n}N}{4}} e^{2^{-n} t \delta \|c_{2^n}^\psi\|_u} w_{2^n}^\psi(t).$$

Notice that, for any $\delta \in (0, 1]$ (cf. (4.1.8)),

$$\|c_1^\psi\|_u \le \|c\|_u + \frac{AB^2}{\delta} + \frac{1 + \delta}{2} D(\psi)^2$$

and, for $n \ge 1$,

$$\|c_{2^n}^\psi\|_u \le \|c\|_u + 5AB^2 + 4D(\psi)^2.$$

Thus, applying induction and passing to the limit as $n \to \infty$, we find that

$$\|\mathbf{Q}_t^\psi \varphi\|_u \le K_N \frac{e^{\frac{t(2\|c\|_u + 6AB^2)}{\delta}}}{(\delta t)^{\frac{N}{4}}} e^{\frac{t(1 + 5\delta) D(\psi)^2}{2}} \|\varphi\|_{L^2(\mathbb{R}^N; \mathbb{R})},$$

for some universal $K_N < \infty$, and therefore, since $(\mathbf{Q}_t^\psi)^\top$ satisfies the same estimate, we can repeat the argument following Lemma 4.1.7 to get

$$(4.4.5) \qquad q(t, x, y) \le \frac{2^{\frac{N}{2}} K_N e^{\frac{t(2\|c\|_u + 6AB^2)}{\delta}}}{(\delta t)^{\frac{N}{2}}} \exp\left(-\frac{d(x, y)^2}{2(1 + 5\delta)t}\right).$$

THEOREM 4.4.6.   *If $p(t, x, y)$ is the transition probability corresponding to the operator L given in (4.4.1), where*

$$\epsilon I \le a(x) \le AI \quad and \quad \|b\|_u \le B$$

for some $0 < \epsilon \leq A$, then, for each $\delta \in (0,1]$ and all $t \in (0,\infty)$,

$$p(t,x,y) \leq \frac{K_N}{(\epsilon\delta t)^{\frac{N}{2}}} \exp\left(\frac{30AB^2t}{\delta} - \frac{d(x,y)^2}{2(1+\delta)t}\right),$$

where $K_N < \infty$ is universal. In particular,

$$p(t,x,y) \leq \frac{K_N}{\left(\epsilon(t \wedge 1)\right)^{\frac{N}{2}}} \exp\left(30AB^2t - \frac{|y-x|^2}{4At} \wedge (30AB^2)\right).$$

PROOF: First observe that $\check{b} = 0$ and $c = 0$. Second, note that $p(\epsilon^{-1}t, x, y)$ corresponds to the operator $\epsilon^{-1}L$, which is the operator obtained by replacing a by $\epsilon^{-1}a$. Hence, the initial estimate follows from the above. In particular, by taking $\delta = 1$ and using $d(x,y)^2 \geq A^{-1}|y-x|^2$, one gets

$$p(t,x,y) \leq \frac{K_N}{(\epsilon t)^{\frac{N}{2}}} \exp\left(30AB^2t - \frac{|y-x|^2}{4At}\right).$$

On the other hand, because $\int p(t,x,y)\,dy = 1$, when $t > 1$

$$p(t,x,y) = \int p(t-1,x,\xi)p(1,\xi,y)\,d\xi \leq \frac{K_N}{\epsilon^{\frac{N}{2}}} e^{30AB^2},$$

and so the second estimate follows when one puts this together with the preceding. $\square$

REMARK 4.4.7. Although when $t \in (0,1]$ the estimates in Lemma 4.4.6 resemble our earlier results, readers may be wondering why the estimate for large t's is so different. Specifically, they may be wondering whether $p(t,x,y)$ tends to 0 as $t \to \infty$. In order to see that it may not, take $N = 1$, $a = 1$, and $b \in C_b^\infty(\mathbb{R};\mathbb{R})$, and assume that $\int e^{2U(x)}\,dx = 1$ for some indefinite integral U of b. Because $L\varphi = \frac{1}{2}e^{-2U}\partial(e^{2U}\partial\varphi)$, we can repeat the reasoning in §4.3 to check that $\mathbf{P}_t$ is symmetric in $L^2(\lambda^U;\mathbb{R})$, where $\lambda^U(dx) = e^{2U(x)}\,dx$. Equivalently, $e^{2U(x)}p(t,x,y) = e^{2U(y)}p(t,x,y)$. In particular, $\langle \mathbf{P}_t\varphi, \lambda^U \rangle = \langle \varphi, \lambda^U \rangle$ and $\mathbf{P}_t$ is a contraction on $L^2(\lambda^U;\mathbb{R})$. Now let $\varphi \in C_b(\mathbb{R}^N;\mathbb{R})$ be given. Then

$$t \rightsquigarrow \langle \varphi, \mathbf{P}_t\varphi \rangle = \|\mathbf{P}_{\frac{t}{2}}\varphi\|_{L^2(\lambda^U;\mathbb{R})}^2$$

is non-increasing and so $\lim_{t\to\infty} (\varphi, \mathbf{P}_t\varphi)_{L^2(\lambda^U;\mathbb{R})}$ exists. But, for $s \leq t$,

$$\|\mathbf{P}_t\varphi - \mathbf{P}_s\varphi\|_{L^2(\lambda^U;\mathbb{R})}^2$$
$$= (\varphi, \mathbf{P}_{2t}\varphi)_{L^2(\lambda^U;\mathbb{R})} - 2(\varphi, \mathbf{P}_{s+t}\varphi)_{L^2(\lambda^U;\mathbb{R})} + (\varphi, \mathbf{P}_{2s}\varphi)_{L^2(\lambda^U;\mathbb{R})}$$
$$\leq 2(\varphi, \mathbf{P}_{2s}\varphi)_{L^2(\lambda^U;\mathbb{R})} - 2(\varphi, \mathbf{P}_{s+t}\varphi)_{L^2(\lambda^U;\mathbb{R})},$$

and therefore $\tilde{\varphi} \equiv \lim_{t \to \infty} \mathbf{P}_t \varphi$ exists in $L^2(\lambda^U; \mathbb{R})$. Clearly, $\tilde{\varphi} = \mathbf{P}_t \tilde{\varphi}$ for all $t > 0$, and so, since we know that $\mathbf{P}_1$ maps $L^2(\lambda^U; \mathbb{R})$ into $C_b(\mathbb{R}; \mathbb{R})$ and $C_b(\mathbb{R}; \mathbb{R})$ into $C_b^\infty(\mathbb{R}; \mathbb{R})$, $\tilde{\varphi} \in C_b^\infty(\mathbb{R}; \mathbb{R})$. Furthermore, from $\tilde{\varphi} = \mathbf{P}_t \tilde{\varphi}$, it is clear that $L\tilde{\varphi} = 0$, and therefore that

$$\tilde{\varphi}(x) = A \int_0^x e^{-2U(\xi)} \, d\xi + B$$

for some $A, B \in \mathbb{R}$. But, because $\int e^{2U(x)} \, dx < \infty$ and therefore

$$\int_0^x e^{-2U(\xi)} \, d\xi \longrightarrow \pm\infty \quad \text{as } x \to \pm\infty,$$

A must be 0. That is, $\tilde{\varphi}$ is constant, and, since $\langle \tilde{\varphi}, \lambda^U \rangle = \lim_{t \to \infty} \langle \mathbf{P}_t \varphi, \lambda^U \rangle = \langle \varphi, \lambda^U \rangle$, we conclude first that $\mathbf{P}_t \varphi \longrightarrow \langle \varphi, \lambda^U \rangle$ in $L^2(\lambda^U; \mathbb{R})$, and then that

$$\lim_{t \to \infty} \|\mathbf{P}_t \varphi - \langle \varphi, \lambda^U \rangle\|_u \le \|\mathbf{P}_1\|_{2 \to \infty} \lim_{t \to \infty} \|\mathbf{P}_{t-1} \varphi - \langle \varphi, \lambda^U \rangle\|_{L^2(\lambda^U; \mathbb{R})} = 0.$$

Finally, take $\varphi(x) = p(1, x, y)$. Then the preceding says that

$$\lim_{t \to \infty} p(t, x, y) = \lim_{t \to \infty} \left[\mathbf{P}_{t-1} p(1, \cdot, y) \right](x)$$

$$= \int e^{2Ux)} p(1, x, y) \, dx = e^{2U(y)} \int p(1, y, x) \, dx = e^{2U(y)}$$

uniformly for $x \in \mathbb{R}$. In particular, $p(t, x, y)$ does not tend to 0. For the reader who is concerned that this conclusion contradicts (4.3.5), notice that the U here is not bounded and therefore (4.3.5) does not apply.

4.4.2. The Lower Bound: *Throughout this subsection we will be assuming that L has the form in (4.4.2) with $c = 0$, and, until further notice, that $I \le a(x) \le AI$ and $\|b\|_u + \|\check{b}\|_u \le B$.*

The proof of our lower bound for $q(t, x, y)$ will follow the same general strategy that we used in §§ 4.2.2 and 4.2.3. However, there is one nasty issue here which did not arise there. Namely, in the proof of Lemma 4.2.3, we made essential use of the fact that $\int p(t, x, y) \, dy = 1 = \int p(t, x, y) \, dx$. In fact, one might say that this was the property from which all the positivity of $p(t, x, y)$ derived, but it is a property which $q(t, x, y)$ does not share. Thus, to get a lower bound on the x and y integrals of $q(t, x, y)$ which does not rely on the smoothness properties of the coefficients, we must give a separate argument.

Set $\eta(\xi) = (1 + |\xi|^2)^{\frac{1}{2}} + \kappa_N$, where κ_N is chosen so that $\int e^{-\eta(x)} \, dx = 1$, and define

$$G_\beta(t, y) = \int e^{-\eta(\xi)} q(t, y + \xi, y)^\beta \, d\xi \ \& \ \check{G}_\beta(t, x) = \int e^{-\eta(\xi)} q(t, x, x + \xi)^\beta \, d\xi$$

for $\beta \in (0,1]$. We want to show that there is an $\alpha = \alpha(A,B) > 0$ such that $G_1(t,y) \wedge \check{G}_1(t,x) \geq \alpha$ for all $(t,x,y) \in (0,1] \times \mathbb{R}^N \times \mathbb{R}^N$, and, because $\check{G}_1(t,x)$ is the same as $G_1(t,x)$ for $L^\top$, we need only deal with $G_1(t,y)$. In fact, by translation, it suffices to handle $G_1(t,0)$. Thus, set $u_\theta(t,\xi) = (1-\theta)q(t,\xi,0) + \theta$, $\nu(d\xi) = e^{-\eta(\xi)}\,d\xi$, and $w_{\beta,\theta}(t) = \langle u_\theta(t)^\beta, \nu \rangle$. When $\beta \in (0,1)$, $\dot{w}_{\beta,\theta}(t)$ equals β times

$$
\frac{1-\beta}{2}\big(\nabla u_\theta(t), au_\theta^{\beta-2}(t)\nabla u_\theta(t)\big)_{L^2(\nu;\mathbb{R}^N)}
$$
$$
+ \Big(b + (\beta-1)\check{b}_\theta + \tfrac{1}{2}\nabla\eta, u_\theta(t)^{\beta-1}a\nabla u_\theta(t)\Big)_{L^2(\nu;\mathbb{R}^N)}
$$
$$
- \big(\nabla\eta, u_\theta(t)^\beta a\check{b}_\theta\big)_{L^2(\nu;\mathbb{R}^N)}
$$
$$
\geq \frac{1-\beta}{4}\big(\nabla u_\theta(t), u_\theta(t)^{\beta-2}a\nabla u_\theta(t)\big)_{L^2(\nu;\mathbb{R}^N)} - \frac{2A(B+1)^2}{1-\beta}w_{\beta,\theta}(t),
$$

where $\check{b}_\theta = \frac{(1-\theta)u_0}{u_\theta}\check{b}$. At the same time, for any $\beta \in (0,1)$ and $t \in (0,1]$,

$$
\dot{w}_{1,\theta}(t) = \big(b + \tfrac{1}{2}\nabla\eta, a\nabla u_\theta(t)\big)_{L^2(\nu;\mathbb{R}^N)} - \big(\nabla\eta, u_\theta(t)a\check{b}_\theta\big)_{L^2(\nu;\mathbb{R}^N)}
$$
$$
\geq -\frac{\beta(1-\beta)}{4}\big(\nabla u_\theta(t), u_\theta(t)^{\beta-2}a\nabla u_\theta(t)\big)_{L^2(\nu;\mathbb{R}^N)}
$$
$$
- \frac{1}{\beta(1-\beta)}\big(b + \tfrac{1}{2}\nabla\eta, u_\theta(t)^{2-\beta}a(b + \tfrac{1}{2}\nabla\eta)\big)_{L^2(\nu;\mathbb{R}^N)}
$$
$$
- \big(\nabla\eta, u_\theta(t)a\check{b}_\theta\big)_{L^2(\nu;\mathbb{R}^N)}
$$
$$
\geq -\frac{\beta(1-\beta)}{4}\big(\nabla u_\theta(t), u_\theta(t)^{\beta-2}a\nabla u_\theta(t)\big)_{L^2(\nu;\mathbb{R}^N)}
$$
$$
- \frac{A(B+1)^2}{\beta(1-\beta)}\big(1 + K_N(A,B)^{1-\beta}t^{-\frac{N(1-\beta)}{2}}\big)w_{1,\theta}(t),
$$

where, in the passage to the last line, we have used the arithmetic-geometric mean inequality, (4.4.5) with $\delta = 1$, and taken $K_N(A,B) = 2^{\frac{N}{2}}K_N e^{AB^2}$. Now, taking $\beta = 1 - \frac{1}{2N}$ and adding $\dot{w}_{\beta,\theta}(t)$ to $\dot{w}_{1,\theta}(t)$, we obtain

$$
\frac{d}{dt}\big(w_{\beta,\theta}(t) + w_{1,\theta}(t)\big)
$$
$$
\geq -4AN(B+1)^2\big(1 + K_N(A,B)^{\frac{1}{2N}}t^{-\frac{1}{4}}\big)\big(w_{\beta,\theta}(t) + w_{1,\theta}(t)\big)
$$

for $t \in (0,1]$. Hence, since $\varliminf_{t \searrow 0} w_{1,\theta}(t) \geq e^{-1-\kappa_N}$ and, by Jensen's inequality, $w_{\beta,\theta}(t) \leq w_{1,\theta}(t)^\beta$, we can conclude that $w_{1,\theta}(t)$ is bounded below by an $\alpha(A,B) > 0$ which depends only on N, A, and B, and, in view of our earlier comments, this means that

$$
\left(\int e^{-\eta(\xi)}q(t,y+\xi,y)\,d\xi\right) \wedge \left(\int e^{-\eta(\xi)}q(t,x,x+\xi)\,d\xi\right) \geq \alpha(A,B)
$$

for all $t \in (0,1]$. Finally, by again applying the second estimate in Lemma 4.4.6, one can use this to find an $R = R(A,B) < \infty$ and an slightly different $\alpha = \alpha(A,B) > 0$ such that

$$(4.4.8) \qquad \left(\int_{B(0,R)} q(t, y+\xi, y)\, d\xi \right) \wedge \left(\int_{B(0,R)} q(t, x, x+\xi)\, d\xi \right) \geq \alpha$$

for $t \in (0,1]$.

    Given $(4.4.8)$, we can more or less repeat the argument in §§4.2.2 and 4.3, and we begin with the analog of Lemma 4.2.3. Set $q_\theta = (1-\theta)q + \theta$. We want to show that there is a $\kappa = \kappa(A,B) < \infty$ for which

$$(4.4.9) \quad \left(\int \log q_\theta(1, y+\xi, y)\, \gamma(d\xi) \right) \wedge \left(\int \log q_\theta(t, x, x+\xi)\, \gamma(d\xi) \right) \geq -\kappa,$$

and, for the usual reasons, it suffices to handle the first integral and to do so when $y = 0$. Thus, take $u_\theta(t,\xi) = q_\theta(t,\xi,0)$ and set $w_\theta(t) = \langle \log u_\theta(t), \gamma \rangle$ for $\theta \in \left(0, \frac{1}{2}\right]$. Then

$$\dot{w}_\theta(t) = \tfrac{1}{2}\big(\nabla \log u_\theta(t), a\nabla \log u_\theta(t)\big)_{L^2(\gamma;\mathbb{R}^N)}$$
$$+ \big(\tfrac{1}{2}\xi + b - \check{b}_\theta, a\nabla \log u_\theta(t)\big)_{L^2(\gamma;\mathbb{R}^N)} - \big(\xi, a\check{b}_\theta\big)_{L^2(\gamma;\mathbb{R}^N)}$$
$$\geq \tfrac{1}{4}\big(\nabla \log u_\theta(t), a\nabla \log u_\theta(t)\big)_{L^2(\gamma;\mathbb{R}^N)}$$
$$- \big(\tfrac{1}{2}\xi + b - \check{b}_\theta, a(\tfrac{1}{2}\xi + b - \check{b}_\theta)\big)_{L^2(\gamma;\mathbb{R}^N)} - \big(\xi, a\check{b}\big)_{L^2(\gamma;\mathbb{R}^N)}$$
$$\geq \tfrac{1}{4}\| \log u_\theta(t) - w_\theta(t)\|^2_{L^2(\gamma;\mathbb{R})} - C,$$

where again $\check{b}_\theta = \frac{(1-\theta)u_0}{u_\theta}b$, $C = C(A,B) < \infty$, and, at the last step, we have used Lemma $(4.2.2)$. From here one proceeds in exactly the same way as we did at the analogous place in the derivation of $(4.3.7)$. Namely, if $w_\theta(t) \geq 0$ for some $t \in (0,1]$, then $w_\theta(1) \geq -C$. If $w_\theta(t) \leq 0$ for all $t \in (0,1]$, then one can use $(4.4.8)$ to conclude that, apart from an additive constant, for $t \in \left[\frac{1}{2}, 1\right]$, $\| \log u_\theta(t) - w_\theta(t)\|^2_{L^2(\gamma;\mathbb{R})}$ dominates a strictly positive constant times $w_\theta(t)^2$ and that both these constants depend only on A and B. Once one knows this, the existence of κ comes from exactly the same line of reasoning that we used earlier.

    The next step is to prove the analog of Lemma 4.2.5, and this is easy. In fact, set $u_\theta(t,\xi) = q_\theta\big(t, \xi + \pi(t), 0\big)$, $a(t,\xi) = a\big(\xi + \pi(t)\big)$, $b(t,\xi) = b\big(\xi + \pi(t)\big)$, and $\check{b}(t,\xi) = \check{b}\big(\xi + \pi(t)\big)$, and using the notation in Lemma

4.2.5, check that[2]

$$\frac{d}{dt} \int \rho(\xi) \log u_\theta(t, \xi) \, d\xi$$

$$= \frac{1}{2} \left(\nabla \log u_\theta(t), \rho a(t) \nabla \log u_\theta(t) \right)_{L^2(\mathbb{R}^N; \mathbb{R})}$$

$$+ \left((a(t)^{-1} \dot{\pi}(t) + b(t) - \check{b}(t)_\theta - \tfrac{1}{2} \nabla \log \rho), \rho a(t) \nabla \log u_\theta(t) \right)_{L^2(\mathbb{R}^N; \mathbb{R})}$$

$$\geq -\frac{1}{2} \left((a(t)^{-1} \dot{\pi}(t) + b(t) - \check{b}(t)_\theta - \tfrac{1}{2} \nabla \log \rho), \right.$$

$$\left. \times \rho a(t) \big(a(t)^{-1} \dot{\pi}(t) + b(t) - \check{b}(t)_\theta - \tfrac{1}{2} \nabla \log \rho \big) \right)_{L^2(\mathbb{R}^N; \mathbb{R})}$$

$$\geq -\frac{1+\delta}{2} \left(\dot{\pi}(t), \rho a(t)^{-1} \dot{\pi}(t) \right)_{L^2(\mathbb{R}^N; \mathbb{R})}$$

$$- \frac{1}{\delta} \left((b(t) - \check{b}(t)_\theta - \tfrac{1}{2} \nabla \log \rho), \rho a(t) \big(b(t) - \check{b}(t)_\theta - \tfrac{1}{2} \nabla \log \rho \big) \right)_{L^2(\mathbb{R}^N; \mathbb{R})}$$

$$\geq -\frac{1+\delta}{2} \left(\dot{\pi}(t), \rho a(t)^{-1} \dot{\pi}(t) \right)_{L^2(\mathbb{R}^N; \mathbb{R}^N)} - \frac{3A(H(\rho) + B^2)}{\delta},$$

where $\check{b}(t)_\theta = \frac{(1-\theta)u_0}{u_\theta} \check{b}(t)$. *Hence, just as in Lemma 4.2.5, we find that (cf.* *(4.2.4))*

(4.4.10)
$$\left(\int \rho(\xi) \log \frac{q(\tau + t, \xi + \pi(t), y)}{q(\tau, y + \xi, y)} \, d\xi \right)$$
$$\wedge \left(\int \rho(\xi) \log \frac{q(\tau + t, x, \xi + \pi(t))}{q(\tau, x, x + \xi)} \, d\xi \right)$$
$$\geq -\frac{1+\delta}{2} E_\rho(t, \pi) - \frac{3A(H(\rho) + B^2)t}{\delta}$$

whenever, in the first case, $\pi(0) = y$ *and, in the second,* $\pi(0) = x$. *Starting from here and using (4.4.9), the same argument that we used at the beginning of § 4.2.3 leads to*

$$q(3, x, y) \geq e^{-2\kappa - A(3N + B^2) - A|y-x|^2}.$$

Next, because $q(t, x, y) = \lambda^{-\frac{N}{2}} q(\lambda^{-1} t, \lambda^{-\frac{1}{2}} x, \lambda^{-\frac{1}{2}} y)$, *where* q_λ *corresponds to the coefficients* $a(\lambda^{\frac{1}{2}} \cdot)$, $\lambda^{\frac{1}{2}} b(\lambda^{\frac{1}{2}} \cdot)$, *and* $\lambda^{\frac{1}{2}} \check{b}(\lambda^{\frac{1}{2}} \cdot)$, *this gives*

$$q(t, x, y) \geq \left(\frac{3}{t} \right)^{\frac{N}{2}} e^{-2\kappa - A(3N + tB^2) - \frac{3A|y-x|^2}{t}} \quad \textit{for } t \in (0, 3].$$

---

[2] Just as in the proof of Lemma 4.2.5, it is easy to reduce to the case when $\rho > 0$ everywhere, and so we will implicitly assume that $\rho > 0$ in the following calculation.

Now, applying (4.4.10) in the same way as we did in the discussion preceding Theorem 4.2.9, we arrive at (cf. (4.2.8))

$$(4.4.11) \quad q(t, x, y) \geq \frac{\beta(A, B)}{(\delta^2 t)^{\frac{N}{2}}} \exp\left(-\frac{A(\kappa(\rho) + 5B^2)}{\delta} - \frac{(1+\delta)d_{t,\rho}(x, y)^2}{2t} \right)$$

for $t \in (0, 3]$, where $\beta(A, B) \in (0, 1]$ depends only on N, A, and B and $\kappa(\rho) = 12S(\rho) + 5H(\rho)$.

THEOREM 4.4.12.   *Let L be given by (4.4.1), and assume that $\epsilon I \leq a(x) \leq (\epsilon \Lambda) I$ for some $\epsilon > 0$ and $\Lambda \in [1, \infty)$ and that $\|b\|_u \leq B$. Then there exists a $\beta \in (0, 1]$, depending only on Λ and B, such that, for each $\delta \in (0, 1]$ and all $t \in (0, 1]$ (cf. (4.2.8)),*

$$p(t, x, y) \geq \frac{\beta}{(\epsilon \delta^2 t)^{\frac{N}{2}}} \exp\left(-\frac{\Lambda(\kappa(\rho) + 5B^2)}{\delta} - \frac{(1+\delta)d_{\epsilon t, \rho}(x, y)^2}{2t} \right),$$

where $\kappa(\rho) = 12S(\rho) + 3H(\rho)$. In particular,

$$\lim_{t \searrow 0} t \log p(t, x, y) = -\frac{d(x, y)^2}{2}$$

continues to hold, and, if $\beta' = \beta e^{-2\Lambda(15N + B^2)}$, then

$$p(t, x, y) \geq \frac{\beta'}{(\epsilon t)^{\frac{N}{2}}} e^{-\frac{2|y-x|^2}{\epsilon t}} \qquad \text{for } t \in (0, 1].$$

Finally, there exists an M, depending only on ϵ, Λ, and B, such that

$$p(t, x, y) \geq \frac{1}{Mt^{\frac{N}{2}}} \exp\left(-Mt - \frac{M|y-x|^2}{t} \right) \qquad \text{for } t \in (1, \infty).$$

PROOF: The first estimate follows from rescaling to take ϵ into account and applying (4.4.11). Given the first estimate here and the one in Theorem 4.4.6, the logarithmic limit is obvious. Also, the second estimate is an easy application of the first. Finally, to prove the last assertion, start with the second estimate and apply the second part of Lemma 4.3.8. □

REMARK 4.4.13.   Again the reader might be wondering about the precision of the preceding estimate when t is large. To see that it is correct, at least qualitatively, one need only consider the constant coefficient operator $L = \frac{1}{2}\Delta + b \cdot \nabla$, for which

$$p(t, 0, y) = (2\pi t)^{-\frac{N}{2}} e^{-\frac{|y-tb|^2}{2t}} \leq (2\pi t)^{-\frac{N}{2}} e^{-|b|^2 t - \frac{|y|^2}{2t}} \qquad \text{when } (y, b)_{\mathbb{R}^N} \leq 0.$$

4.5 Historical Notes and Commentary

The theory presented in this section has its origins in the famous paper [43] by J. Nash. Prior to Nash, no one suspected that one could prove estimates which depended only on ellipticity and bounds on the coefficients, and it is still surprising that one can. For analysts, the importance of such estimates is that they provide a starting point for the analysis of non-linear equations.

The history of this topic can be summarized as follows. The goal of Nash's paper was the proof of an *a priori* continuity result (cf. § 5.3.4 below) for the solutions of elliptic and parabolic equations. Shortly after Nash's paper appeared, E. Di Georgi [11] proved a Harnack principle (cf. § 5.3.5 below) for non-negative solutions to $Lu = 0$, again getting estimates which depend only on the ellipticity and bounds on the coefficients. Following Di Georgi's work, J. Moser published [41], in which he reproved Nash's and Di Georgi's results in a spectacularly original and powerful way. In fact, Moser's method, now known as the *Moser iteration scheme*, has become a basic tool in the analysis of elliptic and parabolic partial differential equations, where it has been used in a myriad of applications. A little later, in [42], Moser extended his results to the parabolic setting. In recent years, Moser's methodology has been honed and abstracted. See [51] for a particularly elegant treatment of these developments.

Definitive applications of Moser's theory to estimates on the fundamental solution appeared for the first time in D. Aronson's paper [4]. What Aronson proved is that $p(t, x, y)$ can be bounded above and below by appropriate Gaussians. Many years later, E. Fabes and I realized that Nash's paper contains techniques which lead directly to Aronson's estimates and that, given Aronson's estimates, the results of Nash, Di Georgi, and Moser are relatively easy corollaries. The treatment given in this chapter as well as that in §§ 5.3.4 and 5.3.5 below are derived from the article [16].

The sharpening of Aronson's estimates to give estimates in which Euclidean distance is replaced by Riemannian distance began with an article [10] by E.B. Davies, who introduced the idea of looking at (cf. § 4.1.2) $p^\psi(t, x, y)$ in order to get off-diagonal upper bounds. Somewhat later, J. Norris and I introduced in [45] the method used here to get the complementary lower bound in terms of Riemannian distance.

As is pointed out in Theorem 4.2.9 and again in Theorem 4.4.12, these sharpened upper and lower bounds can be combined to prove S.R.S. Varadhan's result,

$$\lim_{t \searrow 0} t \log p(t, x, y) = -\frac{d(x, y)^2}{2},$$

which appears in [56] and is the starting place for his derivation of the large deviation theory for diffusions in short time. In this connection, it should be mentioned that one cannot take $\delta = 0$ in the exponential term of the estimates in Theorems 4.2.9 and 4.4.12. The reason for this is best understood from a geometric standpoint. To be precise, consider of the L in (4.3.1) with $U = -\log(\det a)$. Then L is the Laplacian for the Riemannian metric a^{-1}. A famous theorem of Pleijel shows that, as $t \searrow 0$, $p(t, x, y)$ is asymptotic to $\left((2\pi t)^N \det a(x)\right)^{-\frac{1}{2}} e^{-\frac{d(x,y)^2}{2t}}$ as long as y is not in the cut locus of x. However, when y is in the cut locus of x, the asymptotics change. The interested reader might want to consult Molchanov's article [40] or my article [33] with S. Kusuoka. What Varadhan's result shows is that these changes are sub-exponential and therefore invisible after one takes logarithms.

In this chapter we have assumed that L is presented with its second order part in divergence form. Of course, as long as the coefficients are once continuously differentiable, this is simply a matter of notation, since then one can transform L's given by (1.1.8) into ones given by (4.4.1) and vice verse. However, when the coefficients are less than differentiable, the distinction between the properties of those in (1.1.8) and (4.4.1) can be profound. Using (cf. § 1.4) Levi's parametrix method, Pogorzelski [48] and Aronson [3] showed that a Hölder continuous fundamental solution satisfying Gaussian upper bounds exists for operators L given by (1.1.8) when a is uniformly elliptic and all the coefficients are uniformly Hölder continuous and bounded. An alternative, and in some ways more powerful, approach to studying such operators is based on (cf. [31]) Schauder estimates. Be that as it may, as far as I know, none of the existing approaches yields lower bounds comparable to the ones which we have obtained here for L's with divergence form second order parts. Worse, when L is given by (1.1.8) with a uniformly elliptic and a and b bounded but merely continuous, the fundamental solution need not be even bounded, much less continuous, at positive times, although it will be integrable to all orders. Such operators are handled using perturbative techniques in which estimates coming from the theory of singular integral operators play the role which Schauder estimates play in the Hölder continuous case. See, for example, E. Fabes and N. Riviere [15]. Finally, when the continuity hypothesis is dropped, matters are much more complicated. For example, although there is a fundamental solution, it will not be integrable to all orders. Essentially everything that is known in this case derives from a beautiful idea in convex analysis of A.D. Alexandrov [2], whose idea was ingeniously developed by N. Krylov. For applications specifically to the fundamental solution, see [17].

CHAPTER 5

Localization

Thus far, all our results have been about parabolic equations in the whole of Euclidean space, and, particularly in Chapter 4, we took consistent advantage of that fact. However, in many applications it is important to have localized versions of these results, and the purpose of this chapter is to develop some of them.

Because probability theory provides an elegant and ubiquitous localization procedure, we will begin by summarizing a few of the well-known facts about the Markov process determined by an operator L. We will then use that process to obtain a very useful perturbation formula, known as Duhamel's formal. Armed with Duhamel's formula, it will be relatively easy to get localized statements of the global results which we already have, and we will then apply these to prove Nash's Continuity Theorem and the Harnack principle of Di Georgi and Moser.

5.1 Diffusion Processes on $\mathbb{R}^N$

Throughout, we will be assuming that a and b are smooth functions with bounded derivatives of all orders and that $a \geq \epsilon I$ for some $\epsilon > 0$. Given such a and b, L will be one of the associated operators given by (1.1.8), (4.4.1), or, when appropriate (4.3.1). Of course, under the hypotheses made about a and b, the choice between using (1.1.8) or (4.4.1) is a simple matter of notation, whereas the ability to write it as in (4.3.1) imposes special conditions of the relationship between b and a.

There are a lot of ways to pass from an operator of the form in (1.1.8) to the associated Markov family $\{\mathbb{P}_x : x \in \mathbb{R}^N\}$ of Borel probability measures on the path space $\Omega = \Omega(\mathbb{R}^N) \equiv C\big([0, \infty); \mathbb{R}^N\big)$. One approach is to start with Brownian motion and use Itô's stochastic integral equations to transform the Brownian paths into the paths of the desired diffusion. A second approach is to follow Kolmogorov and start with the transition probability function $(t, x) \leadsto P(t, x)$ determined by L and use it to give a direct construction of the $\mathbb{P}_x$'s. Whichever method one chooses, one will end up with a measure $\mathbb{P}_x \in \mathbf{M}_1(\Omega)$ which solves the *martingale problem for L starting at x*. That is,

$$(5.1.1) \qquad \left(\varphi\big(\omega(t)\big) - \varphi(x) - \int_0^t L\varphi\big(\omega(\tau)\big) \, d\tau, \mathcal{B}_t, \mathbb{P}_x \right)$$

will be a mean-zero martingale for every $\varphi \in C^\infty(\mathbb{R}^N; \mathbb{C})$, where we have introduced the notation $\mathcal{B}_t$ to denote the σ-algebra $\sigma(\{\omega(\tau) : \tau \in [0, t]\})$ which is generated by the path $\omega \upharpoonright [0, t]$. Indeed, when thinking *a la* Itô, this is just the observation that the stochastic integral term in Itô's formula is a martingale. When following Kolmogorov, it is an encoding of his forward equation. Namely, for Kolmogorov, $\mathbb{P}_x$ is determined by

$$\mathbb{P}_x\big(\omega(t_m) \in \Gamma_m \text{ for } 0 \leq m \leq n\big)$$

$$= \mathbf{1}_{\Gamma_0}(x) \int \cdots \int_{\Gamma_1 \times \cdots \times \Gamma_n} P(t_1, x, dy_1) P(t_2 - t_1, y_1, dy_2) \cdots P(t_n - t_{n-1}, y_{n-1}, dy_n)$$

for $n \geq 1$, $0 = t_0 < \cdots < t_n$, and $\Gamma_0, \ldots, \Gamma_m \in \mathcal{B}_{\mathbb{R}^N}$, or, equivalently,

$$(*) \qquad \begin{aligned} \mathbb{P}_x\big(\omega(0) = x\big) &= 1 \quad \text{and} \\ \mathbb{P}_x\big(\omega(s + t) \in \Gamma \,\big|\, \mathcal{B}_s\big) &= P\big(t, \omega(s), \Gamma\big) \quad \text{(a.s., } \mathbb{P}_x\text{)}. \end{aligned}$$

To see that this description of $\mathbb{P}_x$ leads to a solution of the martingale problem, let $\varphi \in C_c^\infty(\mathbb{R}^N; \mathbb{C})$, and verify that

$$\begin{aligned} \mathbb{E}_x^{\mathbb{P}}\big[\varphi\big(\omega(s + t)\big) - \varphi\big(\omega(s)\big) \,\big|\, \mathcal{B}_s\big] &= \big\langle \varphi, P\big(t, \omega(s)\big)\big\rangle - \varphi\big(\omega(s)\big)\big\rangle \\ &= \int_0^t \big\langle L\varphi, P\big(\tau, \omega(s)\big)\big\rangle \, d\tau = \mathbb{E}^{\mathbb{P}_x}\left[\int_s^{s+t} L\varphi\big(\omega(\tau)\big) \, d\tau \,\bigg|\, \mathcal{B}_s\right], \end{aligned}$$

which is tantamount to the required martingale property.

It will be useful to know that, under the assumptions we have made about a and b, the measure $\mathbb{P}_x$ is uniquely determined by the martingale problem which it solves. To see this, suppose that $\mathbb{P} \in \mathbf{M}_1(\Omega)$ and that (5.1.1) holds when $\mathbb{P}_x = \mathbb{P}$. We want to show that $\mathbb{P}$ must be obtainable from $(t, x) \rightsquigarrow P(t, x)$ by Kolmogorov's prescription. That $\mathbb{P}\big(\omega(0) = x\big) = 1$ is trivial, since (5.1.1) implies that $\mathbb{E}^{\mathbb{P}}\big[\varphi\big(\omega(0)\big)\big] = \varphi(x)$ for all $\varphi \in C_c^\infty(\mathbb{R}^N; \mathbb{C})$. In order to check the second line of $(*)$, we first need to extend the class of functions for which the martingale property holds. Specifically, we need to know that if $u \in C_b^{1,2}\big([0, T] \times \mathbb{R}^N; \mathbb{C}\big)$, then

$$(5.1.2) \qquad \left(u\big(t \wedge T, \omega(t \wedge T)\big) - \int_0^{t \wedge T} (\partial_\tau + L)u\big(\tau, \omega(\tau)\big) \, d\tau, \mathcal{B}_t, \mathbb{P}\right)$$

is a martingale. When u is smooth and has compact support, this comes down to the observation that (5.1.1) for $\mathbb{P}$ implies

$$\frac{d}{dt}\mathbb{E}^{\mathbb{P}}\big[u\big(s + t, \omega(s + t)\big), A\big] = \mathbb{E}^{\mathbb{P}}\big[(\partial_t + L)u\big(s + t, \omega(s + t)\big), A\big]$$

for $s \in [0, T)$, $0 < t < T - s$, and $A \in \mathcal{B}_s$. One can extend this first to $u \in C_c^{1,2}\big([0, T] \times \mathbb{R}^N; \mathbb{C}\big)$ by an obvious mollification procedure and can then complete the extension by using bump functions. Knowing (5.1.2), the proof of (*) is easy. Given $\varphi \in C_b^2(\mathbb{R}^N; \mathbb{C})$, take u_φ as in (2.0.2), and apply (5.1.2) with $u(t, y) = u_\varphi(T - t, y)$. Because $(\partial_t + L)u = 0$, one concludes that

$$(5.1.3) \qquad \left(u_\varphi\big(T - t \wedge T, \omega(t \wedge T)\big), \mathcal{B}_t, \mathbb{P} \right)$$

is a martingale, and so, by taking $T = s + t$, we see that

$$\mathbb{E}^{\mathbb{P}}\big[\varphi\big(\omega(s + t)\big) \,\big|\, \mathcal{B}_s\big] = u_\varphi\big(t, \omega(s)\big) = \langle \varphi, P\big(t, \omega(s)\big)\rangle \quad (\text{a.s.}, \mathbb{P}),$$

which, together with $P\big(\omega(0) = x\big) = 1$, leads quickly to (*).

Finally, recall that the Markov property for $\{\mathbb{P}_x \,:\, x \in \mathbb{R}^N\}$ is best expressed in terms of the time shift maps $\Sigma_s \,:\, \Omega \longrightarrow \Omega$ determined by $[\Sigma_x(\omega)](t) = \omega(s + t)$. Namely, for a Borel measurable Φ on Ω which is bounded below,

$$(5.1.4) \qquad \mathbb{E}^{\mathbb{P}_x}\big[\Phi \circ \Sigma_s \,\big|\, \mathcal{B}_s\big](\omega) = \mathbb{E}^{\mathbb{P}_{\omega(s)}}[\Phi] \quad (\text{a.s.}, \mathbb{P}_x).$$

5.2 Duhamel's Formula

Let G be a connected open subset of $\mathbb{R}^N$, take $\zeta^G(\omega) \equiv \inf\{t \geq 0 \,:\, \omega(t) \notin G\}$ to be the first time that ω leaves G, note that ζ^G is a *stopping time* relative to $\{\mathcal{B}_t \,:\, t \geq 0\}$,[1] and define the operator $\mathbf{P}_t^G$ by

$$(5.2.1) \qquad \mathbf{P}_t^G\varphi(x) = \mathbb{E}^{\mathbb{P}_x}\big[\varphi\big(\omega(t)\big), \zeta_G(\omega) > t\big]$$

for $\mathcal{B}_{\mathbb{R}^N}$-measurable φ's which are bounded below. As an application of (5.1.4), we see that

$$\begin{aligned} \mathbf{P}_{s+t}^G\varphi(x) &= \mathbb{E}^{\mathbb{P}_x}\big[\varphi\big(\Sigma_s\omega(t)\big), \zeta^G \circ \Sigma_s(\omega) > t \ \& \ \zeta^G(\omega) > s\big] \\ &= \mathbb{E}^{\mathbb{P}_x}\big[\mathbf{P}_t^G\varphi\big(\omega(s)\big), \zeta^G(\omega) > s\big] = \mathbf{P}_s^G \circ \mathbf{P}_t^G\varphi(x). \end{aligned}$$

That is, $\{\mathbf{P}_t^G \,:\, t \geq 0\}$ has the semigroup property $\mathbf{P}_{s+t}^G = \mathbf{P}_s^G \circ \mathbf{P}_t^G$. In addition, it is obvious that $\mathbf{P}_t^G$ is dominated by $\mathbf{P}_t$ in the sense that $\mathbf{P}_t^G\varphi \leq \mathbf{P}_t\varphi$ when $\varphi \geq 0$. In particular, since, by Lemma 3.3.3, we know that $P(t, x, dy) = p(t, x, y)\,dy$, we now know that

$$\mathbf{P}_t^G(x) = \int_G p^G(t, x, y)\varphi(y)\,dy,$$

where $0 \leq p^G(t, x, y) \leq \mathbf{1}_G(y)p(t, x, y)\,dy$. Finally, note that, by Doob's Stopping Time Theorem (cf. Theorem 7.1.6 in [53]),

$$\mathbf{P}_t^G\varphi(x) - \varphi(x) = \mathbb{E}^{\mathbb{P}_x}\left[\int_0^{t \wedge \zeta^G(\omega)} L\varphi\big(\omega(\tau)\big)\,d\tau\right],$$

---

[1] Recall that a stopping time relative to $\{\mathcal{B}_t \,:\, t \geq 0\}$ is a function $\zeta : \Omega \longrightarrow [0, \infty]$ with the property that $\{\zeta \leq t\} \in \mathcal{B}_t$ for all $t \geq 0$

and therefore

$$(5.2.2) \qquad \mathbf{P}_t^G \varphi(x) - \varphi(x) = \int_0^t \mathbf{P}_\tau^G L\varphi(x)\, d\tau, \quad (t, x) \in (0, \infty) \times G,$$

for $\varphi \in C_c^2(G; \mathbb{R})$.

5.2.1. The Basic Formula: The goal of this subsection is to develop another expression for $p^G(t, x, \cdot)$, one which will allow us to transfer to $p^G(t, x, \cdot)$ properties that we already know $p(t, x, \cdot)$ possesses. For this purpose, let $\varphi \in C_b^2(\mathbb{R}^N; \mathbb{R})$ be given, and set $u_\varphi(t, x) = \mathbf{P}_t \varphi(x)$. By (5.1.3) and Doob's Stopping Time Theorem, we know that

$$\begin{aligned}
\mathbf{P}_t \varphi(x) = u_\varphi(t, x) &= \mathbb{E}^{\mathbb{P}_x}\left[u_\varphi\big(t - t \wedge \zeta^G(\omega), \omega(t \wedge \zeta^G)\big)\right] \\
&= \mathbb{E}^{\mathbb{P}_x}\left[\varphi\big(\omega(t)\big),\, \zeta^G(\omega) > t\right] + \mathbb{E}^{\mathbb{P}_x}\left[u_\varphi\big(t - \zeta^G(\omega), \omega(\zeta^G)\big),\, \zeta_G(\omega) \le t\right] \\
&= \mathbf{P}_t^G \varphi(x) + \mathbb{E}^{\mathbb{P}_x}\left[\mathbf{P}_{t-\zeta^G(\omega)} \varphi\big(\omega(\zeta^G)\big),\, \zeta^G(\omega) \le t\right].
\end{aligned}$$

Thus, for all Borel measurable φ's which are bounded below,

$$(5.2.3) \qquad \mathbf{P}_t \varphi(x) = \mathbf{P}_t^G \varphi(x) + \mathbb{E}^{\mathbb{P}_x}\left[\mathbf{P}_{t-\zeta^G} \varphi\big(\omega(\zeta^G)\big),\, \zeta^G \le t\right],$$

which is the starting point for everything that follows.

Our first application of (5.2.3) is to the following crude version of the *reflection principle*.

LEMMA 5.2.4. *There is a $C < \infty$, depending only on the bounds on $\|a\|_{\mathrm{op}}$ and b, such that*

$$\mathbb{P}_x(\zeta^G \le t) \le C \exp\left(Ct - \frac{|x - G\complement|^2}{Ct}\right) \quad \text{for } (t, x) \in (0, \infty) \times G.$$

PROOF: Given $x \in G$, set $R = \frac{1}{2}|x - G\complement|$, $B_1 = B(x, R)$, and $B_2 = B(x, 2R)$. Clearly, $\mathbb{P}_x(\zeta^G \le t) \le \mathbb{P}_x(\zeta^{B_2} \le t)$. Now take $\varphi = \mathbf{1}_{B_1\complement}$ in (5.2.3), and conclude that

$$\begin{aligned}
P(t, x, B_1\complement) &\ge \mathbb{E}^{\mathbb{P}_x}\left[P\big(t - \zeta^{B_2}(\omega), \omega(\zeta^{B_2}), B_1\complement\big),\, \zeta^{B_2}(\omega) \le t\right] \\
&\ge \left(\inf_{(\tau, \xi) \in (0, t] \times B_2\complement} \mathbb{P}_\xi(\tau, \xi, B_1\complement)\right) \mathbb{P}_x(\zeta^G \le t).
\end{aligned}$$

By the second estimate in Lemma 3.3.12, $P(t, x, B_1\complement) \le C e^{Ct - \frac{R^2}{Ct}}$ and

$$P(\tau, \xi, B_1\complement) \ge P(\tau, \xi, B(\xi, R)) = 1 - P\big(\tau, \xi, B(\xi, R)\complement\big) \ge 1 - C e^{Ct - \frac{R^2}{Ct}}$$

for $(\tau, \xi) \in [0, t] \times B_2\complement$, where C has the required dependence. The desired conclusion is easy from these plus $\mathbb{P}_x(\zeta^G \le t) \le 1$. $\square$

Our second application of (5.2.3) is to the derivation of *Duhamel's formula* in the form

$$
(5.2.5) \quad p^G(t,x,y) = p(t,x,y) - \mathbb{E}^{\mathbb{P}_x}\left[p\big(t - \zeta^G(\omega), \omega(\zeta^G), y\big), \ \zeta^G(\omega) < t\right]
$$
$$
\text{for } (t,x,y) \in (0,\infty) \times G \times G,
$$

and again this is an essentially immediate. Namely, for any $\varphi \in C_c(G;\mathbb{R})$, (5.2.3) says that $\mathbf{P}_t^G \varphi(x)$ is equal to

$$
\int_G \varphi(y) p(t,x,y)\, dy - \mathbb{E}^{\mathbb{P}_x}\left[\int_G \varphi(y) p\big(t - \zeta^G(\omega), \omega(\zeta^G), y\big)\, dy, \ \zeta^G(\omega) < t\right],
$$

where the replacement of $\zeta^G \le t$ by $\zeta^G < t$ is justified by the observation that $\mathbf{P}_{t-\zeta^G(\omega)} \varphi\big(\omega(\zeta_G)\big) = \varphi\big(\omega(\zeta^G)\big) = 0$ if $\zeta^G = t$. Hence, the right-hand side of (5.2.5) is indeed a kernel for the operator $\mathbf{P}_t^G$.

Starting from (5.2.5), it is easy to check that, for each $(t,x) \in (0,\infty) \times G$, $y \in G \longmapsto p^G(t,x,y)$ is smooth and that, for each $y \in G$, $(t,x) \in (0,\infty) \times G \longmapsto p^G(t,x,y)$ is Borel measurable. The only concern comes from the fact that $t - \zeta^G(\omega)$ can be arbitrarily small. However, since $y \in G$ and $\omega(\zeta^G) \in \partial G$, the estimates in Theorem 3.3.11 shows that there is nothing to worry about. Hence, $(t,x,y) \in (0,\infty) \times G \times G \longmapsto p^G(t,x,y)$ is Borel measurable, and so the semigroup property for $\{\mathbf{P}_t^G : t \ge 0\}$ leads to the Chapman–Kolmogorov equation

$$
(5.2.6) \quad p^G(s+t,x,y) = \int_G p^G(t,x',y) p^G(s,x,x')\, dx'
$$

for $(s,x), (t,y) \in (0,\infty) \times G$.

5.2.2. Application of Duhamel's Formula to Regularity: Here we will show that $p^G(t,x,y)$ is a smooth function on $(0,\infty) \times G \times G$. However, our estimates will blow up as x and y tend to the boundary ∂G. Thus, the estimates developed here are what a specialist in partial differential equations would call *interior estimates*.

As we have already noted, for each $(t,x) \in (0,\infty) \times G$, $y \leadsto p^G(t,x,y)$ is smooth on G. In fact,

$$
(5.2.7) \quad \partial_y^\beta p^G(t,x,y) = \partial_y^\beta p(t,x,y) - \mathbb{E}^{\mathbb{P}_x}\left[\partial_y^\beta p\big(t - \zeta^G, \omega(\zeta^G), y\big), \ \zeta_G < t\right],
$$

where once again one can use that estimates in Theorem 3.3.11 to justify the differentiation of under the expectation. At the same time, with very little further effort, one sees that there is a constant A_n, with the same dependence as the one in Theorem 3.3.11, such that, for $\|\beta\| \le n$,

$$
(*) \quad \sup_{(\tau,\xi,y) \in (0,t] \times \partial G \times G} \left|\partial_y^\beta p(\tau,\xi,y)\right|
$$
$$
\le \frac{A_n}{|y - G^\complement|^{\|\beta\|+N}} \exp\left(A_n t - \frac{|y - G^\complement|^2}{A_n t}\right).
$$

THEOREM 5.2.8. $p^G(t, x, y)$ is a smooth function on $(0, \infty) \times G \times G$ and, for each $n \geq 1$, there exists an $A_n < \infty$, depending only on ϵ and the bounds on (cf. (2.5.2)) a, b, $b^\top$, $V^\top$, and their derivatives through order $(n+1)$, such that

$$
\left| \partial_t^m \partial_x^\alpha \partial_y^\beta p^G(t, x, y) \right| \leq \frac{A_n e^{A_n t}}{(t^{\frac{1}{2}} \wedge |x - G\complement| \wedge |y - G\complement|)^{2m + \|\alpha\| + \|\beta\| + N}}
$$
$$
\times \exp\left(-\frac{\left(|y - x| \wedge (|x - G\complement| + |y - G\complement|) \right)^2}{A_n t} \right)
$$

for $(m, \alpha, \beta) \in \mathbb{N} \times \mathbb{N}^N \times \mathbb{N}^N$ with $2m + \|\alpha\| + \|\beta\| \leq n$. Moreover, $\partial_t^m p^G(t, x, y) = [L^m p^G(t, \cdot, y)](x)$.

PROOF: We have already proved the existence of $\partial_y^\beta p^G(t, x, y)$, and Theorem 3.3.11 combined with Lemma 5.2.4 and (*) give the required estimates on it.

Proving the smoothness as a function of (t, x) is a bit more challenging. The idea is to differentiate (5.2.7) with respect to x. Because of the estimates in Theorem 3.3.11, the first term on the right is no problem. To handle the second, we want to take advantage of the fact that, since $x \in G$, the probability of ζ^G being close to 0 is small. To capitalize on this observation, define $\zeta_s(\omega) = \inf\{t \geq s : \omega(t) \in \partial G\}$. Clearly, $\zeta_0 = \zeta^G$ if $\omega(0) \in G$, $\zeta_{s+t} = s + \zeta_t \circ \Sigma_s$, and $\zeta_s = \zeta_t$ if $s \leq t \leq \zeta_s$. In particular, $\omega(0) \in G \implies \zeta_s(\omega) \searrow \zeta_0(\omega)$ as $s \searrow 0$, and so, for any $x \in G$ and bounded, measurable $f : [0, t] \times \partial G \longrightarrow \mathbb{R}$,

$$
\mathbb{E}^{\mathbb{P}_x}\left[f\left(t - \zeta_0, \omega(\zeta_0)\right), \zeta_0 < t \right]
$$
$$
= \sum_{n=0}^\infty \mathbb{E}^{\mathbb{P}_x}\left[f\left(t - \zeta_{t_{n+1}}, \omega(\zeta_{t_{n+1}})\right) \mathbf{1}_{[0,t)}(\zeta_{t_{n+1}}) - f\left(t - \zeta_{t_n}, \omega(\zeta_{t_n})\right) \mathbf{1}_{[0,t)}(\zeta_{t_n}) \right]
$$
$$
= \sum_{n=0}^\infty \mathbb{E}^{\mathbb{P}_x}\left[F_{n+1} \circ \Sigma_{t_{n+1}} \right] = \sum_{n=1}^\infty \mathbf{P}_{t_n} f_n(x),
$$

where $t_n = 2^{-n} t$,

$$
F_n = \Big[f\left(t - t_n - \zeta_0, \omega(\zeta_0)\right)
$$
$$
- f\left(t - t_n - \zeta_{t_n}, \omega(\zeta_{t_n})\right) \mathbf{1}_{[0, t - t_n)}(\zeta_{t_n}) \Big] \mathbf{1}_{[0, t_n)}(\zeta_0),
$$

and $f_n(\xi) = \mathbb{E}^{\mathbb{P}_\xi}[F_n]$. Next, observe that (cf. Lemmas 5.2.4)

$$
|f_n(\xi)| \leq 2\|f\|_u \mathbb{P}_\xi(\zeta_0 < t_n) \leq 2\|f\|_u C \exp\left(Ct_n - \frac{|\xi - G\complement|^2}{Ct_n} \right)
$$
$$
\leq 2\|f\|_u C e^{Ct} e^{-\frac{2^n |\xi - G\complement|^2}{Ct}}.
$$

We now apply Theorem 3.3.11 to see that, for each $\alpha \in \mathbb{N}^N$, there is an $A < \infty$ such that $|\partial_x^\alpha \mathbf{P}_{t_n} f_n(x)|$ is dominated by

$$A\|f\|_u \left(\frac{2^n}{t}\right)^{\frac{\|\alpha\|+N}{2}} e^{At} \int \exp\left(-\frac{2^n(|x-\xi|^2 + |\xi - G\mathbb{C}|^2)}{At}\right) d\xi$$

$$\leq A\|f\|_u \left(\frac{2^n}{t}\right)^{\frac{\|\alpha\|+N}{2}} e^{At - \frac{2^n|x-G\mathbb{C}|^2}{4At}} \int \exp\left(-\frac{2^n|x-\xi|^2}{4At}\right) d\xi$$

$$= \frac{A(4\pi A)^{\frac{N}{2}} e^{At} \|f\|_u}{t^{\frac{\|\alpha\|}{2}}} 2^{\frac{n\|\alpha\|}{2}} e^{-\frac{2^n|x-G\mathbb{C}|^2}{4At}},$$

where, in the passage to the second line, we have used $|x-\xi|^2 + |\xi - G\mathbb{C}|^2 \geq \frac{1}{4}\left(|x-\xi|^2 + |x - G\mathbb{C}|^2\right)$. Summing over $n \geq 1$ and using

$$\sum_{n=1}^{\infty} 2^{\frac{n\|\alpha\|}{2}} e^{-\frac{2^n|x-G\mathbb{C}|^2}{4At}} \leq \sum_{n=1}^{\infty} \int_{2^n}^{2^{n+1}} \rho^{\frac{\|\alpha\|}{2}-1} e^{-\frac{\rho|x-G\mathbb{C}|^2}{8At}} d\rho$$

$$\leq \frac{(8At)^{\frac{\|\alpha\|}{2}}}{|x-G\mathbb{C}|^{\|\alpha\|}} \int_0^{\infty} \rho^{\frac{\|\alpha\|}{2}-1} e^{-\rho} d\rho = \frac{(8At)^{\frac{\|\alpha\|}{2}}}{|x-G\mathbb{C}|^{\|\alpha\|}} \Gamma\left(\frac{\|\alpha\|}{2}\right)$$

we conclude that, for a slightly adjusted $A < \infty$ with the required dependence,

$$(5.2.9) \quad \left|\partial_x^\alpha \mathbb{E}^{\mathbb{P}_x}\left[f\left(t - \zeta^G(\omega), \omega(\zeta^G)\right), \zeta^G(\omega) < t\right]\right| \leq \frac{Ae^{At}\|f\|_u}{|x-G\mathbb{C}|^{\|\alpha\|}} e^{-\frac{|x-G\mathbb{C}|^2}{At}}.$$

Now let $\alpha, \beta \in \mathbb{N}^N$ with $\|\alpha\| \geq 1$ and $(t,y) \in (0,\infty) \times G$ be given, and take $f(\tau, \xi) = \partial_y^\beta p(\tau, \xi, y)$ in the preceding. Then, by combining the preceding with $(*)$ above, we get the asserted estimate when $m = 0$. Thus, to complete the proof, all that we need to do is check the final assertion. For this purpose, let $(t,x) \in (0,\infty) \times G$ be given, choose $\eta \in C_c^\infty(G; [0,1])$ so that $\eta = 1$ on $\overline{B(x,r)} \subset\subset G$ for some $r > 0$, and set $\varphi_m(\xi) = \eta(\xi)[L^m p^G(t, \cdot, y)](\xi)$. Because

$$\frac{d}{ds}\mathbf{P}_s\varphi_0(x)\bigg|_{s=0} = L\varphi_0(x) = [Lp^G(t, \cdot, y)](x),$$

and, by (5.2.6) and Lemma 5.2.4,

$$s^{-1}\left|p^G(s+t, x, y) - \mathbf{P}_s\varphi_0(x)\right| \leq 2 \sup_{\xi \in \mathbb{R}^N} p(t, x, \xi) s^{-1} \mathbb{P}_x\left(\zeta^{B(x,r)} \leq s\right) \longrightarrow 0$$

as $s \searrow 0$, we see that $\partial_t p^G(t, x, y) = [Lp^G(t, \cdot, y)](x)$. Finally, assume that $\partial_t^m p^G(t, x, y) = [L^m p^G(t, \cdot, y)](x)$, and apply the preceding line of reasoning with φ_m in place of φ_0. $\square$

5.2.3. Application of Duhamel's Formula to Positivity: Here we will show how to use (5.2.5) to get lower bounds on $p^G(t, x, y)$. The basic result is the following.

LEMMA 5.2.10. *Assume that L is given by either (4.3.1) or (4.4.1). Then, there is a non-decreasing map $\theta \in (0,1) \longmapsto M_\theta \in [1,\infty)$, depending only on ϵ and the bounds on the coefficients (but not their derivatives), such that*

$$p^{B(\xi,1)}(t,x,y) \geq \frac{1}{M_\theta t^{\frac{N}{2}}} e^{-\mu t - \frac{M_\theta |y-x|^2}{t}}$$

for all $\xi \in \mathbb{R}^N$, $t \in (0,\infty)$, and $x,y \in B(\xi,\theta)$,

where $\mu \in (0,\infty)$ depends only on ϵ and the bounds on the coefficients but not on θ.

PROOF: Without loss in generality, we will assume that $\xi = 0$. Also, we may and will assume that $\theta \geq \frac{1}{2}$.

Set $B = B(0,1)$, and, for $\theta \in \left[\frac{1}{2}, 1\right)$, take $B_\theta = B(0,\theta)$. From Theorem 4.3.9 or Theorems 4.4.6 and 4.4.12 combined with (5.2.5) and Lemma 5.2.4, there is an $M \in [1,\infty)$, depending only on ϵ and the bounds on the coefficients, such that

$$p^B(t,x,y) \geq \frac{1}{Mt^{\frac{N}{2}}} e^{-\frac{M|y-x|^2}{t}} - \sup_{\tau \in (0,t]} \frac{M}{\tau^{\frac{N}{2}}} e^{-\frac{(1-\theta)^2}{M\tau} - \frac{(1-\theta)^2}{Mt}}$$

$$\geq \frac{1}{Mt^{\frac{N}{2}}} e^{-\frac{M|y-x|^2}{t}} - M \left(\frac{MN}{2e(1-\theta)^2}\right)^{\frac{N}{2}} e^{-\frac{(1-\theta)^2}{Mt}},$$

for $(t,x,y) \in (0,1] \times B_\theta \times B_\theta$. Thus, if $\alpha = \left(\frac{1}{M} \wedge \sqrt{\frac{2^{1-\frac{2}{N}}e}{M^{1+\frac{4}{N}}N}}\right)(1-\theta)$, then

(*)
$$p^B(t,x,y) \geq \frac{1}{2Mt^{\frac{N}{2}}} e^{-\frac{M|y-x|^2}{t}}$$

for $(t,x,y) \in (0,\alpha^2] \times B_\theta \times B_\theta$ with $|y-x| \leq \alpha$.

Next, suppose that $(t,x,y) \in (0,\infty) \times B_\theta \times B_\theta$ with $0 < t \leq |y-x|^2$. Take n to be the smallest integer dominating $\frac{4|y-x|^2}{\alpha^2 t}$, and set $\tau = \frac{t}{n}$, $r = \frac{|y-x|}{2n}$, and, for $0 \leq m \leq n$, $z_m = x + \frac{m}{n}(y-x)$ and $\Gamma_m = B_\theta \cap B(z_m, r)$. Then, since $t \leq |y-x|^2 \leq 4$, $n \geq \frac{4}{\alpha^2}$ and so $\tau \leq \alpha^2$. Also, since $\frac{n}{2} \leq n-1 \leq \frac{4|y-x|^2}{\alpha^2 t} \leq n$ and $\frac{r^2}{\tau} = \frac{|y-x|^2}{4nt}$, $\frac{\alpha^2}{32} \leq \frac{r^2}{\tau} \leq \frac{\alpha^2}{16}$. In particular, if $\xi \in \Gamma_m$ and $\xi' \in \Gamma_{m+1}$, then $|\xi' - \xi|^2 \leq 16r^2 \leq \alpha^2 \tau \leq \alpha^2$. Finally, since $r \leq \theta$, there exists a dimensional constant $\beta \in (0,1)$ such that $|\Gamma_m| \geq \beta r^N \geq \beta \left(\frac{\alpha^2 \tau}{32}\right)^{\frac{N}{2}}$. Now, proceeding in the same way as we did in the proof of Lemma 4.3.8 and applying (*), we see from (5.2.6) that $p^B(t, x, y)$ dominates

$$\int \cdots \int_{\Gamma_1 \times \cdots \times \Gamma_{n-1}} p^B(\tau, x, \xi_1) p^B(\tau, \xi_1, \xi_2) \cdots p^B(\tau, \xi_{n-1}, y) \, d\xi_1 \cdots d\xi_{n-1}$$

$$\geq \left(\frac{e^{-M\alpha^2}}{2M\tau^{\frac{N}{2}}} \right)^n (\beta r^N)^{n-1} \geq \frac{e^{-M}}{2Mt^{\frac{N}{2}}} \left(\frac{\beta \alpha^N e^{-M}}{2(32)^{\frac{N}{2}} M} \right)^{n-1}$$

$$\geq \frac{e^{-M}}{2Mt^{\frac{N}{2}}} \exp\left(-\frac{\lambda |y - x|^2}{t} \right),$$

where $\lambda = -\frac{4}{\alpha^2} \log \frac{\beta \alpha^N e^{-M}}{2(32)^{\frac{N}{2}} M}$.

So far we have proved our estimate with $\mu = 0$ for $(t, x, y) \in (0, \infty) \times B_\theta \times B_\theta$ when either $|y - x|^2 \vee t \leq \alpha^2$ or $t \leq |y - x|^2$. We now want to deal with $(t, x, y) \in [\alpha^2, \infty) \times B_\theta \times B_\theta$ with $t \geq |y - x|^2$. For this purpose, take n to be the smallest integer dominating $\frac{3t}{\alpha^2}$, and set $\tau = \frac{t}{n}$, $r = \frac{\alpha}{3}$, and, for $0 \leq m \leq n$, $z_m = x + \frac{m}{n}(y - x)$ and $\Gamma_m = B_\theta \cap B(z_m, r)$. Then $n \geq 3$, $\frac{\alpha^2}{6} \leq \tau \leq \frac{\alpha^2}{3}$, $\frac{|y-x|}{n} \leq \frac{t^{\frac{1}{2}}}{n} \leq \frac{\alpha}{3}$, and so $|\xi' - \xi|^2 \leq \alpha^2 \leq 6\tau$ if $\xi \in \Gamma_m$ and $\xi' \in \Gamma_{m+1}$. Hence, by the same sort of chaining argument used above, $p^B(t, x, y)$ dominates

$$\left(\frac{e^{-6M}}{2M\tau^{\frac{N}{2}}} \right)^n (\beta r^N)^{n-1} \geq \frac{3^{\frac{N}{2}} e^{-6M}}{2M\alpha^N} \left(\frac{\beta 2^{\frac{N}{2}} e^{-6M}}{2M3^{\frac{N}{2}}} \right)^{\frac{3t}{\alpha^2}}.$$

After combining this with the preceding, we have proved the desired estimate, but with a μ which depends on θ.

Finally, to rid μ of its dependence on θ, let μ be the one for $\theta = \frac{1}{2}$. Given $\theta \in \left[\frac{1}{2}, 1 \right)$, x, $y \in B_\theta$, and $t \geq 3$, observe that $p^B(t, x, y)$ dominates

$$\iint_{B_{\frac{1}{2}} \times B_{\frac{1}{2}}} p^B(1, x, \xi) p^B(t - 2, \xi, \eta) p^B(1, \eta, y) \, d\xi d\eta$$

$$\geq \Omega_N^2 4^{-N} \inf\{ p^B(1, x, \xi) p^B(1, \eta, y) : (\xi, \eta) \in B_{\frac{1}{2}}^2 \}$$

$$\times \inf\{ p^B(t - 2, \xi, \eta) : (\xi, \eta) \in B_{\frac{1}{2}}^2 \} \geq \rho_\theta e^{-\mu t}$$

for some $\rho_\theta > 0$. $\square$

THEOREM 5.2.11.   *Again assume that L is given by either (4.3.1) or (4.4.1). Then, for each $R \geq 1$, there exist a $\mu(R)$ and a non-decreasing $\theta \in (0, 1) \longmapsto M_\theta(R) \in [1, \infty)$, depending only on ϵ and the bounds on the coefficients, such that, for all $r \in (0, R]$ and $\xi \in \mathbb{R}^N$,*

$$p^{B(\xi, r)}(t, x, y) \geq \frac{1}{M_\theta(R) t^{\frac{N}{2}}} \exp\left(-\frac{\mu(R) t}{r^2} - \frac{M_\theta(R) |y - x|^2}{t} \right),$$

for $(t, x, y) \in (0, \infty) \times B(\xi, \theta r) \times B(\xi, \theta r)$.

Moreover, when L is given by (4.3.1), μ and M can be chosen to be independent of R.

PROOF: Throughout, without loss in generality, we may and will take ξ to be the origin.

Given $\lambda > 0$, set $a_\lambda(x) = a(\lambda x)$, $U_\lambda(x) = U(\lambda x)$, $b_\lambda(x) = \lambda b(\lambda x)$, and, depending on whether L is given by (4.3.1) or (4.4.1), L_λ so that

$$L_\lambda \varphi = \tfrac{1}{2} e^{-2U_\lambda} \nabla \cdot \left(e^{2U_\lambda} a_\lambda \nabla \varphi \right) \quad \text{or} \quad L_\lambda \varphi = \tfrac{1}{2} \nabla \cdot \left(a_\lambda \nabla \varphi \right) + b_\lambda \cdot \nabla \varphi.$$

At the same time, define $\omega \in \Omega \longmapsto \omega_\lambda \in \Omega$ so that $\omega_\lambda(t) = \lambda^{-1} \omega(\lambda^2 t)$. Then the distribution of $\omega \leadsto \omega_\lambda$ under $\mathbb{P}_{\lambda x}$ is $(\mathbb{P}_\lambda)_x$, where $\{(\mathbb{P}_\lambda)_x : x \in \mathbb{R}^N\}$ is a Markov family determined by L_λ. To see this, we use the martingale problem characterization of $(\mathbb{P}_\lambda)_x$. Let $\varphi \in C_c^2(\mathbb{R}^N; \mathbb{R})$ and $\varphi_\lambda(x) = \varphi(\lambda^{-1} x)$, then $L_\lambda \varphi(\lambda^{-1} x) = \lambda^2 L \varphi_\lambda(x)$. Hence

$$\varphi(\omega_\lambda(t)) - \int_0^t L_\lambda \varphi(\omega_\lambda(\tau))\, d\tau = \varphi_\lambda(\omega(\lambda^2 t)) - \lambda^2 \int_0^t L\varphi_\lambda(\omega(\lambda^2 \tau))\, d\tau$$

$$= \varphi_\lambda(\omega(\lambda^2 t)) - \int_0^{\lambda^2 t} L\varphi_\lambda(\omega(\tau))\, d\tau,$$

and so

$$\left(\varphi(\omega_\lambda(t)) - \varphi(x) - \int_0^t L_\lambda \varphi(\omega_\lambda(\tau))\, d\tau, \mathcal{B}_{\lambda^2 t}, \mathbb{P}_{\lambda x} \right)$$

is a mean-zero martingale. Since $\mathcal{B}_{\lambda^2 t} = \sigma(\{\omega_\lambda(\tau) : \tau \in [0, t]\})$, this proves that the distribution of $\omega \leadsto \omega_\lambda$ under $\mathbb{P}_{\lambda x}$ solves the martingale problem for L_λ starting from x and is therefore equal to $(\mathbb{P}_\lambda)_x$. Next set $B_R = B(0, R)$ for $R \in (0, \infty)$, and observe that $\zeta^{B_R}(\omega_\lambda) = \lambda^2 \zeta^{B_{\lambda R}}(\omega)$. Thus,

$$\mathbb{P}_{\lambda x}(\omega(\lambda^2 t) \in \lambda \Gamma \ \& \ \zeta^{B_{\lambda R}}(\omega) > \lambda^2 t) = \mathbb{P}_{\lambda x}(\omega_\lambda(t) \in \Gamma \ \& \ \zeta^{B_R}(\omega_\lambda) > t)$$

$$= (\mathbb{P}_\lambda)_x(\omega(t) \in \Gamma \ \& \ \zeta^{B_R}(\omega) > t).$$

Equivalently,

$$\lambda^N \int_\Gamma p^{B_{\lambda R}}(\lambda^2 t, \lambda x, \lambda y)\, dy = \int_{\lambda \Gamma} p^{B_{\lambda R}}(\lambda^2 t, \lambda x, y)\, dy = \int_\Gamma p_\lambda^{B_R}(t, x, y)\, dy,$$

where $p_\lambda^{B_R}(t, x, y)$ is the density associated with L_λ. In other words,

$$(*) \qquad p^{B_{\lambda R}}(t, x, y) = \lambda^{-N} p_\lambda^{B_R}(\lambda^{-2} t, \lambda^{-1} x, \lambda^{-1} y).$$

Given $(*)$, the rest is easy. Namely, if L is given by (4.3.1), then, by Lemma 5.2.10, there exists a $\mu \in (0, \infty)$ and, for each $\theta \in (0, 1)$, an $M = M_\theta$ such that

$$\lambda^{-N} p_\lambda^{B_1}(\lambda^{-2} t, \lambda^{-1} x, \lambda^{-1} y) \geq \frac{1}{M t^{\frac{N}{2}}} \exp\left(-\frac{\mu t}{\lambda^2} - \frac{M|y - x|^2}{t} \right)$$

for $(t, x, y) \in (0, \infty) \times B_{\theta \lambda} \times B_{\theta \lambda}$. Hence, by taking $R = 1$ and $\lambda = r$ in $(*)$, one gets the asserted result in this case. When L is given by (4.4.1) and $b \not\equiv 0$, the same reasoning holds only for $\lambda \in (0, 1]$. On the other hand, if $R \geq 1$, then Lemma 5.2.10 can be used to find $\mu(R)$ and $\theta \leadsto M_\theta(R)$ for $p_{\lambda R}^{B_1}(t, x, y)$ with $\lambda \in (0, 1]$, after which one can apply $(*)$ with $\lambda = \frac{r}{R}$ to get the result for $r \leq R$. $\square$

COROLLARY 5.2.12. *Let G be a non-empty, connected, open subset of $\mathbb{R}^N$. Under the conditions in Theorem 5.2.11, $p^G(t,x,y) > 0$ for all $(t,x,y) \in (0,\infty) \times G \times G$. Moreover, there exists a $\mu_G \in [0,\infty)$ such that*

$$\varliminf_{t \to \infty} \inf_{(x,y) \in K^2} t^{-1} \log p^G(t,x,y) \geq -\mu_G$$

for all $K \subset\subset G$. In fact, if L is given by (4.3.1), then there is a $\mu > 0$ such that $\mu_G \leq \frac{\mu}{R^2}$ where $R = \sup\{r > 0 : \exists x\ B(x,r) \subseteq G\}$ is the inner radius of G.

PROOF: Let $(t,x) \in (0,\infty) \times G$ be given, and set $H = \{y \in G : p^G(t,x,y) > 0\}$. Obviously, H is open. In addition, if $0 < r < |x - G\complement|$, then, by Theorem 5.2.11, $p^G(t,x,y) \geq p^{B(x,r)}(t,x,y) > 0$ for $y \in B(x,r)$. Hence $H \neq \emptyset$. Thus, we will know that $H = G$ once we show that $G \setminus H$ is open. To this end, suppose that $y \in G \setminus H$ and choose $0 < r < |y - G\complement|$. Then, again by Theorem 5.2.11, $p^G(\tau,\xi,y) > 0$ for all $(\tau,\xi) \in (0,\infty) \times B(y,r)$. Hence, since

$$0 = p^G(t,x,y) = \int_G p^G(t-\tau,x,\xi) p^G(\tau,\xi,y)\,d\xi \quad \text{for } \tau \in (0,t),$$

$p^G(t,x,\xi) = \lim_{\tau \searrow 0} p^G(t-\tau,x,\xi) = 0$ for all $\xi \in B(y,r)$.

To prove the last part, choose $\xi \in G$ and $R > 0$ so that $\overline{B(\xi,2R)} \subset\subset G$ and set $B = B(\xi,R)$. Then $p^G(t,x,y)$, for $t \geq 3$, dominates

$$\iint\limits_{B \times B} p^G(1,x,x') p^{B(\xi,2R)}(t-2,x',y') p^G(1,y',y)\,dx'dy'.$$

Since $p^G(1,x,x') p^G(1,y',y)$ is uniformly positive for $(x,y) \in K^2$ and $(x',y') \in B^2$, the desired result follows from the estimate in Theorem 5.2.11 applied to $p^{B(\xi,2R)}(t-2,x',y')$. Finally, suppose the L is given by (4.3.1), let μ be the one in Theorem 5.2.11, and conclude from that theorem that $\mu_G \geq \frac{\mu}{R^2}$ if R is the inner radius of G. $\square$

5.2.4. A Refinement: Suppose that L is given by (4.4.1). Given a connected, open $G \neq \emptyset$, we will show here that Corollary 5.2.12 can be sharpened. The basic result is contained in the following lemma, in which we use the notation introduced in § 4.2.2.

LEMMA 5.2.13. *Given a continuously differentiable $\pi : [0,t] \longrightarrow G$, set $r(\pi) = \inf\{|\pi(\tau) - G\complement| : \tau \in [0,t]\} \wedge 1$. If $t \in \big(0, r(\pi)^2\big]$ and $\rho \in C^2\big(\mathbb{R}^N;[0,\infty)\big)$ is supported in $B\big(0,\frac{1}{2}r(\pi)\big)$ and has total integral 1, then, for each $\delta \in (0,1]$,*

$$p^G\big((1+\delta)t, \pi(0), \pi(t)\big)$$
$$\geq \frac{\alpha}{\|\rho\|_{\mathrm{u}}(\delta t)^N} \exp\left(-\frac{S(\rho)}{\alpha\delta t} - \frac{H(\rho)t}{\alpha\delta} - \frac{(1+\delta)E_\rho(t,\pi)}{2}\right),$$

where $\alpha \in (0,1]$ depends only on ϵ and the bounds on a and b.

PROOF: The argument is essentially the same as the one with which we derived the first estimate in Theorem 4.4.12. Namely, set $r = r(\pi)$ and, using the same reasoning as we used to derive (4.4.10), only here taking into account the fact that $|(\xi + \pi(\tau)) - G\mathbb{C}| \geq \frac{r}{2}$ for all $(\tau, \xi) \in [0, t] \times \mathrm{supp}(\rho)$, check that

$$\int \rho(\xi) \log \frac{p^G(\tau + t, \pi(0), \xi + \pi(t))}{p^G(\tau, \pi(0), \xi + \pi(0))} \, d\xi$$

and

$$\int \rho(\xi) \log \frac{p^G(\tau + t, \xi + \pi(t), \pi(0))}{p^G(\tau, \xi + \pi(0), \pi(0))} \, d\xi$$

both dominate

$$-\frac{3A(H(\rho) + B^2)t}{\delta} - \frac{(1 + \delta)E_\rho(t, \pi)}{2},$$

where A and B are the bounds on a and b. Hence, by the estimate in Theorem 5.2.11 with $R = 1$ and $\theta \in = \frac{1}{2}$,

$$\int \rho(\xi) \log p^G\big((1 + \delta)\tfrac{t}{2}, \pi(0), \xi + z\big) \, d\xi$$

$$\geq \int \rho(\xi) \log p^{B(\pi(0), r)}\big(\tfrac{\delta t}{2}, \pi(0), \xi + \pi(0)\big) \, d\xi$$

$$-\frac{3A(H(\rho) + B^2)t}{\delta} - \frac{(1 + \delta)E_\rho(t, \pi)}{2}$$

$$\geq -\log \frac{2^{\frac{N}{2}}}{M(\delta t)^{\frac{N}{2}}} - \frac{\mu \delta t}{2r^2} - \frac{2MS(\rho)}{\delta t} - \frac{3A(H(\rho) + B^2)t}{\delta} - \frac{(1 + \delta)E_\rho(\tfrac{t}{2}, \pi)}{2},$$

where $z = \pi\big(\tfrac{t}{2}\big)$. Similarly, with $\check{\pi}(\tau) = \pi(t - \tau)$ replacing π,

$$\int \rho(\xi) \log p^G\big((1 + \delta)\tfrac{t}{2}, \xi + z, \pi(t)\big) \, d\xi$$

$$\geq -\log \frac{2^{\frac{N}{2}}}{M(\delta t)^{\frac{N}{2}}} - \frac{\mu \delta t}{2r^2} - \frac{2MS(\rho)}{\delta t} - \frac{3A(H(\rho) + B^2)t}{\delta} - \frac{(1 + \delta)E_\rho(\tfrac{t}{2}, \check{\pi})}{2}.$$

Next,

$$\log p^G\big((1 + \delta)t, \pi(0), \pi(t)\big)$$

$$= \log \int_{G-z} p^G\big((1 + \delta)\tfrac{t}{2}, \pi(0), \xi + z\big) p^G\big((1 + \delta)\tfrac{t}{2}, \xi + z, \pi(t)\big) \, d\xi$$

$$\geq \log \int_{B(z, r)} p^G\big((1 + \delta)\tfrac{t}{2}, \pi(0), \xi + z\big) p^G\big((1 + \delta)\tfrac{t}{2}, \xi + z, \pi(t)\big) \, d\xi$$

$$\geq \log \frac{1}{\|\rho\|_{\mathrm{u}}} \int \rho(\xi) p^G\big((1 + \delta)\tfrac{t}{2}, \pi(0), \xi + z\big) p^G\big((1 + \delta)\tfrac{t}{2}, \xi + z, \pi(t)\big) \, d\xi$$

$$\geq -\log \|\rho\|_{\mathrm{u}} + \int \rho(\xi) \log p^G\big((1 + \delta)\tfrac{t}{2}, \pi(0), \xi + z\big) \, d\xi$$

$$+ \int \rho(\xi) \log p^G\big((1 + \delta)\tfrac{t}{2}, \xi + z, \pi(t)\big) \, d\xi.$$

Hence, when we plug in the preceding estimates and remember that $t \leq r^2$, we get the asserted estimate. $\square$

Before stating the main conclusion, set $G_{(t)} = \{x \in G : |x - G\mathbb{C}| \geq t^{\frac{1}{2}}\}$, and define

$$d_{t,\rho}^G(x,y) = \inf\left\{ \sqrt{E_{\rho_t}(1,\pi)} : \pi \in C^1\big([0,1]; G_{(t)}\big) \right.$$

$$\left. \text{with } \pi(0) = x \ \& \ \pi(1) = y \right\},$$

where the infimum over the empty set is taken to be ∞ and $\rho_t(\xi) = t^{-\frac{N}{2}}\rho(t^{-\frac{1}{2}}\xi)$.

The following statement is an immediate consequence of Lemma 5.2.13 and (4.2.7).

THEOREM 5.2.14. Let $\rho \in C_c^\infty\big(B(0,\tfrac{1}{2}); [0,\infty)\big)$. There is a $\alpha \in (0,1)$, depending only on ρ, ϵ, and the bounds on a and b, such that, for each $\delta \in (0,1)$,

$$p^G(t,x,y) \geq \frac{\alpha}{t^{\frac{N}{2}}} \exp\left(-\frac{1}{\alpha\delta} - \frac{(1+\delta)d_{t,\rho}^G(x,y)^2}{2t} \right)$$

for all $(t,x,y) \in (0,1] \times G \times G$. In particular,

$$\varliminf_{t\searrow 0} t \log p^G(t,x,y) \geq -\frac{d^G(x,y)^2}{2}, \quad x, y \in G,$$

where

$$d^G(x,y) = \inf\left\{ \sqrt{\int_0^1 \left(\dot\pi(\tau), a\big(\pi(\tau)\big)^{-1}\dot\pi(\tau) \right)_{\mathbb{R}^N} d\tau} : \right.$$

$$\left. \pi \in C^1\big([0,1]; G\big) \text{ with } \pi(0) = x \ \& \ \pi(1) = y \right\}.$$

REMARK 5.2.15. One should ask whether there is an upper bound to complement the lower bound in Theorem 5.2.14. In particular, what happens if one attempts to mimic the procedure which we used in § 4.4.1? From the outset, there are technical difficulties which have to be confronted. Specifically, because we have not discussed the behavior of $p^G(t,x,y)$ near the boundary ∂G, there is a question about the validity of the integration by parts on which that procedure is based. It turns out that smoothness of ∂G suffices and that ∂G has to be pretty bad before this problem becomes insurmountable. In any case, assuming that one has justified the integration by parts, one can proceed as we did in § 4.4.1 to show that, for each $\delta \in (0,1]$ and $\psi \in C_b^1(\mathbb{R}^N; \mathbb{R})$,

$$p^G(t,x,y) \leq \frac{K}{(\delta t)^{\frac{N}{2}}} \exp\left(\frac{Kt}{\delta} + \psi(x) - \psi(y) + \frac{t(1+\delta)D^G(\psi)^2}{2} \right),$$

where
$$D^G(\psi)^2 = \sup_{x \in \bar{G}} \big(\nabla\psi(x), a(x)\nabla\psi(x)\big)_{\mathbb{R}^N}.$$

Thus, one gets an upper bound in terms of
$$D^G(x,y)^2 = \sup\{\psi(y) - \psi(x) - \tfrac{1}{2}D^G(\psi)^2\}.$$

At this point, there remains the question of relating $D^G(x,y)$ to $d^G(x,y)$. By repeating the argument in §4.1.1 to check that $D(x,y) = d(x,y)$, one can show that $D^G(x,y) = d^{\bar{G}}(x,y)$, where $d^{\bar{G}}(x,y)$ is defined by the same prescription as $d^G(x,y)$ only with G replaced by $\bar{G}$. Indeed, the inequality $D^G(x,y) \leq d^{\bar{G}}(x,y)$ is easy. To prove the opposite inequality, one can first check that, as $r \searrow 0$, $d^{G^{(r)}}(x,y) \searrow d^{\bar{G}}(x,y)$, where $G^{(r)} = \{x \in \mathbb{R}^N : |x - G| < r\}$. One can then show that, for each $r > 0$, $D^G(x,y) \leq d^{G^{(r)}}(x,y)$.[2] Finally, one has to determine when $d^{\bar{G}}(x,y) = d^G(x,y)$. Again, smoothness of ∂G is a sufficient. On the other hand, it is easy to construct G's for which $d^{\bar{G}}(x,y) < d^G(x,y)$. For example, take $N = 2$ and $G = B(0,2) \setminus \overline{B(c,1)}$, where $c = (0,1)$. If $x \in G$ lies close to $(0,2)$ and $y = (-x_1, x_2)$, then, when $a \equiv I$, $d^{\bar{G}}(x,y) \longrightarrow 0$, whereas $d^G(x,y)$ tends to 2π as $x \to c$. Since it is clear on probabilistic grounds that in this, and most situations, d^G is more likely to give the correct estimate than is $d^{\bar{G}}$, one can try an approximation procedure. For instance, even if ∂G is not smooth, it may be possible (as it is in our example) to write G as the union of an increasing sequence of G_n's, with $\bar{G}_n \subset G$, each of which is sufficiently nice that one can apply the argument in §4.4.1. Then, because $d^{\bar{G}_n}(x,y) \searrow d^G(x,y)$ and $p^{G_n}(t,x,y) \nearrow p^G(t,x,y)$, one gets the upper bound with d^G.

5.3 Minimum Principles

The estimates in §5.2 provide a powerful tool with which to prove various minimum principles.

5.3.1. The Weak Minimum Principle Revisited: In §2.4.1 we already discussed one minimum principle, which (cf. Lemma 2.4.1) said that if u is bounded below and $(L-\partial_t)u \leq 0$ in $(0,T] \times \mathbb{R}^N$ satisfies $\underline{\lim}_{t \searrow 0} u(t, \cdot) \geq 0$ uniformly on compacts, then $u \geq 0$. In particular, this means that for any solution to $\partial_t u = Lu$ in $(0,T] \times \mathbb{R}^N$ which is bounded below, $\inf_{x \in \mathbb{R}^N} u(s, \cdot) \leq \inf_{x \in \mathbb{R}^N} u(t, \cdot)$ whenever $0 < s \leq t \leq T$. This is a *weak minimum principle* because it does not rule out the possibility of equality for non-constant u.

For purposes of comparison, we will now state and prove a quite general form of the weak minimum principle. In its statement, $\mathfrak{G}$ is an open subset of $\mathbb{R} \times \mathbb{R}^N$, and, for $(t,y) \in \mathbb{R} \times \mathbb{R}^N$,

$$\zeta^{(t,\mathfrak{G})}(\omega) \equiv \inf\big\{\tau \geq 0 : \big(t-\tau, \omega(\tau)\big) \notin \mathfrak{G}\big\}.$$

---

[2] The crucial point is that one needs a little extra room when carrying out the mollification step used in the argument preceding Lemma 4.1.11.

THEOREM 5.3.1.   *Let L be given by (1.1.8) with bounded, smooth coefficients, and let $\{\mathbb{P}_x : x \in \mathbb{R}^N\}$ be a Markov family determined by L. Assume that (t, y) is a point in $\mathfrak{G}$ for which $\mathbb{P}_y(\zeta^{(t,\mathfrak{G})} < \infty) = 1$ and that $u \in C^{1,2}(\mathfrak{G}; \mathbb{R})$ is bounded below and satisfies*

$$(L - \partial_\tau)u \leq 0 \text{ in } \mathfrak{G} \quad \text{and} \quad \mathbb{P}_y\left(\lim_{\tau \nearrow \zeta^{(t,\mathfrak{G})}(\omega)} u(t - \tau, \omega(\tau)) < 0\right) = 0.$$

Then $u(t, y) \geq 0$.[3] In particular, if $u \in C^{1,2}(\mathfrak{G}; \mathbb{R}) \cap C(\overline{\mathfrak{G}}; \mathbb{R})$ is bounded below and satisfies $(L - \partial_\tau)u \leq 0$ in $\mathfrak{G}$, then

$$u(t, y) \geq \inf\{u(\tau, \xi) : \tau < t \text{ and } (\tau, \xi) \in \partial\mathfrak{G}\}$$

whenever $\mathbb{P}_y(\zeta^{(t,\mathfrak{G})} < \infty) = 1$.

PROOF: Choose a non-decreasing sequence $\{\mathfrak{G}_n : n \geq 1\}$ of open sets so that $(t, y) \in \mathfrak{G}_n$, $\overline{\mathfrak{G}}_n \subset\subset \mathfrak{G}$ for each $n \geq 1$, and $\mathfrak{G} = \bigcup_1^\infty \mathfrak{G}_n$. Clearly, $\zeta_n = \zeta^{(t,\mathfrak{G}_n)}(\omega) \nearrow \zeta^{(t,\mathfrak{G})}(\omega)$. Thus, if we show that $u(t, y) \geq \mathbb{E}^{\mathbb{P}_y}[u(t - \zeta_n, \omega(\zeta_n))]$ for each n, then, by Fatou's Lemma, it will follow that $u(t, y) \geq 0$. But, by using a smooth bump function, we can construct a $u_n \in C_c^{1,2}(\mathbb{R}^N; \mathbb{R})$ so that $u_n = u$ on $\overline{\mathfrak{G}}_n$, and therefore, by Doob's Stopping Time Theorem (cf. the discussion at the end of §5.1),

$$u(t, y) = u_n(t, y) = \mathbb{E}^{\mathbb{P}_y}\left[u_n(t - \zeta_n, \omega(\zeta_n))\right]$$

$$- \mathbb{E}^{\mathbb{P}_y}\left[\int_0^{\zeta_n} (L - \partial_\tau)u_n(t - \tau, \omega(\tau))\, d\tau\right]$$

$$\geq \mathbb{E}^{\mathbb{P}_y}\left[u(t - \zeta_n, \omega(\zeta_n))\right].$$

The final statement comes from applying the earlier one to the difference between u and the constant on right-hand side of the inequality. □

REMARK 5.3.2. It is the final statement in Theorem 5.3.1 which is usually called the weak minimum principle, and it most commonly applied to situations in which $\mathbb{P}_y(\zeta^{(t,\mathfrak{G})} < \infty) = 1$ for all $(t, y) \in \mathfrak{G}$. The term "weak" has a pejorative connotation which is not entirely justified here. Indeed, although the weak minimum principle is less dramatic than the statements which we will prove below, it applies to situations (e.g., when L is degenerate) where the stronger statements fail and sometimes gives more information even when seemingly more refined versions hold. On the other hand,

---

[3] Neither the boundedness nor the smoothness of the coefficients is really needed here. All that one needs is the existence, for each $x \in \mathbb{R}^N$, of a $\mathbb{P}_x$ for which (5.1.2) holds whenever $\varphi \in C_c^{1,2}([0, \infty) \times \mathbb{R}^N; \mathbb{R})$. In §6.1 of [52] it is shown that $\mathbb{P}_x$ exists whenever a and b are bounded and continuous. When they are merely continuous, the only problem comes from the possibility of *explosion*. However, even if explosion occurs, one can simply "kill" the process when it does.

there is a subtlety in its statement. Namely, before applying it, one needs
to have checked that $\mathbb{P}_y(\zeta^{(t,\mathfrak{G})} < \infty) = 1$, which is more or less equivalent
to showing that the weak minimum principle holds when u is a bounded.
Fortunately, there are criteria which allow one to check this condition in
lots of situations. For example, suppose that one can find a $u \in C^{1,2}(\mathfrak{G};\mathbb{R})$
with the properties that u is bounded above and $(L - \partial_\tau)u \geq \alpha > 0$. Then,
using the reasoning and notation of the preceding proof, we have

$$\sup_{\mathfrak{G}} u - u(t,y) \geq \mathbb{E}^{\mathbb{P}_y}\left[u\left(t - \zeta_n, \omega(\zeta_n)\right)\right] - u(t,y) \geq \alpha \mathbb{E}^{\mathbb{P}_y}\left[\zeta_n\right]$$

for all $n \geq 0$, and so, after letting $n \to \infty$, we get $\mathbb{E}^{\mathbb{P}_y}\left[\zeta^{(t,\mathfrak{G})}\right] < \infty$. Of
course, this criterion is useless in cases when $\zeta^{(t,\mathfrak{G})}$ is $\mathbb{P}_y$-almost surely
finite but has infinite expectation value. For instance, it cannot be used to
check (cf. Corollary 7.2.14 in [53]) that, for $L = \frac{1}{2}\Delta$, $r > 0$ and $|y| > r$,
$\mathbb{P}_y(\zeta^{\mathbb{R} \times B(0,r)^\complement} < \infty) = 1$ when $N \in \{1,2\}$ but not when $N \geq 3$.

5.3.2. A Mean Value Property: All the other forms of the minimum
principles which we will prove derive from the following *mean value property*. In its statement, and below,

(5.3.3) $$Q\big((t,y),r\big) = (t - r^2, t) \times B(y,r)$$

and

$$\fint_\Gamma \psi(\xi)\, d\xi = \frac{1}{|\Gamma|}\int_\Gamma \psi(\xi)\, d\xi,$$

the average of ψ over Γ.

THEOREM 5.3.4. *Let L be given by either (4.3.1) or (4.4.1). Then, for
each $R \in [1,\infty)$, there exists a non-increasing $\theta \in (0,1) \longmapsto \alpha(R,\theta) \in
(0,1)$, depending only on the bounds on the coefficients and ϵ, such that*

$$u(t,y) \geq \alpha e^{-\frac{r^2}{\alpha s}} \fint_{B(\xi,\theta r)} u(t - s, \eta)\, d\eta$$

*for all $r \in (0,R]$, $s \in (0,r^2]$, $y \in B(\xi,\theta r)$, and $u \in C_b\big(\overline{Q((t,\xi),r)}; [0,\infty)\big) \cap
C^{1,2}\big(Q((t,\xi),r); [0,\infty)\big)$ satisfying $(L - \partial_\tau)u \leq 0$. Furthermore, when L is
given by (4.3.1), α can be taken independent of R.*

PROOF: We begin by observing that there is no loss in generality in as-
suming that $r < R$ and that $u \in C_b^{1,2}(\mathbb{R} \times \mathbb{R}^N; [0,\infty))$. Indeed, just as
in the proof of Theorem 5.3.1, the general case can be obtained from this
one by an easy limit procedure and the use of bump functions. Thus, we
will proceed under these assumptions, in which case the asserted result is
an easy application of Theorem 5.2.11. Namely, by Doob's Stopping Time
Theorem,

$$u(t,y) = \mathbb{E}^{\mathbb{P}_y}\left[u\big(t - s \wedge \zeta^{B(\xi,r)}, \omega(s \wedge \zeta^{B(\xi,r)})\big)\right]$$

$$- \mathbb{E}^{\mathbb{P}_y}\left[\int_0^{s \wedge \zeta^{B(\xi,r)}} (L + \partial_\tau)u\big(t - \tau, \omega(\tau)\big)\, d\tau\right].$$

Hence, because $u \geq 0$ and $(L - \partial_\tau)u \leq 0$ on $Q((t, \xi), r)$,

$$u(t, y) \geq \int_{B(\xi, \theta r)} p^{B(\xi, r)}(s, y, \eta) u(t - s, y, \eta) \, d\eta,$$

and so the desired estimate follows from the one in Theorem 5.2.11. □

5.3.3. The Strong Minimum Principle: The purpose of this section is to prove the *strong minimum principle*. In its statement, $\mathfrak{G}$ is an open subset of $\mathbb{R} \times \mathbb{R}^N$ and, for $(t, y) \in \mathfrak{G}$, $\mathfrak{G}(t, y)$ denotes the set

$$\left\{ \big(t - \tau, \omega(\tau)\big) : \omega \in \Omega \text{ with } \omega(0) = y \text{ and } 0 < \tau < \zeta^{(t, \mathfrak{G})}(\omega) \right\}$$

of points which are *backwards* to (t, y) in $\mathfrak{G}$.

THEOREM 5.3.5. *Assume that L is given by (1.1.8), where a and b are smooth and $a(x)$ is strictly positive definite for all $x \in \mathfrak{G}$. If $u \in C^{1,2}(\mathfrak{G}; \mathbb{R})$ satisfies $(L - \partial_\tau)u \leq 0$ and u achieves its minimum value at $(t, y) \in \mathfrak{G}$, then $u = u(t, y)$ on $\mathfrak{G}(t, y)$.*

PROOF: We begin by showing that if $u \in C^{1,2}(\mathfrak{G}; [0, \infty))$ satisfies $(L - \partial_\tau)u \leq 0$ and $u(t, y) = 0$ for some $(t, y) \in \mathfrak{G}$, then there exists an $r > 0$ such that $u = 0$ on $Q((t, y), r)$. To this end, choose $r > 0$ so that $\overline{Q((t, y), 2r)} \subset\subset \mathfrak{G}$. Then one can easily construct smooth coefficients a' and b' with bounded derivatives of all orders so that $a' \geq \epsilon I$ for some $\epsilon > 0$ and $L\psi = \frac{1}{2}\nabla \cdot (a'\nabla\psi) + b' \cdot \nabla\psi$ on $\overline{Q((t, y), 2r)}$. Thus, by Theorem 5.3.4,

$$0 = u(t, y) \geq \alpha e^{-\frac{r^2}{\alpha s}} \fint_{B(y, r)} u(t - s, \xi) \, d\xi, \quad s \in (0, r^2),$$

and so u must vanish on $\overline{Q((t, y), r)}$.

To complete the proof, let u and (t, y) be as in the statement. Without loss in generality, we will assume that $u(t, y) = 0$. Given $\omega \in \Omega$ with $\omega(0) = y$, set $s = \sup\{\tau \in [0, \zeta^{(t, \mathfrak{G})}(\omega)) : u(t - \tau, \omega(\tau)) = 0\}$. Then, by the preceding, s must be equal to $\zeta^{(t, \mathfrak{G})}(\omega)$. □

REMARK 5.3.6. It should be recognized that the preceding is a slightly ridiculous derivation of the strong minimum principle. Indeed there are (cf. Chapter 2 of [19]) far more elementary proofs, based on sharpened versions of the ideas used to prove Lemma 2.4.1, and those proofs require nothing but continuity and ellipticity of the coefficients.

5.3.4. Nash's Continuity Theorem: Another, and much more profound, conclusion which can be drawn from Theorem 5.3.4 is Nash's celebrated continuity theorem. We begin with the following lemma, in which (cf. (5.3.3))

$$\text{Osc}(u; (t, \xi), r) \equiv \sup\{u(t'x') - u(t, x) : (t, x), (t'x') \in Q((t, \xi), r)\}$$

is the *oscillation* of u on $Q((t, \xi), r)$.

LEMMA 5.3.7. *Let L be as in Theorem 5.3.4. Then, for each $R \in [1, \infty)$, there is a non-decreasing function $\theta \in (0,1) \longmapsto \rho(R, \theta) \in (0,1)$ with the property that, for all $r \in (0, R]$, $\theta \in (0,1)$, $(t, \xi) \in \mathbb{R} \times \mathbb{R}^N$, and $u \in C^{1,2}(Q(t, \xi), r); \mathbb{R}) \cap C_b(\overline{Q((t, \xi), r)}; \mathbb{R})$ satisfying $(L - \partial_\tau) u = 0$,*

$$\mathrm{Osc}(u; (t, \xi), \theta r) \leq \rho \, \mathrm{Osc}(u; (t, \xi), r).$$

Moreover, when L is given by (4.3.1), ρ can be taken independent of R.

PROOF: For $\delta \in \{\theta, 1\}$, set

$$M_\delta = \sup\{u(\tau, \eta) : (\tau, \eta) \in Q((t, \xi), \delta r)\},$$
$$m_\delta = \inf\{u(\tau, \eta) : (\tau, \eta) \in Q((t, \xi), \delta r)\},$$

and take

$$\Gamma = \left\{ \eta \in B(\xi, \theta r) : u(t - r^2, \eta) \geq \frac{M_1 + m_1}{2} \right\}.$$

Clearly, either $|\Gamma| \geq \frac{1}{2}|B(\xi, \theta r)|$ or $|\Gamma| \leq \frac{1}{2}|B(\xi, \theta r)|$. In the first case, we apply Theorem 5.3.4 to $u - m_1$ and conclude that, for all $(\tau, \eta) \in Q((t, \xi), \theta r)$,

$$u(\tau, \eta) - m_1 \geq \beta \fint_{B(\xi, \theta r)} \big(u(t - r^2, \eta') - m_1\big) \, d\eta' \geq \frac{\beta}{4}(M_1 - m_1),$$

where $\beta = \alpha e^{-\frac{1}{\alpha(1 - \theta^2)}}$. Hence, $m_\theta - m_1 \geq \frac{\beta}{4}(M_1 - m_1)$; and so, if $\rho = 1 - \frac{\beta}{4}$, then

$$\mathrm{Osc}(u; (t, \xi), \theta r) = M_\theta - m_\theta \leq M_1 - m_\theta \leq \rho(M_1 - m_1) = \rho \, \mathrm{Osc}(u; (t, \xi), r).$$

In the case when $|\Gamma| \leq \frac{1}{2}|B(\xi, \theta r)|$, we work in the same way with $M_1 - u$ to reach the same conclusion. Finally, observe that the ρ we have found has the same dependence properties as α and is therefore acceptable. □

THEOREM 5.3.8. *Let L be given by either (4.3.1) or (4.4.1). Then, for each $R \in (0, \infty)$, there exists a function, depending only on ϵ and the bounds on the coefficients (but not their derivatives), $\theta \in (0,1) \longmapsto \nu(R, \theta) \in (0,1]$ such that, if $(T, \Xi) \in \mathbb{R} \times \mathbb{R}^N$ and $u \in C^{1,2}(Q((T, \Xi), R); \mathbb{R})$ satisfies $(L - \partial_\tau) u = 0$, then*

$$|u(s', x') - u(s, x)| \leq 2 \left(\frac{|s' - s|^{\frac{1}{2}} \vee |x' - x|}{(1 - \theta) R} \right)^{\nu(R, \theta)} \|u\|_u$$

for all (s, x), $(s', x') \in Q((t, y), \theta R)$. Moreover, when L is given by (4.3.1), ν can be taken independent of R. Hence, if L is given by (4.3.1), then, for each $T \in \mathbb{R}$, any bounded $u \in C^{1,2}((-\infty, T) \times \mathbb{R}^N; \mathbb{R})$ satisfying $(L - \partial_\tau) u = 0$ is constant.

PROOF: Set $\delta = 1 - \theta$, and, given (s,x), $(s',x') \in Q\big((T,\Xi),\theta R\big)$, set $r \equiv |s' - s|^{\frac{1}{2}} \vee |x' - x|$, $t = s \vee s'$, and $\xi = \frac{x+x'}{2}$.

If $r \geq R$, then, since $r \leq R$,

$$|u(s',x') - u(s,x)| \leq 2 \left(\frac{r}{\delta R}\right)^{\nu} \|u\|_{\mathrm{u}}$$

for any $\nu \in (0,1]$. Now assume that $r \leq \delta R$, and determine $n \in \mathbb{N}$ by $\delta^{n+1} r \leq R < \delta^n r$. Then (s,x), $(s',x') \in Q\big((t,\xi),\delta^n R\big) \subseteq Q\big((T,\Xi),R\big)$. Therefore, by Lemma 5.3.7,

$$|u(s',x') - u(s,x)| \leq \mathrm{Osc}\big(u;(t,\xi),\delta^n R\big) \leq \rho^n \mathrm{Osc}\big(u;(t,\xi),R\big) \leq 2\rho^n \|u\|_{\mathrm{u}},$$

where $\rho = \rho(R,\delta)$. Thus if $\nu \in (0,1]$ is chosen so that $\nu = 1$ when $\rho \leq \delta$ and $\rho = \delta^{\nu}$ when $\rho \geq \delta$, then, since $\delta^n \leq \frac{r}{\delta R}$, we get

$$|u(s',x') - u(s,x)| \leq 2 \left(\frac{r}{\delta R}\right)^{\nu} \|u\|_{\mathrm{u}}.$$

Finally, because ρ can be chosen independent of R when L is given by (4.3.1), the same is true about ν. Hence, when L is given by (4.3.1) and u is a bounded solution to $(L - \partial_\tau)u = 0$ on $(-\infty, T) \times \mathbb{R}^N$, the last part of the theorem follows when one lets $R \nearrow \infty$. $\square$

REMARK 5.3.9. Under the given smoothness hypotheses, the basic conclusion drawn in Theorem 5.3.8 looks less than striking. Indeed, Theorems 3.4.1 tells us that u is not only Hölder continuous but also has continuous derivatives of all orders. Thus, the interest in Theorem 5.3.8 is the independence of the estimates from any regularity properties of the coefficients. This independence is important for applications to non-linear equations as well as equations with rough coefficients. For an example of the second sort, consider the problem of constructing solutions to $\partial_t u = Lu$ in $(0,\infty) \times \mathbb{R}^N$ with $u(0, \cdot) = f \in C_{\mathrm{b}}(\mathbb{R}^N; \mathbb{R})$, where L is given by (4.4.1), $a \geq \epsilon I$, but its coefficients are merely bounded and continuous. One way of going about this would be to construct sequences $\{a_n : n \geq 1\}$ and $\{b_n : n \geq 1\}$ of mollified coefficients which approach a and b uniformly on compacts and then construct the solution u_n to the corresponding problem for the associated operator L_n. Theorem 5.3.8 says that $\{u_n : n \geq 1\}$ is relatively compact in the sense of uniform convergence on compact subsets of $(0,\infty) \times \mathbb{R}^N$. Of course, this does not solve the problem, but at least it gets one started. Finally, it should be pointed out that the dependence of the estimates on size alone is important even when the equation is linear and the coefficients are smooth. Namely, the *Liouville-type* theorem in the final assertion is a consequence of a scaling argument which works only because the estimates do not depend on smoothness properties of the coefficients.

5.3.5. The Harnack Principle of De Giorgi and Moser: There is an interesting and instructive way in which to understand the relationship between the weak and strong minimum principles for L. Namely, for a connected open set $G \subseteq \mathbb{R}^N$ and an $x \in \mathbb{R}^N$, denote by Π_x^G the distribution under $\mathbb{P}_x$ of $\omega \rightsquigarrow \omega(\zeta^G)$ on $\{\zeta^G < \infty\}$. Then Π_x^G is the *harmonic measure* for G based at x, and, for $x \in G$, the weak minimum principle holds for L if and only if Π_x^G is a probability measure. Indeed, $\Pi_x^G(\mathbb{R}^N) = \mathbb{P}_x(\zeta^G < \infty)$. The strong minimum principle deals with a quite different property of the Π_x^G's. Whether or not they are probability measures, it says that $\Pi_{x'}^G \ll \Pi_x^G$ for all $x, x' \in G$. To see this, let Γ be a Borel subset of ∂G, and set $u(x) = \Pi_x^G(\Gamma)$. Then, for any $\varphi \in C_c^\infty(G; \mathbb{R})$, one has

$$\int_G L^\top \varphi(x) u(x)\, dx = \lim_{t \searrow 0} \frac{1}{t} \int_G \varphi(x)\big(\mathbf{P}_t u(x) - u(x)\big)\, dx.$$

At the same time (cf. the proof of Theorem 5.2.8),

$$\big|\mathbf{P}_t u(x) - u(x)\big| = \Big|\mathbb{P}_x\big(\omega(\zeta_t^G) \in \Gamma,\, \zeta_t^G < \infty\big) - \mathbb{P}_x\big(\omega(\zeta^G) \in \Gamma,\, \zeta^G < \infty\big)\Big|$$
$$\leq \mathbb{P}_x\big(\zeta^G \leq t\big),$$

where $\zeta_t^G(\omega) = \inf\{\tau \geq t : \omega(\tau) \notin G\}$. Hence, by Lemma 5.2.4, as $t \searrow 0$, $t^{-1}\big|\mathbf{P}_t u(x) - u(x)\big| \longrightarrow 0$ uniformly on compact subsets of G, and so, as a Schwartz distribution, $Lu = 0$ on G. But, by Theorem 3.4.1, this means that u is a smooth solution to $Lu = 0$ in G, and therefore, since $u : G \longrightarrow [0,1]$, the strong minimum principle says that it is either always positive or identically 0. That is, $\Pi_x^G(\Gamma) = 0$ for some $x \in G$ implies $\Pi_x^G(\Gamma) = 0$ for all $x \in G$. Similarly, by considering $1 - \Pi_x^G(\partial G)$, one sees that, when it holds, the strong maximum guarantees that Π_x^G is a probability measure either for all or for no $x \in G$.

The preceding digression provides a context for *Harnack's principle*. In terms of the Π_x^G's, it is the statement (cf. Remark 5.3.11 below) that $\Pi_{x'}^G$ is not only absolutely continuous with respect to Π_x^G but also that, so long as x and x' range over a compact subset of G, the associated Radon–Nikodym derivatives are bounded above and below by positive constants.

THEOREM 5.3.10. *Assume that L is given by $(4.3.1)$ or $(4.4.1)$. Then, for each $R \in (0, \infty)$ there is a non-decreasing $\theta \in \left[\frac{1}{2}, 1\right) \longmapsto \kappa(R, \theta) \in [1, \infty)$, depending only on the bounds on the coefficients and ϵ, such that, for any $(T, \Xi) \in \mathbb{R} \times \mathbb{R}^N$ and non-negative $u \in C^{1,2}\big(Q(T, \Xi), R); \mathbb{R}\big) \cap C\big(\overline{Q((T, \Xi), R)}; \mathbb{R}\big)$ satisfying $(L - \partial_\tau)u = 0$,*

$$u(s, x) \leq \kappa u(T, y) \quad \text{when } (x, y) \in B(\Xi, \theta R) \text{ and } 1 - \theta^2 \leq \frac{T - s}{R^2} \leq \theta^2.$$

Moreover, when L is given by $(4.3.1)$, κ can be chosen independent of R. In particular, if L is given by $(4.3.1)$, then any non-negative $u \in C^2(\mathbb{R}^N; \mathbb{R})$ satisfying $Lu = 0$ is constant.

PROOF: If $u(T, y) = 0$, then, by the strong minimum principle, $u = 0$ on $Q((T, \Xi), R)$. Thus we will assume that $u(T, y) = 1$ and will show that there is a $\kappa < \infty$, with the required dependence, such that $u(s, x) \le \kappa$ for all $(s, x) \in [T - \theta^2 R^2, T - (1 - \theta^2)R^2] \times \overline{B(\Xi, \theta R)}$.

Set $\delta = \frac{1+\theta}{2}$, and use Theorem 5.3.4 to find a $\beta \in (0, 1)$, with the required dependence, such that

$$(*) \qquad u(T, y) \ge \beta \fint_{B(\Xi, \delta R)} u(\tau, \eta) \, d\eta$$
$$\text{for } T - R^2 \le \tau \le T - (1 - \theta^2)R^2.$$

Referring to Lemma 5.3.7, set $\rho = \rho(R, \frac{1}{2})$, $\mu = \frac{1-\rho}{2}$, and $\lambda = \frac{1-\mu}{\rho} = \frac{1+\rho}{2\rho} > 1$. Then, from $(*)$ and $u(T, y) = 1$, for any $M > 0$ and $T - R^2 \le \tau \le T - (1 - \theta^2)R^2$,

$$\left| \left\{ \eta \in B(\Xi, \delta R) : u(\tau, \eta) > \mu M \right\} \right| < \Omega_N r(M)^N,$$
$$\text{where } r(M) = \frac{\delta R}{(\mu \beta)^{\frac{1}{N}}} M^{-\frac{1}{N}}.$$

Now set $\mathcal{R} = (T - R^2, T - (1 - 2\theta^2)R^2)$, and suppose that $(s, x) \in \mathcal{R}$ with $u(s, x) \ge M$, where M is large enough that $\overline{Q((s, x), 2r(M))} \subset\subset \mathcal{R}$. Then, by the preceding, $B(x, r(M))$ must contain a ξ for which $u(s, \xi) \le \mu M$, and so, by Lemma 5.3.7, there exists an $(s', x') \in \overline{Q((s, x), 2r(M))}$ such that

$$u(x', x') \ge \mathrm{Osc}(u; (s, x), 2r(M)) \ge \rho^{-1} \mathrm{Osc}(u; (s, x), r(M)) \ge \lambda M.$$

That is, we now know that

$$(**) \qquad \begin{aligned} &\overline{Q((s, x), 2r(M))} \subset\subset \mathcal{R} \text{ and } u(s, x) \ge M \\ &\implies \exists (s', x') \in \overline{Q((s, x), 2r(M))} \; u(s', x') \ge \lambda M. \end{aligned}$$

Finally, take

$$\kappa = \frac{4^N \lambda}{\beta \mu (1 - \theta)^N (\lambda^{\frac{1}{N}} - 1)^N}.$$

Then

$$\theta + \frac{2}{(\beta \mu \kappa)^{\frac{1}{N}}} \sum_{n=0}^{\infty} \lambda^{-\frac{n}{N}} = \delta,$$

and therefore

$$\theta R + 2 \sum_{n=0}^{\infty} r(\lambda^n \kappa) = \delta R \quad \text{and} \quad \theta^2 R^2 + 4 \sum_{n=0}^{\infty} r(\lambda^n \kappa)^2 \le \delta^2 R^2.$$

Hence, if $u(s,x) \geq \kappa$ for some $(s,x) \in [T-\theta^2 R^2, T-(1-\theta^2)R^2] \times B(\Xi, \theta R)$, then we could use (**) to inductively produce a sequence $\{(s_n, x_n) : n \geq 0\} \subseteq \mathcal{R}$ such that $(s_0, x_0) = (s,x)$, $(s_{n+1}, x_{n+1}) \in \overline{Q\big((s_n, x_n), 2r(\lambda^n M)\big)}$, and $u(s_n, x_n) \geq \lambda^n \kappa$, which would lead to the contradiction that u is unbounded on $Q\big((T, \Xi), R\big)$.

Because both β and ρ have the required dependence properties, so does κ. In particular, when L is given by (4.3.1), κ can be taken independent of R. Thus, if $u \in C^2\big(\mathbb{R}^N; [0, \infty)\big)$ satisfies $Lu = 0$, then $0 \leq u(x) \leq \kappa u(0)$ for all $x \in \mathbb{R}^N$. But this means that u is bounded, and therefore, by the last part of Theorem 5.3.8, u is constant. $\square$

REMARK 5.3.11. Returning to the discussion at the beginning of this subsection, suppose that $\overline{B(x, 2r)} \subset\subset G$. Then, for any $\Gamma \in \mathcal{B}_{\partial G}$, Theorem 5.3.10 applied to $y \rightsquigarrow \Pi_y^G(\Gamma)$ says that $\Pi_y^G(\Gamma) \leq \kappa \Pi_x^G(\Gamma)$ for all $y \in \overline{B(x,r)}$, from which the assertion made earlier about Radon–Nikodym derivatives is an easy consequence. In a different direction, one might wonder whether the last part of Theorem 5.3.10 cannot be extended to non-negative solutions of $(L - \partial_\tau)u = 0$ in $\mathbb{R} \times \mathbb{R}^N$. However, this is not the case. Indeed, take $N = 1$, $L = \frac{1}{2}\partial_x^2$, and consider $(\tau, x) \rightsquigarrow \exp\big(x + \frac{\tau}{2}\big)$.

5.4 Historical Notes and Commentary

From the purely analytic standpoint, Duhamel's formula is a simple application of the familiar procedure for solving a boundary value problem by starting with a solution to the interior equation and then correcting it so that it satisfies the boundary condition. The implementation of this procedure with stopping times goes back to J.L. Doob [12], who also initiated the application of stopping times to the derivation of estimates of the sort in § 5.2.2. The strong minimum principle in Theorem 5.3.5 is due to L. Nirenberg [44].

Whether or not it did so for the first time, the application of stopping times to the derivation of the mean value property in § 5.2.3 appears in [16], where, following ideas of N. Krylov and M. Safonov in [32], it was used, in the same way as it is here, to prove the oscillation result in Lemma 5.3.7. Given Lemma 5.3.7, the derivations given in [16] and here of Nash's Continuity Theorem and the Di Georgi–Moser Harnack principle are essentially the same as those in Moser's [41] and [42]. Again the interested reader should consult [51] to see interesting extensions of these results.

In connection with the discussion in the final part of § 4.5, it should be mentioned that Lemma 5.3.7 and, as a consequence, both the Nash Continuity Theorem as well as the Di Georgi–Moser Harnack principle are true for uniformly elliptic L's given by (1.1.8), even if the coefficients are only bounded and measurable. In view of the discouraging remarks made in § 4.5 about such operators, this information may come as something of a surprise. Indeed, when they became known, these results shocked experts who had devoted years of time to such questions. Nonetheless, in their brilliant paper [31], Krylov and Safonov showed how one could parlay Alexandrov's sort of ideas into a proof of them.

Finally, I would be remiss were I to not mention an entirely different approach introduced by S.T. Yau to prove Harnack inequalities for the Laplacian on a Riemannian manifold. His goal was to control the constants in terms of geometric quantities, and he showed that they can be controlled in terms of lower bounds on the Ricci curvature. In their famous sequel [36], P. Li and Yau proved versions of these results for the associated heat operator.

CHAPTER 6

On a Manifold

The purpose of this chapter is to show how one can transfer the results obtained earlier, particularly those in Chapters 3 and 4, from Euclidean space to a differentiable manifold.

As we have already seen in the Euclidean setting, it is important to distinguish between local (short time) and global (long time) aspects of the theory. When dealing with differentiable manifolds, this distinction becomes even more important. Indeed, by definition, all smooth manifolds are locally Euclidean, and so one should expect that, aside from a few technical details, there is no problem about transferring the local theory. When the manifold is compact, global theory is relatively easy. Namely, because it has nowhere to spread, the heat flow quickly equilibrates and so the fundamental solution to a non-degenerate (i.e., a is elliptic) Kolmogorov equation tends rapidly to its stationary state. On the other hand, when the manifold is non-compact, the global theory reflects the geometric growth properties of the particular manifold under consideration, and so the Euclidean case cannot be used to predict long time behavior.

For the reason just given, we will restrict our attention to the local theory. In fact, in order to avoid annoying questions about possible "explosion," we will restrict our attention to compact manifolds, a restriction which, for the local theory, is convenient but inessential.

6.1 Diffusions on a Compact Riemannian Manifold

Throughout, M will be a compact, connected, C^∞-manifold of dimension N. For our purposes, the most convenient way to introduce an operator on M is to give M a Riemannian structure and to define L on $C^2(M;\mathbb{R})$ by

$$(6.1.1) \qquad L\varphi = \tfrac{1}{2}\Delta\varphi + B\varphi,$$

where Δ is the standard (i.e., the one corresponding to the Levi-Civita connection) Laplacian on M determined by the Riemannian structure, and B is a vector field on M. By analogy with the expressions in (1.1.8) or (4.4.1), one might have thought that we would write L in the form

$$L\varphi = \tfrac{1}{2}\text{Trace}\big(a\text{Hess}\varphi\big) + B\varphi \quad \text{or} \quad L\varphi = \text{div}\big(a\text{grad}\varphi\big) + B\varphi,$$

where $x \in M \longmapsto a(x) \in T_x(M)^* \otimes T_x(M)^*$ is symmetric and positive definite with respect to the Riemann metric. However, as long as the coefficients are smooth, the first of these can always be rewritten in the form given in the second, and, by incorporating a^{-1} into the Riemannian metric, operators in the second form can be rewritten in the form of (6.1.1). (Of course, the vector field B will depend on the form used.) Thus, nothing essential is lost by assuming from the outset that L has the form given in (6.1.1).

Our first goal is to contract a continuous transition probability function $(t, x) \in [0, \infty) \times M \longmapsto P(t, x) \in \mathbf{M}_1(M)$ with the property that

$$(6.1.2) \qquad \langle \varphi, P(t, x) \rangle - \varphi(x) = \int_0^t \langle L\varphi, P(\tau, x) \rangle \, d\tau, \quad \varphi \in C^2(M; \mathbb{R}).$$

Because it provides an effective way to analyze $P(t, x)$, we will employ a a pathspace approach which will enable us to "lift" the construction from $\mathbb{R}^N$ to M. Thus, we begin by characterizing the measures on pathspace associated with L as solutions to a martingale problem. In the process, we will show that, at least locally, these measures are obtained by lifting solutions to martingale problems for elliptic operators on $\mathbb{R}^N$, and, in turn, this will give us a way of lifting to the manifold setting the properties which we proved earlier in the Euclidean setting.

6.1.1. Splicing Measures: Given a Polish space (i.e., a complete, separable, metric space) X, let $\Omega(X)$ denote the Polish space $C([0, \infty); X)$ with the topology of uniform convergence on compacts. If $x \in X \longmapsto \mathbb{Q}'_x \in \mathbf{M}_1(\Omega(X))$ is a (Borel) measurable map with the property that $\mathbb{Q}'_x(\omega(0) = x) = 1$ for each $x \in X$, define the measurable map

$$(t, \omega) \in [0, \infty) \times \Omega(X) \longmapsto \delta_\omega \underset{t}{\otimes} \mathbb{Q}'. \in \mathbf{M}_1(\Omega(X))$$

so that, for all $n \geq 1$, $0 \leq \tau_0 < \cdots < \tau_m \leq t < \tau_{m+1} < \cdots < \tau_n$, and $\Gamma_0, \ldots, \Gamma_n \in \mathcal{B}_X$,

$$\delta_\omega \underset{t}{\otimes} \mathbb{Q}'. \left(\omega'(\tau_\ell) \in \Gamma_\ell \text{ for all } 0 \leq \ell \leq n \right)$$

$$= \left(\prod_{\ell=0}^m \mathbf{1}_{\Gamma_\ell}(\omega(\tau_\ell)) \right) \mathbb{Q}_{\omega(t)} \left(\{ \omega' : \omega'(\tau_\ell - t) \in \Gamma_\ell \text{ for all } m < \ell \leq n \} \right).$$

Next, given $\mathbb{Q} \in \mathbf{M}_1(\Omega(X))$ and a $\{\mathcal{B}_t : t \geq 0\}$-stopping time $\zeta : \Omega(X) \longrightarrow [0, \infty]$, the *splice of* $\mathbb{Q}$ *to* $\{\mathbb{Q}'_x : x \in X\}$ *at time* ζ is the measure such that

$$\mathbb{Q} \underset{\zeta}{\otimes} \mathbb{Q}'. (B) = \int_{\{\zeta(\omega) < \infty\}} \delta_\omega \underset{\zeta(\omega)}{\otimes} \mathbb{Q}'. (B) \, \mathbb{Q}(d\omega) + \mathbb{Q}(B \cap \{\zeta = \infty\})$$

for $B \in \mathcal{B}_{\Omega(X)}$. The crucial fact which we need to know about $\mathbb{Q} \underset{\zeta}{\otimes} \mathbb{Q}'.$ is contained in the following lemma.

LEMMA 6.1.3.   Let ζ be a stopping time, and assume that $\Xi : [0, \infty) \times \Omega(X) \longrightarrow \mathbb{R}$ and $(t, x, \omega) \in [0, \infty) \times X \times \Omega(X) \longmapsto \Xi'_x(t, \omega) \in \mathbb{R}$ are measurable functions which, as functions of (t, ω), are progressively measurable[1] with respect to $\{\mathcal{B}_t : t \geq 0\}$ and, as functions of t, are continuous. Further, assume that $\Xi(0, \omega) = \Xi'_x(0, \omega) \equiv 0$,

$$\mathbb{E}^{\mathbb{Q}}\left[\sup_{\tau \in [0,t]} |\Xi(\tau)|\right] \vee \sup_{x \in X} \mathbb{E}^{\mathbb{Q}'_x}\left[\sup_{\tau \in [0,t]} |\Xi'_x(t)|\right] < \infty \quad \text{for all } t \in [0, \infty),$$

and that

$$\left(\Xi(t \wedge \zeta), \mathcal{B}_t, \mathbb{Q}\right) \quad \text{and, for each } x \in X, \quad \left(\Xi'_x(t), \mathcal{B}_t, \mathbb{Q}'_x\right)$$

are martingales. Finally, define

$$\bar{\Xi}(t, \omega) = \begin{cases} \Xi(t, \omega) & \text{if } 0 \leq t < \zeta(\omega) \\ \Xi\big(\zeta(\omega), \omega\big) + \Xi'_{\omega(\zeta)}\big(t - \zeta(\omega), \Sigma_{\zeta(\omega)}\omega\big) & \text{if } t \geq \zeta(\omega), \end{cases}$$

where Σ_τ is the time-shift map on $\Omega(X)$ given by $\Sigma_\tau \omega(t) = \omega(\tau + t)$. Then

$$\left(\bar{\Xi}(t), \mathcal{B}_t, \mathbb{Q} \underset{\zeta}{\otimes} \mathbb{Q}'.\right)$$

is a martingale.

PROOF: Let $0 \leq t_1 < t_2$ and $0 = \tau_0 < \cdots < \tau_n = t_1$, $\Gamma_0, \ldots, \Gamma_n \in \mathcal{B}_X$ be given, and set $A = \{\omega : \omega(\tau_\ell) \in \Gamma_\ell, \ 0 \leq \ell \leq n\}$. We need to check that

$$\mathbb{E}^{\mathbb{Q} \otimes \mathbb{Q}'.}_{\zeta}\left[\bar{\Xi}(t_2), A\right] = \mathbb{E}^{\mathbb{Q} \otimes \mathbb{Q}'.}_{\zeta}\left[\bar{\Xi}(t_1), A\right].$$

By Doob's Stopping Time Theorem, we know that

$$\mathbb{E}^{\mathbb{Q} \otimes \mathbb{Q}'.}_{\zeta}\left[\bar{\Xi}(t_2 \wedge \zeta), A\right] = \mathbb{E}^{\mathbb{Q}}\left[\Xi(t_2 \wedge \zeta), A\right]$$

$$= \mathbb{E}^{\mathbb{Q}}\left[\Xi(t_1 \wedge \zeta), A\right] = \mathbb{E}^{\mathbb{Q} \otimes \mathbb{Q}'.}_{\zeta}\left[\bar{\Xi}(t_2 \wedge \zeta), A\right].$$

Thus, if $\tilde{\Xi}(t, \omega) = \bar{\Xi}(t, \omega) - \Xi\big(t \wedge \zeta(\omega), \omega\big)$, then what remains to be shown is that

$$\mathbb{E}^{\mathbb{Q} \otimes \mathbb{Q}'.}_{\zeta}\left[\tilde{\Xi}(t_2), A\right] = \mathbb{E}^{\mathbb{Q} \otimes \mathbb{Q}'.}_{\zeta}\left[\tilde{\Xi}(t_1), A\right].$$

For this purpose, set

$$A_0 = A \cap \{\zeta = 0\}, \ C = A \cap \{\zeta > t_1\}$$
$$\text{and } A_m = A \cap \{\tau_{m-1} < \zeta \leq \tau_m\} \text{ for } 1 \leq m \leq n.$$

---

[1] A function on $[0, \infty) \times \Omega$ is progressively measurable if its restriction to $[0, t] \times \Omega$ is $\mathcal{B}_{[0,t]} \times \mathcal{B}_t$-measurable for each $t \geq 0$.

Next, for $1 \leq m \leq n$, define

$$B_m = \{\omega : \tau_{m-1} < \zeta(\omega) \leq \tau_m \ \& \ \omega(\tau_\ell) \in \Gamma_\ell, \ 0 \leq \ell \leq m-1\} \quad \text{and}$$
$$B'_m(\omega) = \{\omega' : \omega'(\tau_\ell - \zeta(\omega)) \in \Gamma_\ell, \ m \leq \ell \leq n\} \quad \text{if } \tau_{m-1} < \zeta(\omega) \leq \tau_m.$$

Then, for $i \in \{1, 2\}$,

$$\mathbb{E}^{\overset{\mathbb{Q} \otimes \mathbb{Q}'}{\varsigma}} \big[\tilde{\Xi}(t_i), \, A\big] = \mathbb{E}^{\overset{\mathbb{Q} \otimes \mathbb{Q}'}{\varsigma}} \big[\tilde{\Xi}(t_i), \, A_0\big]$$
$$+ \sum_{m=1}^{n} \mathbb{E}^{\overset{\mathbb{Q} \otimes \mathbb{Q}'}{\varsigma}} \big[\tilde{\Xi}(t_i), \, A_m\big] + \mathbb{E}^{\overset{\mathbb{Q} \otimes \mathbb{Q}'}{\varsigma}} \big[\tilde{\Xi}(t_i), \, C\big].$$

Clearly,

$$\mathbb{E}^{\overset{\mathbb{Q} \otimes \mathbb{Q}'}{\varsigma}} \big[\tilde{\Xi}(t_2), \, A_0\big] = \int_{A_0} \mathbb{E}^{\mathbb{Q}'_{\omega(0)}} \big[\Xi'_{\omega(0)}(t_2), \, A_0\big] \, \mathbb{Q}(d\omega)$$
$$= \int_{A_0} \mathbb{E}^{\mathbb{Q}'_{\omega(0)}} \big[\Xi'_{\omega(0)}(t_1), \, A_0\big] \, \mathbb{Q}(d\omega) = \mathbb{E}^{\overset{\mathbb{Q} \otimes \mathbb{Q}'}{\varsigma}} \big[\tilde{\Xi}(t_1), \, A_0\big]$$

and

$$\mathbb{E}^{\overset{\mathbb{Q} \otimes \mathbb{Q}'}{\varsigma}} \big[\tilde{\Xi}(t_2), \, C\big] = \mathbb{E}^{\overset{\mathbb{Q} \otimes \mathbb{Q}'}{\varsigma}} \big[\tilde{\Xi}(t_2) - \Xi(\zeta), \, C \cap \{\zeta(\omega) < t_2\}\big]$$
$$= \int_{C \cap \{\zeta < t_2\}} \mathbb{E}^{\mathbb{Q}'_{\omega(\zeta)}} \big[\Xi'_{\omega(\zeta)}(t_2 - \zeta(\omega))\big] \, \mathbb{Q}(d\omega) = 0 = \mathbb{E}^{\overset{\mathbb{Q} \otimes \mathbb{Q}'}{\varsigma}} \big[\tilde{\Xi}(t_1), \, C\big].$$

Finally, for each $1 \leq m \leq n$,

$$\mathbb{E}^{\overset{\mathbb{Q} \otimes \mathbb{Q}'}{\varsigma}} \big[\tilde{\Xi}(t_2), \, A_m\big] = \int_{B_m} \mathbb{E}^{\mathbb{Q}'_{\omega(\zeta)}} \big[\Xi'_{\omega(\zeta)}(t_2 - \zeta(\omega)), \, B'_m(\omega)\big] \, \mathbb{Q}(d\omega)$$

$$= \int_{B_m} \mathbb{E}^{\mathbb{Q}'_{\omega(\zeta)}} \big[\Xi'_{\omega(\zeta)}(t_1 - \zeta(\omega)), \, B'_m(\omega)\big] \, \mathbb{Q}(d\omega) = \mathbb{E}^{\overset{\mathbb{Q} \otimes \mathbb{Q}'}{\varsigma}} \big[\tilde{\Xi}(t_1), \, A_m\big]. \quad \square$$

6.1.2. Existence of Solutions: In this subsection we will show that there is a measurable map $x \in M \longmapsto \mathbb{P}_x \in \mathbf{M}_1\big(\Omega(M)\big)$ such that $\mathbb{P}_x$ solves the martingale problem for L starting from x. That is,

$$(6.1.4) \qquad \left(\varphi\big(\omega(t)\big) - \varphi(x) - \int_0^t L\varphi\big(\omega(\tau)\big) \, d\tau, \mathcal{B}_t, \mathbb{P}_x \right)$$

is a mean-zero martingale for each $\varphi \in C^2(M; \mathbb{R})$.

In order to distinguish points and paths in M from points and paths on $\mathbb{R}^N$, we will use x and ω to denote, respectively, generic elements of M and $\Omega(M)$, and ξ and η to denote, respectively, generic elements of

$\mathbb{R}^N$ and $\Omega(\mathbb{R}^N)$. Similarly, $\mathbb{P}$ and $\mathbb{Q}$ will denote elements of, respectively, $\mathbf{M}_1\big(\Omega(M)\big)$ and $\mathbf{M}_1\big(\Omega(\mathbb{R}^N)\big)$. Finally, $B_{\mathbb{R}^N}$ will denote Euclidean balls in $\mathbb{R}^N$, and B_M will denote Riemannian balls in M.

Because M is compact, we can choose open covers $\{U_k : 1 \le k \le K\}$, $\{V_k : 1 \le k \le K\}$, and $\{W_k : 1 \le k \le K\}$ of M so that, for each $1 \le k \le K$, there is a diffeomorphism ψ_k from an open neighborhood of $\overline{W_k}$ onto an open neighborhood of $\overline{B_{\mathbb{R}^N}(0,3)}$ with the properties that $U_k = \psi_k^{-1}\big(B_{\mathbb{R}^N}(0,1)\big)$, $V_k = \psi_k^{-1}\big(B_{\mathbb{R}^N}(0,2)\big)$, $W_k = \psi_k^{-1}\big(B_{\mathbb{R}^N}(0,3)\big)$, and, whenever $W_k \cap W_{k'} \ne \emptyset$, $\psi_k \circ \psi_{k'}^{-1}$ has bounded derivatives of all orders there. It is an easy matter to check that there exists a $\theta \in (0,1)$ such that

(6.1.5)
$$\theta d_M(x,y) \le |\psi_k(y) - \psi_k(x)| \le \theta^{-1} d_M(x,y)$$
$$\text{for all } 1 \le k \le K \text{ and } x,\, y \in \overline{W_k},$$

where $d_M(x,y)$ is the Riemannian distance between x and y. Further, for each $1 \le k \le K$, we can choose $a^k \in C_{\mathrm{b}}^\infty\big(\mathbb{R}^N; \mathrm{Hom}(\mathbb{R}^N;\mathbb{R}^N)\big)$ and $b^k \in C_{\mathrm{b}}^\infty(\mathbb{R}^N;\mathbb{R}^N)$ with the properties that[2]

(6.1.6)
$$[L\varphi](x) \equiv \tfrac{1}{2}\nabla \cdot \big(a^k \nabla(\varphi \circ \psi_k^{-1})\big)\big(\psi_k(x)\big) + b^k \cdot \nabla(\varphi \circ \psi_k^{-1})\big(\psi_k(x)\big)$$
$$= \big[L_k(\varphi \circ \psi_k^{-1})\big]\big(\psi_k(x)\big)$$

whenever $x \in \overline{W_k}$ and, for some $\epsilon \in (0,1)$, $a^k \ge \epsilon I$ and $\|a^k\|_{\mathrm{u}} \vee \|b^k\|_{\mathrm{u}} \le \epsilon^{-1}$.

For each $1 \le k \le K$, let $\{\mathbb{Q}_\xi^k : \xi \in \mathbb{R}^N\}$ be the Markov family of probability measures on $\Omega(\mathbb{R}^N)$ determined by the martingale problem for L_k. We want to build our $\mathbb{P}_x$'s by *splicing* these families together. With this in mind, choose and fix a point $x_0 \in M$, and, for each $1 \le k \le K$, define $\Phi_k : \Omega(\mathbb{R}^N) \longrightarrow \Omega(M)$ by

(6.1.7)
$$[\Phi_k(\eta)](t) = \begin{cases} \psi_k^{-1}\big(\eta(t \wedge \zeta^{B_{\mathbb{R}^N}(0,3)})\big) & \text{if } \eta(0) \in \overline{B_{\mathbb{R}^N}(0,3)} \\ x_0 & \text{if } \eta(0) \notin \overline{B_{\mathbb{R}^N}(0,3)}. \end{cases}$$

Given $x \in M$, take $k(x) = \min\{k : x \in V_k\}$ and set

(6.1.8)
$$\mathbb{P}_x^0 = \mathbb{Q}_{\psi_{k(x)}(x)}^{k(x)} \circ \Phi_{k(x)}^{-1}.$$

That is, $\mathbb{P}_x^0$ is the distribution of $\eta \rightsquigarrow \Phi_{k(x)}(\eta)$ under $\mathbb{Q}_{\psi_{k(x)}(x)}^{k(x)}$. Finally, set $\zeta_0(\omega) = \zeta^{W_{k(\omega(0))}}(\omega)$ and

$$\zeta_n(\omega) = \begin{cases} \infty & \text{if } \zeta_{n-1}(\omega) = \infty \\ \zeta_{n-1}(\omega) + \zeta_0\big(\Sigma_{\zeta_{n-1}(\omega)}\omega\big) & \text{if } \zeta_{n-1}(\omega) < \infty \end{cases} \quad \text{for } n \ge 1.$$

---

[2] The use of ∇ in this chapter is a little ambiguous. When applied to functions on $\mathbb{R}^N$, as in the line which follows, it is the Euclidean gradient. When applied to functions on M, it is the gradient determined by the Riemannian structure on M.

With the preceding notation in place, we are ready to splice the families $\{\mathbb{Q}^k_\xi : \xi \in \mathbb{R}^N\}$ together. Namely, given $x \in M$, set $\mathbb{P}^n_x = \mathbb{P}^{n-1}_x \underset{\zeta_{n-1}}{\otimes} \mathbb{P}^0_\cdot$ for $n \geq 1$. We want to show that, for each $n \geq 0$ and $\varphi \in C^2(M;\mathbb{R})$,

$$\left(\varphi\big(\omega(t \wedge \zeta_n)\big) - \varphi(x) - \int_0^{t \wedge \zeta_n} L\varphi\big(\omega(\tau)\big) \, d\tau, \mathcal{B}_t, \mathbb{P}^n_x \right)$$

is a mean-zero martingale. For this purpose, set

$$\Xi'_x(t, \omega) = \varphi\big(\omega(t \wedge \zeta^{W_{k(x)}})\big) - \varphi(x) - \int_0^{t \wedge \zeta^{V_{k(x)}}(\omega)} L\varphi\big(\omega(\tau)\big) \, d\tau.$$

Then, because the distribution of $\omega \leadsto \Xi'_x(t, \omega)$ under $\mathbb{P}^0_x$ is the same as the distribution of

$$\eta \leadsto \varphi \circ \psi^{-1}_{k(x)}\big(\eta(t \wedge \zeta^{B_{\mathbb{R}^N}(0,3)})\big) - \varphi(x) - \int_0^{t \wedge \zeta^{B_{\mathbb{R}^N}(0,3)}(\eta)} L_{k(x)} \big(\varphi \circ \psi^{-1}_{k(x)}\big)\big(\eta(\tau)\big) \, d\tau$$

under $\mathbb{Q}^{k(x)}_{\psi_k(x)}$, we know that $\big(\Xi'(t, x), \mathcal{B}_t, \mathbb{P}^0_x\big)$ is a mean-zero martingale, which completes the case when $n = 0$. Now assume the result for n, set

$$\Xi_n(t, \omega) = \varphi\big(\omega(t \wedge \zeta_n)\big) - \varphi(x) - \int_0^{t \wedge \zeta_n} L\varphi\big(\omega(\tau)\big) \, d\tau,$$

and observe that $\Xi_{n+1}(t, \omega) = \bar{\Xi}(t, \omega)$ when $\bar{\Xi}(t, \omega)$ is constructed from $\zeta_n(\omega)$, $\Xi_n(t, \omega)$, and $\Xi'_\cdot(t, \omega)$ by the prescription in Lemma 6.1.3. Hence, that lemma allows us to complete the induction.

In view of the preceding, what we would like to do is take $\mathbb{P}_x$ so that[3] $\mathbb{P}_x \restriction \mathcal{B}_{\zeta_n} = \mathbb{P}^n_x \restriction \mathcal{B}_{\zeta_n}$ for all $n \geq 0$. However, in order to know that this will complete our construction, we must check that

$$\lim_{n \to \infty} \mathbb{P}^n_x(\zeta_n \leq t) = 0 \quad \text{for all } t \in (0, \infty).$$

To do so, we will use the fact that

(6.1.9) $$\rho \equiv \max_{1 \leq k \leq K} \sup_{x \in \bar{V}_k} \mathbb{E}^{\mathbb{P}^0_x}\big[e^{-\zeta^{W_k}}\big] < 1.$$

To check this, set

$$r = \min\big\{ d_M(x, y) : x \in \overline{V_k} \ \& \ y \notin W_k \text{ for some } 1 \leq k \leq K \big\}.$$

---

[3] Recall that if ζ is a stopping time relative to $\{\mathcal{B}_t : t \geq 0\}$, then $\mathcal{B}_\zeta$ is the σ-algebra of $A \subseteq \Omega$ such that $A \cap \{\zeta \leq t\} \in \mathcal{B}_t$ for every $t \geq 0$.

Then, $x \in \overline{V_k}$ implies that $B_M(x,r) \subseteq W_k$ and therefore, by (6.1.5), that

$$\mathbb{E}^{\mathbb{P}^0_x}\left[e^{-\zeta^{W_k}}\right] \leq \mathbb{E}^{\mathbb{P}^0_x}\left[e^{-\zeta^{B_M(x,r)}}\right] \leq \mathbb{E}^{\mathbb{Q}^{k(x)}_{\xi(x)}}\left[e^{-\zeta^{B_{\mathbb{R}^N}(\xi(x),\theta r)}}\right],$$

where $\xi(x) = \psi_{k(x)}(x)$. Since, by the estimates in Lemma 5.2.4, there is a $\delta \in (0,1)$ such that

$$\max_{1 \leq k \leq K} \sup_{\xi \in \overline{B_{\mathbb{R}^N}(0,2)}} \mathbb{Q}^k_\xi\left(\zeta^{B_{\mathbb{R}^N}(\xi,\theta r)} \leq \delta\right) \leq \delta,$$

we now see that (6.1.9) holds with $\rho \leq (1-\delta)e^{-\delta} + \delta$.

Knowing (6.1.9), we see that, for $n \geq 1$,

$$\mathbb{E}^{\mathbb{P}^n_x}\left[e^{-\zeta_n}\right] = \int e^{-\zeta_{n-1}(\omega)} \mathbb{E}^{\mathbb{P}^0_{\omega(\zeta_{n-1})}}\left[e^{-\zeta_0}\right] \mathbb{P}^{n-1}_x(d\omega)$$

$$\leq \rho\, \mathbb{E}^{\mathbb{P}^{n-1}_x}\left[e^{-\zeta_{n-1}}\right],$$

and therefore that

(6.1.10) $$\sup_{x \in M} \mathbb{E}^{\mathbb{P}^n_x}\left[e^{-\zeta_n}\right] \leq \rho^{n+1} \quad \text{for all } n \geq 0.$$

Armed with (6.1.10), we can now define

(6.1.11) $$\mathbb{P}_x = \lim_{n \to \infty} \mathbb{P}^n_x.$$

In fact, for any $t \geq 0$ and bounded, $\mathcal{B}_t$-measurable $F : \Omega(M) \longrightarrow \mathbb{R}$,

$$\left|\mathbb{E}^{\mathbb{P}^n_x}[F] - \mathbb{E}^{\mathbb{P}^m_x}[F]\right| \leq 2\|F\|_u \mathbb{P}^m_x(\zeta_m \leq t) \leq 2e^t\|F\|_u \rho^{m+1} \quad \text{if } 0 \leq m \leq n.$$

Furthermore, since $\mathbb{P}_x \upharpoonright \mathcal{B}_{\zeta_n} = \mathbb{P}^n_x \upharpoonright \mathcal{B}_{\zeta_n}$ for all n, it is easy to check that $\mathbb{P}_x$ solves the martingale problem for L starting at x. Finally, observe that, by induction on $n \geq 0$, $x \in M \longmapsto \mathbb{P}^n_x \in \mathbf{M}_1(\Omega(M))$ is measurable and therefore, by the preceding, so is $x \in M \longmapsto \mathbb{P}_x \in \mathbf{M}_1(\Omega(M))$.

6.1.3. Uniqueness of Solutions: There are two ingredients in the proof that, for each $x \in M$, the $\mathbb{P}_x$ just constructed is the only solution to the martingale problem for L starting from each x. The first of these is the following local uniqueness statement.

LEMMA 6.1.12. *Suppose that $x \in W_k$ and that $\mathbb{P}$ is a solution to the martingale problem for L starting from x. Then (cf. (6.1.7))*

$$\mathbb{P} \upharpoonright \mathcal{B}_{\zeta^{W_k}} = \left(\mathbb{Q}^k_{\psi_k(x)} \circ \Phi_k^{-1}\right) \upharpoonright \mathcal{B}_{\zeta^{W_k}}.$$

PROOF: Define $\Psi_k : \Omega(M) \longrightarrow \Omega(\mathbb{R}^N)$ so that

$$[\Psi_k(\omega)](t) = \begin{cases} \psi_k\left(\omega(t \wedge \zeta^{W_k})\right) & \text{if } \omega(0) \in W_k \\ 0 & \text{if } \omega(0) \notin W_k. \end{cases}$$

Then $\omega(0) \in W_k \implies \omega(t) = \Phi_k \circ \Psi_k(t)$ for $t \in [0, \zeta^{W_k})$. Hence, the desired conclusion is equivalent to saying that $\mathbb{Q} \equiv \mathbb{P} \circ \Psi_k^{-1}$ equals $\mathbb{Q}_{\psi_k(x)}^k$ on $\mathcal{B}_{\zeta^{B_{\mathbb{R}^N}(0,3)}}$. But, by (6.1.6), for any $\varphi \in C_c^2(\mathbb{R}^N; \mathbb{C})$,

$$[L^k \varphi] \circ \psi_k\big(\omega(\tau)\big) = [L(\varphi \circ \psi_k)]\big(\omega(\tau)\big) \quad \text{for } 0 \le \tau < \zeta^{W_k}(\omega),$$

and therefore

$$\left(\varphi\big(\eta(t \wedge \zeta^{B_{\mathbb{R}^N}(0,3)})\big) - \varphi\big(\psi_k(x)\big) - \int_0^{t \wedge \zeta^{B_{\mathbb{R}^N}(0,3)}} L^k \varphi\big(\eta(\tau)\big)\, d\tau, \mathcal{B}_t, \mathbb{Q} \right)$$

is a mean-zero martingale, which, by Lemma 6.1.3, implies that

$$\left(\varphi\big(\eta(t)\big) - \varphi\big(\psi_k(x)\big) - \int_0^t L^k \varphi\big(\eta(\tau)\big)\, d\tau, \mathcal{B}_t, \mathbb{Q} \underset{\zeta^{B_{\mathbb{R}^N}(0,3)}}{\otimes} \mathbb{Q}. \right)$$

is a mean-zero martingale. Thus, because $\mathbf{Q}_{\psi_k(x)}^k$ is the only solution to the martingale problem for L^k starting from $\psi_k(x)$, we have now shown that $\mathbb{Q} \underset{\zeta^{B_{\mathbb{R}^N}(0,3)}}{\otimes} \mathbb{Q}. = \mathbf{Q}_{\psi_k(x)}^k$ and therefore that $\mathbb{Q} \upharpoonright \mathcal{B}_{\zeta^{B_{\mathbb{R}^N}(0,3)}} = \mathbb{Q}_{\psi_k(x)}^k \upharpoonright \mathcal{B}_{\zeta^{B_{\mathbb{R}^N}(0,3)}}$. $\square$

The second ingredient is a general fact about what happens to martingales under conditioning. First, recall (cf. Theorem 5.1.15 in [53]) that if $\mathbb{P}$ is a probability measure on a Polish space Ω and $\mathcal{F}$ is a countably generated sub-σ-algebra of $\mathcal{B}_\Omega$, then there exists an $\mathcal{F}$-measurable map $\omega \in \Omega \longmapsto \mathbb{P}_\omega \in \mathbf{M}_1(\Omega)$ and a $\Lambda \in \mathcal{F}$ such that $\mathbb{P}(\Lambda) = 0$, $\mathbb{P}_\omega(A) = 1_A(\omega)$ for all $\omega \notin \Lambda$ and $A \in \mathcal{F}$, and

$$\mathbb{P}(A \cap B) = \int_A \mathbb{P}_\omega(B)\, \mathbb{P}(d\omega) \quad \text{for all } A \in \mathcal{F} \text{ and } B \in \mathcal{B}_\Omega.$$

That is, for each $B \in \mathcal{B}_\Omega$, $\omega \rightsquigarrow \mathbb{P}_\omega(B)$ is a conditional probability of B given $\mathcal{F}$, and for this reason $\omega \rightsquigarrow \mathbb{P}_\omega$ is called a *regular conditional distribution* of $\mathbb{P}$ given $\mathcal{F}$. Moreover, it is a simple matter to check that $\omega \rightsquigarrow \mathbb{P}_\omega$ is uniquely determined up to an $\mathcal{F}$-measurable set of $\mathbb{P}$-measure 0. Second, we need to know that when Ω is a pathspace like $\Omega(M)$ and ζ is a stopping time, then $\mathcal{B}_\zeta$ is countably generated.[4]

In order to show how to pass from the local uniqueness result in Lemma 6.1.12 to a global uniqueness statement, we will use the following fact about the relationship between solutions to the martingale problem and regular conditional distributions.

---

[4] To see this, one identifies (cf. Lemma 1.3.3 in [52]) as $\sigma(\{\omega(t \wedge \zeta) : t \ge 0\})$, which is possible only because we have insisted that $\{\zeta \le t\} \in \mathcal{B}_t$, as opposed to $\{\zeta < t\} \in \mathcal{B}_t$, for all $t \ge 0$.

LEMMA 6.1.13. *Suppose that $\mathbb{P}$ solves the martingale problem for L starting from x, let ζ be a stopping time, and choose $\omega \leadsto \mathbb{P}_\omega$ to be a regular conditional probability distribution of $\mathbb{P}$ given $\mathcal{B}_\zeta$. Then there is a $\Lambda \in \mathcal{B}_\zeta$ such that $\mathbb{P}(\Lambda) = 0$ and, for each $\omega \notin \Lambda$ with $\zeta(\omega) < \infty$, $\mathbb{P}_\omega \circ \Sigma_{\zeta(\omega)}^{-1}$ solves the martingale problem for L starting from $\omega(\zeta)$.*

PROOF: Suppose that $\big(\Xi(t), \mathcal{B}_t, \mathbb{P}\big)$ is a continuous martingale with the properties that $\Xi(0) = 0$ and $\sup_{(t,\omega) \in [0,T] \times \Omega} |\Xi(t, \omega)| < \infty$ for every $T \in (0, \infty)$. We want to show that, for $\mathbb{P}$-almost every $\omega \in \{\zeta < \infty\}$, $\mathbb{P}_\omega(A) = \mathbf{1}_A(\omega)$ for all $A \in \mathcal{B}_\zeta$ and

$$\big(\Xi(t) - \Xi(t \wedge \zeta(\omega)), \mathcal{B}_t, \mathbb{P}_\omega\big) \text{ is a martingale,}$$

and, since $\mathcal{B}_\zeta$ is countably generated, the first of these causes no problem. Turning to the second, let $A \in \mathcal{B}_\zeta$ be given, and set $A_T = A \cap \{\zeta \leq T\}$ for $T \in (t_2, \infty)$. Given $0 \leq t_1 < t_2$ and $B \in \mathcal{B}_{t_1}$,

$$
\begin{aligned}
\int_{A_T} & \mathbb{E}^{\mathbb{P}_\omega}\big[\Xi(t_2) - \Xi\big(t_2 \wedge \zeta(\omega)\big), \, B\big] \, \mathbb{P}(d\omega) = \mathbb{E}^{\mathbb{P}}\big[\Xi(t_2) - \Xi(t_2 \wedge \zeta), \, A_T \cap B\big] \\
& = \mathbb{E}^{\mathbb{P}}\big[\Xi(t_2) - \Xi(t_2 \wedge \zeta), \, A_T \cap B \cap \{\zeta \leq t_1\}\big] \\
& \quad + \mathbb{E}^{\mathbb{P}}\big[\Xi(t_2) - \Xi(t_2 \wedge \zeta), \, A_T \cap B \cap \{\zeta > t_1\}\big] \\
& = \mathbb{E}^{\mathbb{P}}\big[\Xi(t_1) - \Xi(t_1 \wedge \zeta), \, A_T \cap B \cap \{\zeta \leq t_1\}\big] \\
& \quad + \mathbb{E}^{\mathbb{P}}\big[\Xi(t_2 \vee \zeta) - \Xi(\zeta), \, A_T \cap B \cap \{\zeta > t_1\}\big] \\
& = \mathbb{E}^{\mathbb{P}}\big[\Xi(t_1) - \Xi(t_1 \wedge \zeta), \, A_T \cap B\big] \\
& = \int_{A_T} \mathbb{E}^{\mathbb{P}_\omega}\big[\Xi(t_1) - \Xi\big(t_1 \wedge \zeta(\omega)\big), \, B\big] \, \mathbb{P}(d\omega),
\end{aligned}
$$

since $A_T \cap B \cap \{\zeta \leq t_1\} \in \mathcal{B}_{t_1}$ and $A_T \cap B \cap \{\zeta > t_1\} \in \mathcal{B}_\zeta$. Because this is true for all $A \in \mathcal{B}_\zeta$ and $T > t_2$, it follows that, for $\mathbb{P}$-almost every $\omega \in \{\zeta < \infty\}$,

$$(*) \quad \mathbb{E}^{\mathbb{P}_\omega}\big[\Xi(t_2) - \Xi\big(t_2 \wedge \zeta(\omega)\big), \, B\big] = \mathbb{E}^{\mathbb{P}_\omega}\big[\Xi(t_1) - \Xi\big(t_1 \wedge \zeta(\omega)\big), \, B\big]$$

for each $t_1 < t_2$ and $B \in \mathcal{B}_{t_1}$. Hence, since $\mathcal{B}_{t_1}$ is countably generated and $t \leadsto \Xi(t)$ is continuous, this means that we can find one $\mathcal{B}_\zeta$-measurable Λ set of $\mathbb{P}$-measure 0 so that $(*)$ holds for all $\omega \notin \Lambda$ with $\zeta(\omega) < \infty$, $t_1 < t_2$, and $B \in \mathcal{B}_{t_1}$, which is tantamount to the asserted martingale property.

To complete the proof, choose a dense sequence $\{\varphi_\ell : \ell \geq 1\}$ in $C^2(M; \mathbb{R})$, and, for each $\ell \geq 1$, set

$$\Xi_\ell(t, \omega) = \varphi_\ell\big(\omega(t)\big) - \varphi_\ell\big(\omega(0)\big) - \int_0^t L\varphi_\ell\big(\omega(\tau)\big) \, d\tau.$$

Next, for each $\ell \geq 1$, construct Λ_ℓ accordingly for $(\Xi_\ell(t), \mathcal{B}_t, \mathbb{P})$, and take $\Lambda = \bigcup_{\ell=1}^\infty \Lambda_\ell$. Then, for each $\ell \geq 1$ and $\omega \notin \Lambda$ with $\zeta(\omega) < \infty$,

$$\left(\varphi_\ell\big(\omega'(t + \zeta(\omega))\big) - \varphi_\ell\big(\omega(\zeta)\big) - \int_0^t L\varphi_\ell\big(\omega'(\tau + \zeta(\omega))\big)\, d\tau, \mathcal{B}_{\zeta(\omega)+t}, \mathbb{P}_\omega \right)$$

is a martingale with mean value 0. Because $\{\varphi_\ell : \ell \geq 1\}$ is dense in $C^2(M; \mathbb{R})$, it follows that whenever $\omega \notin \Lambda$ and $\zeta(\omega) < \infty$, $\mathbb{P}_\omega \circ \Sigma_{\zeta(\omega)}^{-1}$ solves the martingale problem for L starting from $\omega(\zeta)$. $\quad\square$

THEOREM 6.1.14. *For each $x \in M$, the $\mathbb{P}_x$ in (6.1.11) is the one and only solution to the martingale problem for L starting from x. Moreover, $x \in M \longmapsto \mathbb{P}_x \in \mathbf{M}_1\big(\Omega(M)\big)$ is measurable and $\{\mathbb{P}_x : x \in M\}$ is a Markov family. In fact, for each stopping time ζ, $\omega \leadsto \delta_\omega \underset{\zeta(\omega)}{\otimes} \mathbb{P}.$ is a regular conditional distribution of $\mathbb{P}_x$ given $\mathcal{B}_\zeta$.*

PROOF: We already know that $x \leadsto \mathbb{P}_x$ is measurable and that $\mathbb{P}_x$ is a solution for each $x \in M$. To prove the uniqueness assertion, let $\mathbb{P}$ be a solution starting from x. We need to show (cf. the notation in § 6.1.2) that $\mathbb{P} \upharpoonright \mathcal{B}_{\zeta_n} = \mathbb{P}_x^n \upharpoonright \mathcal{B}_{\zeta_n}$ for all $n \geq 0$. By Lemma 6.1.12, there is nothing to do when $n = 0$. To prove it in general, assume that $\mathbb{P} \upharpoonright \mathcal{B}_{\zeta_{n-1}} = \mathbb{P}_x^{n-1} \upharpoonright \mathcal{B}_{\zeta_{n-1}}$ for some $n \geq 1$, and let $\omega \leadsto \mathbb{P}_\omega$ be a regular conditional distribution of $\mathbb{P}$ given $\mathcal{B}_{\zeta_{n-1}}$. Then, by Lemma 6.1.13, for $\mathbb{P}$-almost every ω with $\zeta_{n-1}(\omega) < \infty$, $\mathbb{P}_\omega \circ \Sigma_{\zeta_{n-1}(\omega)}^{-1}$ is a solution to the martingale problem for L starting from $\omega(\zeta_{n-1})$, and so, Lemma 6.1.12, $\mathbb{P}_\omega \circ \Sigma_{\zeta_{n-1}(\omega)}^{-1} \upharpoonright \mathcal{B}_{\zeta_0} = \mathbb{P}_{\omega(\zeta_{n-1})}^0 \upharpoonright \mathcal{B}_{\zeta_0}$. But, by the inductive hypothesis, this means that, for $B \in \mathcal{B}_{\zeta_n}$,

$$\mathbb{P}(B) = \mathbb{P} \underset{\zeta_{n-1}}{\otimes} \mathbb{P}^0_\cdot(B) = \mathbb{P}_x^{n-1} \underset{\zeta_{n-1}}{\otimes} \mathbb{P}^0_\cdot(B) = \mathbb{P}_x^n(B).$$

Having proved uniqueness, we can prove the final assertion by applying Lemma 6.1.3 to see that $\mathbb{P}_x \underset{\zeta}{\otimes} \mathbb{P}.$ is a solution starting at x, which therefore must be equal to $\mathbb{P}_x$. $\quad\square$

6.2 The Transition Probability Function on M

Referring to § 6.1, define $(t, x) \in [0, \infty) \times M \longmapsto P(t, x) \in \mathbf{M}_1(M)$ so that $P(t, x)$ is the distribution of $\omega \leadsto \omega(t)$ under $\mathbb{P}_x$. Clearly $(t, x) \leadsto P(t, x)$ is measurable. In addition, by the Markov property,

$$P(s + t, x, \Gamma) = \mathbb{E}^{\mathbb{P}_x}\big[P\big(t, \omega(s), \Gamma\big) \big] = \int P(t, y, \Gamma)\, P(s, x, dy),$$

and so $(t, x) \leadsto P(t, x)$ satisfies the Chapman–Kolmogorov equation. Finally,

$$\langle \varphi, P(t, x) \rangle - \varphi(x) = \mathbb{E}^{\mathbb{P}_x}\big[\varphi\big(\omega(t)\big) - \varphi(x) \big]$$
$$= \mathbb{E}^{\mathbb{P}_x}\left[\int_0^t L\varphi\big(\omega(\tau)\big)\, d\tau \right] = \int_0^t \langle L\varphi, P(\tau, x) \rangle\, d\tau,$$

and so

$$\langle \varphi, P(t,x) \rangle = \varphi(x) + \int_0^t \langle L\varphi, P(\tau, x) \rangle \, d\tau$$

for all $\varphi \in C^2(M; \mathbb{C})$. Hence, $(t, x) \rightsquigarrow P(t, x)$ is a measurable transition probability function which, for each $x \in M$, satisfies Kolmogorov's forward equation for L. In fact, it is the only such transition probability function. One way to see this is to suppose that $(t, x) \rightsquigarrow P'(t, x)$ is a second and, following Kolmogorov, construct the Markov family $\{\mathbb{P}'_x : x \in M\}$ for which P' is the transition probability function. Assuming that $\mathbb{P}'_x \in \mathbf{M}_1(\Omega(M))$, it is clear that $\mathbb{P}'_x$ solves the martingale problem for L starting from x and is therefore equal to $\mathbb{P}_x$, which, in turn, means that $P'(t, x) = P(t, x)$. Using any one of a number of standard criteria, checking that the $\mathbb{P}'_x$'s must live on $\Omega(M)$ is not hard. However, rather than go into the details, we will defer the proof of uniqueness until the end of § 6.3.2 (cf. Corollary 6.3.4 below), when we will have enough information to give a more elementary proof.

6.2.1. Local Representation of $P(t,x)$: In this subsection we will develop a formula on which our entire analysis of $P(t, x)$ rests.

For each $1 \leq k \leq K$, define the stopping times $\{\alpha_{n,k} : n \geq 0\}$ and $\{\beta_{n,k} : n \geq 0\}$ so that

$$\alpha_{0,k}(\omega) = \inf\{t \geq 0 : \omega(t) \in \bar{V}_k\}, \quad \beta_{n,k}(\omega) = \inf\{t \geq \alpha_{n,k}(\omega) : \omega(t) \notin W_k\}$$
$$\text{and } \alpha_{n+1,k}(\omega) = \inf\{t \geq \beta_{n,k} : \omega(t) \in \bar{V}_k\},$$

with the understanding that $\beta_{n,k}(\omega) = \infty$ if $\alpha_{n,k}(\omega) = \infty$ and $\alpha_{n+1,k}(\omega) = \infty$ if $\beta_{n,k}(\omega) = \infty$. Then, because

$$\alpha_{n+1,k}(\omega) \geq \beta_{n,k}(\omega) = \alpha_{n,k}(\omega) + \zeta^{W_k}\left(\Sigma_{\alpha_{n,k}(\omega)}\omega\right) \quad \text{if } \alpha_{n,k}(\omega) < \infty,$$

the Markov property can be used to show that $\mathbb{E}^{\mathbb{P}_x}\left[e^{-\alpha_{n+1,k}}\right]$ is dominated by

$$\int_{\{\alpha_{n,k}(\omega) < \infty\}} e^{-\alpha_{n,k}(\omega)} \mathbb{E}^{\mathbb{P}_{\omega(\alpha_{n,k})}}\left[e^{-\zeta^{W_k}}\right] \mathbb{P}_x(d\omega) \leq \rho \mathbb{E}^{\mathbb{P}_x}\left[e^{-\alpha_{n,k}}\right],$$

where $\rho \in (0,1)$ is the one in (6.1.9). Hence,

$$(6.2.1) \qquad \max_{1 \leq k \leq K} \sup_{x \in M} \mathbb{P}_x(\alpha_{n,k} \leq t) \leq e^t \rho^n.$$

Now set $P^{W_k}(t, x, \Gamma) = \mathbb{P}_x\left(\omega(t) \in \Gamma \ \& \ \zeta^{W_k} > t\right)$. Then, because, by (6.2.1), $\alpha_{n,k} \nearrow \infty$ $\mathbb{P}_x$-almost surely, for any $\Gamma \in \mathcal{B}_{V_k}$,

$$P(t, x, \Gamma) = \sum_{n=0}^{\infty} \mathbb{P}_x\left(\omega(t) \in \Gamma \ \& \ \alpha_{n,k}(\omega) \leq t < \beta_{n,k}(\omega)\right),$$

and, by the Markov property in the final part of Theorem 6.1.14,

$$\mathbb{P}_x\big(\omega(t) \in \Gamma \ \& \ \alpha_{n,k}(\omega) \leq t < \beta_{n,k}(\omega)\big)$$
$$= \mathbb{E}^{\mathbb{P}_x}\big[P^{W_k}(t - \alpha_{n,k}(\omega), \omega(\alpha_{n,k}), \Gamma), \ \alpha_{n,k}(\omega) < t\big].$$

Thus, for $\Gamma \in \mathcal{B}_{V_k}$, we have that

$$(6.2.2) \quad P(t, x, \Gamma) = \sum_{n=0}^{\infty} \mathbb{E}^{\mathbb{P}_x}\big[P^{W_k}(t - \alpha_{n,k}(\omega), \omega(\alpha_{n,k}), \Gamma), \ \alpha_{n,k}(\omega) < t\big].$$

6.2.2. The Transition Probability Density: We will now use (6.2.2) to transfer the results in Chapters 3 and 4 from $\mathbb{R}^N$ to M. For this purpose, first use Lemma 6.1.12 to see that

$$P^{W_k}(t, x, \Gamma) = \int_{\psi_k(\Gamma)} (q^k)^{B_{\mathbb{R}^N}(0,3)}\big(t, \psi_k(x), \xi'\big) \, d\eta' \ \text{ for } x \in W_k \text{ and } \Gamma \in \mathcal{B}_{W_k},$$

where (cf. (5.2.5)) $(q^k)^{B_{\mathbb{R}^N}(0,3)}(t, \xi, \cdot)$ is the density of the transition function $(Q^k)^{B_{\mathbb{R}^N}(0,3)}(t, \xi, \Gamma) = \mathbb{Q}_\xi^k(\eta(t) \in \Gamma, \ \zeta^{B_{\mathbb{R}^N}(0,3)} > t)$. Now let λ_M denote the *Riemannian volume measure* on M. That is, if $\Gamma \in \mathcal{B}_{W_k}$, then

$$\lambda_M(\Gamma) = \int_{\psi_k(\Gamma)} v_k(\xi) \, d\xi \quad \text{where } v_k(\xi) \equiv \sqrt{\det\big(g^k(\psi_k^{-1}(\xi))\big)}$$
$$\text{and } g^k\big(\psi_k^{-1}(\xi)\big) \equiv \left(\Big(\big((\psi_k)_*\partial_{\xi_i}, (\psi_k)_*\partial_{\xi_j}\big)_{T_{\psi_k^{-1}(\xi)}}\Big)\right)_{1 \leq i,j \leq N}$$

is the Riemannian metric computed in the coordinates determined by ψ_k. Then the preceding can be rewritten as the statement that, for $x \in W_k$ and $\Gamma \in \mathcal{B}_{W_k}$,

$$P^{W_k}(t, x, \Gamma) = \int_\Gamma p^{W_k}(t, x, y) \, \lambda_M(dy)$$

$(6.2.3)$

$$\text{where } p^{W_k}(t, x, y) \equiv v_k\big(\psi_k(y)\big)^{-1} (q^k)^{B_{\mathbb{R}^N}(0,3)}\big(t, \psi_k(x), \psi_k(y)\big).$$

The next step is to combine (6.2.2) with (6.2.3). For this purpose, recall that $d_M(x, y)$ denotes the Riemannian distance between points $x, y \in M$. Thus, because, because (cf. Theorem 3.3.11 or Theorem 4.4.6)

$$(q^k)^{B_{\mathbb{R}^N}(0,3)}(t, \xi, \xi') \leq \frac{A}{t^{\frac{N}{2}}} \exp\left(At - \frac{|\xi' - \xi|^2}{At}\right)$$

for some $A \in (0, \infty)$, we can use (6.1.5) to find another $A \in (0, \infty)$ for which

$$p^{W_k}(t, x, y) \leq \frac{A}{t^{\frac{N}{2}}} \exp\left(At - \frac{d_M(x, y)^2}{At}\right)$$

whenever $1 \leq k \leq K$ and $(t, x, y) \in (0, \infty) \times W_k \times W_k$. In particular, because $\alpha_{n,k}(\omega) \in (0, \infty) \implies \omega(\alpha_{n,k}) \in \partial V_k$, this means that, for each $T \in (0, \infty)$,

$$\max_{1 \leq k \leq K} \sup_{\substack{n \geq 0 \\ \omega \in \Omega(M) \\ (t,y) \in [0,T] \times \bar{U}_k}} \mathbf{1}_{(0,t)} \big(\alpha_{n,k}(\omega) \big) p^{W_k} \big(t - \alpha_{n,k}(\omega), \omega(\alpha_{n,k}), y \big) < \infty,$$

and so, by (6.2.1), for each $x \in M$, the series

$$(6.2.4) \quad p(t, x, y) \equiv \sum_{n=0}^{\infty} \mathbb{E}^{\mathbb{P}_x} \big[p^{W_k} \big(t - \alpha_{n,k}(\omega), \omega(\alpha_{n,k}), y \big), \, \alpha_{n,k} < t \big], \; y \in \overline{U_k}$$

converges uniformly on compacts to a continuous function in $(t, y) \in (0, \infty) \times \overline{U_k}$. Of course, one might worry that because y can be in more than one $\bar{U}_k$ and therefore that the definition of $p(t, x, y)$ depends on which k is chosen. However, no matter which k with $\overline{U_k} \ni y$ is chosen, when $p(t, x, y)$ is given by (6.2.4), (6.2.2) and (6.2.3) say that,

$$P(t, x, \Gamma) = \int_{\Gamma} p(t, x, y) \, \lambda_M(dy) \quad \text{for } (t, x) \in (0, \infty) \times M \text{ and } \Gamma \in \mathcal{B}_{\overline{U_k}},$$

and so $p(t, x, y)$ is unambiguously defined by (6.2.4) and, in fact,

$$P(t, x, \Gamma) = \int_{\Gamma} p(t, x, y) \, \lambda_M(dy) \quad \text{for } (t, x) \in (0, \infty) \times M \text{ and } \Gamma \in \mathcal{B}_M.$$

That is, $p(t, x, y)$ is the transition probability density for $P(t, x)$.

6.3 Properties of the Transition Probability Density

We will show in this section how to derive smoothness properties for $p(t, x, y)$ from the results which we proved in Chapter 3. However, before getting into the details, it may be helpful to address the question of how one measures "smoothness" of functions on M. For æsthetic reasons, one might choose to measure them in terms of the Riemannian structure, in which case one could define $x \in M \longmapsto \nabla^n \varphi(x) \in (T_x M)^{\otimes n}$ so that, for $\Xi_1, \ldots, \Xi_n \in T_x M$,

$$\big(\nabla^n \varphi(x), \Xi_1 \otimes \cdots \otimes \Xi_n \big) = \frac{\partial^n}{\partial t_1 \cdots \partial t_n} \varphi \Big(\exp_x \big(t_1 \Xi_1 + \cdots t_n \Xi_n \big) \Big) \Big|_{t_1 = \cdots = t_n = 0},$$

where $\exp_x$ denotes the *exponential map* based at x. That is, $\exp_x(\Xi)$ is the position at time 1 of the geodesic starting from x at time 0 with initial velocity $\Xi \in T_x M$. One might then define

$$\|\varphi\|_{C^n(M;\mathbb{R})} = \sup_{x \in M} \left(\sum_{m=0}^{n} |\nabla^m \varphi(x)|^2_{(T_x M)^{\otimes m}} \right)^{\frac{1}{2}}.$$

However, when dealing, as we are, with compact manifolds, such an intrinsic definition is unnecessary and, for our purposes, inconvenient. Thus, instead, we will start with the finite atlas of coordinate charts $\{(U_k, \psi_k) : 1 \leq k \leq K\}$ described in § 6.1.2 and take

$$\|\varphi\|_{C^n(M;\mathbb{R})} = \sup_{x \in M} \left(\sum_{\{k:x \in U_k\}} \sum_{\|\alpha\| \leq n} \left| \partial^\alpha (\varphi \circ \psi_k^{-1}) (\psi_k(x)) \right|^2 \right)^{\frac{1}{2}}.$$

Although, one can, and should, complain that this definition depends on the choice of atlas, all choices lead to definitions which are commensurate with the intrinsic one, and so not a lot is lost by working with this definition.

6.3.1. Smoothness in the Forward Variable: Using (6.2.4), we can show that smoothness of $p(t, x, y)$ in the forward variable y reduces to the smoothness of $p^{W_k}(t, x, y)$ in y. Indeed, by the second line in (6.2.3), (6.1.5), and the estimate in Theorem 5.2.8 applied to $(q^k)^{B_{\mathbb{R}^N}(0,3)}$, for each $n \geq 0$, there exists an $A_n < \infty$

$$(6.3.1) \quad \max_{\|\beta\| \leq n} \left| \partial_\xi^\beta p^{W_k}(t, x, \psi_k^{-1}(\xi)) \right| \leq \frac{A_n}{t^{\frac{N+n}{2}}} \exp\left(A_n t - \frac{d_M(x, \psi_k^{-1}(\xi))^2}{A_n t} \right)$$

$$\text{for } 1 \leq k \leq K, \ x \in \overline{V_k} \text{ and } \xi \in \overline{B_{\mathbb{R}^N}(0, 2)}.$$

Next, by (6.2.4) $p(t, x, \psi_k^{-1}(\xi)))$ equals

$$\mathbf{1}_{W_k}(x) p^{W_k}(t, x, \psi_k^{-1}(\xi))$$
$$+ \sum_{n=1}^\infty \mathbb{E}^{\mathbb{P}_x} \left[p^{W_k}(t - \alpha_{n,k}(\omega), \omega(\alpha_{n,k}), \psi_k^{-1}(\xi)), \ \alpha_{n,k}(\omega) < t \right]$$

for $x \in M$ and $\xi \in B_{\mathbb{R}^N}(0, 1)$. Putting this together with (6.2.1) and (6.3.1), and remembering that M is compact and therefore that d_M is bounded, one concludes that, after a small adjustment in A_n,

$$\max_{\|\beta\| \leq n} \left| \partial_\xi^\beta p(t, x, \psi_k^{-1}(\xi)) \right| \leq \frac{A_n}{t^{\frac{N+n}{2}}} \exp\left(A_n t - \frac{d_M(x, \psi_k^{-1}(\xi))^2}{A_n t} \right).$$

Finally, one can remove the exponential growth in t by applying the Chapman–Kolmogorov equation. Namely, if $t > 1$, then

$$\partial_\xi^\beta p(t, x, \psi_k^{-1}(\xi)) = \int p(t - 1, x, z) \partial_\xi^\beta p(1, z, \psi_k^{-1}(\xi)) \, \lambda_M(dz),$$

and so

$$\max_{\|\beta\| \leq n} \sup_{(t,\xi) \in [1,\infty) \times B_{\mathbb{R}^N}(0,1)} \left| \partial_\xi^\beta p(t, x, \psi_k^{-1}(\xi)) \right| \leq A_n e^{A_n}.$$

Hence, after again taking into account the boundedness of d_M and adjusting A_n, one arrives at

(6.3.2)
$$\max_{\|\beta\| \leq n} \left| \partial_\xi^\beta p\big(t, x, \psi_k^{-1}(\xi)\big) \right| \leq \frac{A_n}{(1 \wedge t)^{\frac{N+n}{2}}} e^{-\frac{d_M(x, \psi_k^{-1}(\xi))^2}{A_n t}}$$
for $1 \leq k \leq K$ and $(t, x, \xi) \in (0, \infty) \times M \times B_{\mathbb{R}^N}(0, 1)$.

6.3.2. Smoothness in all Variables: In this subsection we will prove the following regularity result about $p(t, x, y)$.

THEOREM 6.3.3.   *For each $n \geq 0$ there is an $A_n < \infty$ with the property that, whenever $m + \|\alpha\| + \|\beta\| \leq n$,*

$$\left| \partial_t^m \partial_\xi^\alpha \partial_{\xi'}^\beta p\big(t, \psi_k^{-1}(\xi), \psi_{k'}^{-1}(\xi')\big) \right| \leq \frac{A_n}{(1 \wedge t)^{\frac{N+n}{2}}} e^{-\frac{d_M(\psi_k^{-1}(\xi), \psi_{k'}^{-1}(\xi'))^2}{A_n t}}$$
for $1 \leq k, k' \leq K$, $(t, \xi, \xi') \in (0, \infty) \times B_{\mathbb{R}^N}(0, 1)^2$.

In addition, $\partial_t^m p(t, x, y) = [L^m p(t, \cdot, y)](x)$ for $m \geq 1$.

PROOF: The reasoning is essentially the same as that given in the proof of Theorem 5.2.8 to prove the corresponding result there.

We begin by applying the Markov property in Theorem 6.1.14 to derive the Duhamel formula

$$p(t, x, y) = p^{W_k}(t, x, y) + \mathbb{E}^{\mathbb{P}_x}\left[p\big(t - \zeta^{W_k}(\omega), \omega(\zeta^{W_k}), y\big), \, \zeta^{W_k}(\omega) < t \right].$$

Next, given $1 \leq k \leq K$, set $p_k(t, \xi, \xi') = p\big(t, \psi_k^{-1}(\xi), \psi_k^{-1}(\xi')\big)$ and $\hat{p}_k(t, \xi, \xi')$ $= p^{W_k}\big(t, \psi_k^{-1}(\xi), \psi_k^{-1}(\xi')\big)$ for $(t, \xi, \xi') \in (0, \infty) \times \overline{B_{\mathbb{R}^N}(0, 3)}^2$. Then, by the preceding together with Lemma 6.1.12, $p_k(t, \xi, \xi')$ equals

$$\hat{p}_k(t, \xi, \xi') + \mathbb{E}^{\mathbb{Q}_\xi^k}\left[p_k\big(t - \zeta^{B_{\mathbb{R}^N}(0,3)}(\eta), \eta(\zeta^{B_{\mathbb{R}^N}(0,3)}), \xi'\big), \, \zeta^{B_{\mathbb{R}^N}(0,3)}(\eta) < t \right].$$

Just as in the derivation of (6.3.1), the estimates in Theorem 5.2.8 say that there is an $A_n < \infty$ such that

$$\left| \partial_\xi^\alpha \partial_{\xi'}^\beta \hat{p}_k(t, \xi), \xi') \right| \leq \frac{A_n}{t^{\frac{N+n}{2}}} \exp\left(A_n t - \frac{d_M(\psi_k^{-1}(\xi), \psi_k^{-1}(\xi'))^2}{A_n t} \right)$$
for $(t, \xi, \xi') \in (0, 1] \times \overline{B_{\mathbb{R}^N}(0, 2)}^2$ and $\|\alpha\| + \|\beta\| \leq n$

To handle the second term, we use (6.3.2) to justify writing

$$\partial_{\xi'}^\beta \mathbb{E}^{\mathbb{Q}_\xi^k}\left[p_k\big(t - \zeta^{B_{\mathbb{R}^N}(0,3)}(\eta), \eta(\zeta^{B_{\mathbb{R}^N}(0,3)}), \xi'\big), \, \zeta^{B_{\mathbb{R}^N}(0,3)}(\eta) < t \right]$$
$$= \mathbb{E}^{\mathbb{Q}_\xi^k}\left[\partial_{\xi'}^\beta p_k\big(t - \zeta^{B_{\mathbb{R}^N}(0,3)}(\eta), \eta(\zeta^{B_{\mathbb{R}^N}(0,3)}), \xi'\big), \, \zeta^{B_{\mathbb{R}^N}(0,3)}(\eta) < t \right]$$
.

for $(t, \xi, \xi') \in (0,1] \times \overline{B_{\mathbb{R}^N}(0,2)} \times B_{\mathbb{R}^N}(0,1)$. Thus, we can apply (5.2.9) to see that

$$\partial_\xi^\alpha \partial_{\xi'}^\beta \mathbb{E}^{\mathbb{Q}_\xi^k}\left[p_k\big(t - \zeta^{B_{\mathbb{R}^N}(0,3)}(\eta), \eta(\zeta^{B_{\mathbb{R}^N}(0,3)}, \xi'), \zeta^{B_{\mathbb{R}^N}(0,3)}(\eta) < t\right]$$

exists and that, for a slightly altered $A_n < \infty$, its absolute value is dominated by $A_n e^{-\frac{1}{A_n t}}$ times

$$\sup\{|\partial_{\xi'}^\beta p_k(\tau, \chi, \chi')| : t \in (0,1], \; \chi \in \partial B_{\mathbb{R}^N}(0,3) \text{ and } \chi' \in B_{\mathbb{R}^N}(0,1)\}$$

for $(t, \xi, \xi') \in (0,1] \times \overline{B_{\mathbb{R}^N}(0,2)} \times B_{\mathbb{R}^N}(0,1)$ and $\|\alpha\| + \|\beta\| \leq n$. After combining this with our earlier estimate on the first term and using the fact that d_M is bounded, we conclude that there is an $A_n < \infty$ such that

$$|\partial_\xi^\alpha \partial_{\xi'}^\beta p_k(t, \xi, \xi')| \leq \frac{A_n}{t^{\frac{N+n}{2}}} e^{-\frac{d_M(\psi_k^{-1}(\xi), \psi_k(\xi'))^2}{A_n t}}$$

for $(t, \xi, \xi') \in (0,1] \times \overline{B_{\mathbb{R}^N}(0,2)} \times B_{\mathbb{R}^N}(0,1)$ and $\|\alpha\| + \|\beta\| \leq n$. In particular, this proves the required estimate when $k' = k$ and $t \in (0,1]$.

Now suppose that $k' \neq k$ and $(t, \xi, \xi') \in (0,1] \times B_{\mathbb{R}^N}(0,1)^2$. If $\psi_k(\xi) \in \overline{V_{k'}}$, then we write

$$p\big(t, \psi_k^{-1}(\xi), \psi_{k'}^{-1}(\xi')\big) = p_{k'}\big(t, \psi_{k'} \circ \psi_k^{-1}(\xi), \xi'\big)$$

and apply the preceding. To handle the case when $\psi_k^{-1}(\xi) \notin \overline{V_{k'}}$, set $G = \{\chi \in B_{\mathbb{R}^N}(0,3) : \psi_k^{-1}(\chi) \notin \overline{\tilde{V}_{k'}}\}$, where $\tilde{V}_{k'} \equiv \psi_{k'}^{-1}\big(B_{\mathbb{R}^N}(0, \frac{3}{2})\big)$. Then, by (6.1.5), there is an $r > 0$ such that $|\xi - G\mathbb{C}| > r$ whenever $\xi \in B_{\mathbb{R}^N}(0,1)$ and $\psi_k^{-1}(\xi) \notin \overline{V_{k'}}$. Moreover, another application of the Markov property and Lemma 6.1.12 shows that

$$\partial_{\xi'}^\beta p\big(t, \psi_k^{-1}(\xi), \xi'\big) = \mathbb{E}^{\mathbb{Q}_\xi^k}\left[\partial_{\xi'}^\beta p\big(t - \zeta^G(\eta), \psi_k^{-1}(\eta(\zeta^G)), \psi_{k'}^{-1}(\xi')\big), \; \zeta^G(\eta) < t\right]$$

when $(t, \xi, \xi') \in (0,1] \times B_{\mathbb{R}^N}(0,1)^2$ and $\psi_k^{-1}(\xi) \notin \overline{V_{k'}}$. Thus, once again, we are in a position to apply (5.2.9) and (6.3.2) in order to complete the proof of the required estimate when $t \in (0,1]$.

To handle $t > 1$, apply the Chapman–Kolmogorov equation to write

$$p(t, x, y) = \iint p(\tfrac{1}{3}, x, z) p\big(t - \tfrac{2}{3}, z, z'\big) p(\tfrac{1}{3}, z', y) \, \lambda_M(dz) \lambda_M(dz').$$

By the estimate just proved and the one in (6.3.2), derivatives of the integrand with respect to x and y up to order n are bounded, independent of $t > 1$, and so, since $\lambda_M(M) < \infty$, it follows that x and y-derivatives of $p(t, x, y)$ up to order n are also bounded independent of $t > 1$.

Finally, knowing that $p(t, x, y)$ is smooth in x and y, the derivation of $\partial_t^m p(t, x, y) = [L^m p(t, \cdot, y)](x)$ is similar to, but easier than, the argument given at the end of the proof of Theorem 5.2.8. $\square$

Armed with the preceding regularity result, we can easily prove the uniqueness result mentioned in § 6.2.

COROLLARY 6.3.4.   *For each $x \in M$, the function $t \rightsquigarrow P(t, x)$ described in § 6.3.2 is the one and only continuous map $t \rightsquigarrow \mu(t)$ with the property that*

$$\langle \varphi, \mu(t) \rangle = \varphi(x) + \int_0^t \langle L\varphi, \mu(\tau) \rangle \, d\tau \quad \text{for all } \varphi \in C^2(M; \mathbb{C}).$$

In particular, $(t, x) \rightsquigarrow P(t, x)$ is the only transition function satisfying

$$\langle \varphi, P(t, x) \rangle = \varphi(x) + \int_0^t \langle L\varphi, P(\tau, x) \rangle \, d\tau \quad \text{for all } \varphi \in C^2(M; \mathbb{C}).$$

PROOF: The proof is identical to that of the uniqueness results in § 2.1.1. Namely, let $\varphi \in C^2(M; \mathbb{C})$ be given, and set $u_\varphi(t, x) = \langle \varphi, P(t, x) \rangle$. Then, for any $t > 0$, $\frac{d}{d\tau} \langle u_\varphi(t - \tau, \cdot), \mu(\tau) \rangle = 0$, and so

$$\langle \varphi, \mu(t) \rangle = \langle u_\varphi(t, \cdot), \mu(0) \rangle = u_\varphi(t, x) = \langle \varphi, P(t, x) \rangle. \quad \square$$

6.4 Nash Theory on a Manifold

In this section we will prove the following.

THEOREM 6.4.1.   *For each $\delta \in (0, 1]$ there exists an $K(\delta) \in [1, \infty)$ such that*

$$\frac{1}{K(\delta)(1 \wedge t)^{\frac{N}{2}}} e^{-\frac{(1+\delta) d_M(x,y)^2}{2t}} \leq p(t, x, y) \leq \frac{K(\delta)}{(1 \wedge t)^{\frac{N}{2}}} e^{-\frac{d_M(x,y)^2}{2(1+\delta)t}}.$$

In particular,

$$\lim_{t \searrow 0} t \log p(t, x, y) = -\frac{d_M(x, y)^2}{2} \quad \text{uniformly on } M \times M.$$

6.4.1. The Lower Bound: Given Theorem 5.2.14, the proof of the lower bound is quite easy.

To get started, set

$$R = \inf_{x \in M} \max \{ d_M(x, V_k \complement) : 1 \leq k \leq K \}.$$

By (6.1.5), $R \geq \theta > 0$. Now let $x \in M$ be given, and choose k so that $B_M(x, R) \subseteq V_k$. If $y \in B_M(x, R)$, then there is a minimal geodesic[5] $\Pi : [0, 1] \longrightarrow B_M(x, R)$ with $x = \Pi(0)$ and $y = \Pi(1)$. Starting from Theorem 5.2.14, it is easy to check that, for each $\delta > 0$, there is an $\alpha(\delta) > 0$ with the property that

$$(q^k)^{B_{\mathbb{R}^N}(0,3)} \big(t, \psi_k(x), \psi_k(y) \big)$$

$$\geq \frac{\alpha(\delta)}{t^{\frac{N}{2}}} \exp \left(-\frac{(1 + \delta) d^k \big(\psi_k(x), \psi_k(y) \big)^2}{2t} \right)$$

---

[5] We will always take geodesics to have speed 1. Thus, when a geodesic is minimal, $d_M(\Pi(s), \Pi(t)) = |t - s|$.

for $t \in (0,1]$, where

$$d^k(\xi, \xi') = \inf \left\{ \sqrt{\int_0^1 \left(\dot{\pi}(\tau), a^k \big(\pi(\tau) \big)^{-1} \dot{\pi}(\tau) \right)_{\mathbb{R}^N} d\tau} : \right.$$

$$\left. \pi \in C^1 \big([0,1]; B_{\mathbb{R}^N}(0,2) \big) \text{ with } \pi(0) = \xi \ \& \ \pi(1) = \xi' \right\}.$$

Hence, since $\pi \equiv \psi_k \circ \Pi : [0,1] \longrightarrow B_{\mathbb{R}^N}(0,2)$,

$$d^k \big(\psi_k(x), \psi_k(y) \big)^2 \leq \int_0^1 \left(\dot{\pi}(t), a^k \big(\pi(t) \big)^{-1} \dot{\pi}(t) \right)_{\mathbb{R}^N} dt = d_M(x,y)^2,$$

and, by (6.2.3),

$$p(t,x,y) \geq v_k(y)^{-1} (q^k)^{B_{\mathbb{R}^N}(0,3)} \big(t, \psi_k(x), \psi_k(y) \big),$$

we have now shown that, for a slightly different choice of $\alpha(\delta) > 0$,

$$(*) \quad p(t,x,y) \geq \frac{\alpha(\delta)}{t^{\frac{N}{2}}} e^{-\frac{(1+\delta)d_M(x,y)^2}{2t}} \quad \text{for } t \in (0,1] \text{ and } d_M(x,y) < R.$$

Now assume that $d_M(x,y) \geq R$, and choose $n \in \mathbb{Z}^+$ to be the smallest integer dominating $\frac{3d_M(x,y)}{R}$. Clearly, $3 \leq n \leq \frac{3 \operatorname{diam}(M)}{R}$. Next, let $\Pi : [0,1] \longrightarrow M$ be a minimal geodesic running from x to y, and set $x_m = \Pi\big(\frac{m}{n}\big)$ for $0 \leq m \leq n$. Then $d_M(x_{m-1}, x_m) = \frac{d_M(x,y)}{n} \leq \frac{R}{3}$. Given $t \in (0,1]$, set $\tau = \frac{t}{n}$, $r = n^{-1} R t^{\frac{1}{2}}$, and $B_m = B_M\big(x_m, r\big)$. Then $d_M(z_{m-1}, z_m) \leq \frac{d_M(x,y) + 2Rt^{\frac{1}{2}}}{n} < R$ for $1 \leq m \leq n$, $z_{m-1} \in B_{m-1}$, and $z_m \in B_m$. Hence, by $(*)$ and the Chapman–Kolmogorov equation, $p(t,x,y)$ dominates

$$\int_{B_1 \times \cdots \times B_{n-1}} \cdots \int p(\tau, x, z_1) p(\tau, z_1, z_2) \cdots p(\tau, z_{n-1}, y)\, \lambda_M(dz_1) \cdots \lambda_M(dz_{n-1})$$

$$\geq \left(\frac{\alpha(\delta)}{\tau^{\frac{N}{2}}} \right)^n \left(\prod_1^{n-1} \lambda_M(B_m) \right) \exp \left(-\frac{(1+\delta)\big(d_M(x,y) + 2Rt^{\frac{1}{2}}\big)^2}{2t} \right).$$

Finally, there is a $\beta \in (0, \infty)$ such that $\lambda_M\big(B(z, \rho)\big) \geq \beta \rho^N$ for any $z \in M$ and $\rho \in (0, R)$, and

$$\frac{\big(d_M(x,y) + 2Rt^{\frac{1}{2}}\big)^2}{2t} \leq \frac{(1+\delta)d_M(x,y)^2}{2t} + \frac{4R^2}{\delta}.$$

Thus, after further adjustment of $\alpha(\delta)$, one gets the lower bound in Theorem 6.4.1 for $t \in (0,1]$. Observe that, because d_M is bounded, the lower bound for $t \geq 1$ is tantamount to saying that $p(t,x,y)$ is bounded below by a positive constant uniformly for $t \geq 1$. But, we already know that $p(1,x,y) \geq \alpha(1) e^{-\frac{\operatorname{diam}(M)^2}{\alpha(1)}}$, and therefore, by the Chapman–Kolmogorov equation, the same is true for all $t \geq 1$.

6.4.2. The Upper Bound: Because the upper bound is less amenable to localization, proving it requires us to repeat the argument in § 4.4.1, only now on M.

The first step is to find a replacement for Lemma 4.1.3. For this purpose, let $h(t, x, y)$ be the transition probability density corresponding to $\frac{1}{2}\Delta$, and define $\mathbf{H}_t\varphi(x) = \int h(t, x, y)\varphi(y)\,\lambda_M(dy)$. By the divergence theorem,

$$\left(\varphi, \Delta\psi\right)_{L^2(\lambda_M;\mathbb{R})} = -\left(\nabla\varphi, \nabla\psi\right)_{L^2(\lambda_M;TM)},$$

where

$$\left(X, Y\right)_{L^2(\lambda_M;TM)} \equiv \int_M \left(X(x), Y(x)\right)_{T_x M} \lambda_M(dx)$$

for vector fields X and Y on M. In particular, this means that Δ is formally self-adjoint on $L^2(\lambda_M;\mathbb{R})$ and therefore that

$$\frac{d}{d\tau}\left(\mathbf{H}_{t-\tau}\varphi, \mathbf{H}_\tau\psi\right)_{L^2(\lambda_M;\mathbb{R})} = 0 \quad \text{for } \tau \in [0, t].$$

Thus, $\mathbf{H}_t$ is symmetric in $L^2(\lambda_M;\mathbb{R})$, which is equivalent to saying that $h(t, x, y) = h(t, y, x)$ and, just as in the derivation of (4.1.2), means that $\mathbf{H}_t$ is a self-adjoint contraction on $L^2(\lambda_M;\mathbb{R})$. In addition, one has that

$$\frac{d}{dt}\left\|\nabla\mathbf{H}_t\varphi\right\|^2_{L^2(\mu;TM)} = -\frac{d}{dt}\left(\mathbf{H}_t\varphi, \Delta\mathbf{H}_t\varphi\right)_{L^2(\lambda_M;\mathbb{R})} = -\left\|\Delta\mathbf{H}_t\right\|^2_{L^2(\lambda_M;\mathbb{R})} \leq 0.$$

Finally, observe that, from (6.3.2) with $n = 0$, we know that $\|\mathbf{H}_t\|_{1\to\infty} \leq C_M(1 \wedge t)^{-\frac{N}{2}}$ for some $C_M < \infty$.

With these preliminaries, it might seem that we have everything we need to have $h(t, x, y)$ play the role that the Gauss kernel played in our second proof of Lemma 4.1.3. However, there is one crucial difference: Our estimate for $\|\mathbf{H}_t\|_{1\to\infty}$ is a short time estimate and gives no information about long time behavior. For this reason, we will resort to the following subterfuge. At the cost of a slight alteration in the constant C_M, we replace $(1 \wedge t)^{-\frac{N}{2}}$ on the right hand side by $t^{-\frac{N}{2}}e^t$. Now set $\varphi_t = e^{-t}\mathbf{H}_t\varphi$, use

$$\varphi_t = \varphi - \int_0^t \left(I - \tfrac{1}{2}\Delta\right)\varphi_\tau \, d\tau,$$

and conclude that

$$\|\varphi\|^2_{L^2(\lambda_M;\mathbb{R})} = \left(\varphi, \varphi_t\right)_{L^2(\lambda_M;\mathbb{R})}$$
$$+ \int_0^t \left(\left(\varphi, \varphi_\tau\right)_{L^2(\lambda_M;\mathbb{R})} + \tfrac{1}{2}\left(\nabla\varphi, \nabla\varphi_\tau\right)_{L^2(\lambda_M;TM)}\right) d\tau.$$

Because

$$\left|\left(\varphi, \varphi_t\right)_{L^2(\lambda_M;\mathbb{R})}\right| \leq \|\varphi\|_{L^1(\lambda_M;\mathbb{R})}e^{-t}\|\mathbf{H}_t\varphi\|_{L^\infty(\lambda_M;\mathbb{R})} \leq C_M t^{-\frac{N}{2}}\|\varphi\|^2_{L^1(\lambda_M;\mathbb{R})},$$

$$\left(\varphi, \varphi_\tau\right)_{L^2(\lambda_M;\mathbb{R})} \leq \|\varphi\|_{L^2(\lambda_M;\mathbb{R})} e^{-\tau} \|\mathbf{H}_\tau \varphi\|_{L^2(\lambda_M;\mathbb{R})} \leq \|\varphi\|^2_{L^2(\lambda_M;\mathbb{R})},$$

and

$$\left(\nabla\varphi, \nabla\varphi_\tau\right)_{L^2(\lambda_M;TM)} \leq \|\nabla\varphi\|_{L^2(\lambda_M;TM)} e^{-\tau} \|\nabla\mathbf{H}_\tau \varphi\|_{L^2(\lambda_M;TM)}$$
$$\leq \|\nabla\varphi\|^2_{L^2(\lambda_M;TM)},$$

we obtain

$$\|\varphi\|^2_{L^2(\lambda_M;\mathbb{R})} \leq C_M t^{-\frac{N}{2}} \|\varphi\|^2_{L^1(\lambda_M;\mathbb{R})} + t\left(\|\varphi\|^2_{L^2(\lambda_M;\mathbb{R})} + \tfrac{1}{2}\|\nabla\varphi\|^2_{L^2(\lambda_M;TM)}\right)$$

for all $t > 0$. Thus, after minimizing with respect to t and altering C_M, we arrive at

$$(6.4.2) \quad \|\varphi\|^{2+\frac{4}{N}}_{L^2(\lambda_M;\mathbb{R})} \leq C_M \left(\|\varphi\|^2_{L^2(\lambda_M;\mathbb{R})} + \tfrac{1}{2}\|\nabla\varphi\|^2_{L^2(\lambda_M;TM)}\right)\|\varphi\|^{\frac{4}{N}}_{L^1(\lambda_M;\mathbb{R})}.$$

Given (6.4.2), we have the engine which powers the reasoning in § 4.4.1. In particular, we can now prove the following.

LEMMA 6.4.3.   Define $D(\psi) = \|\nabla\psi\|_u$ for $\psi \in C^\infty(M;\mathbb{R})$ and

$$D_M(x,y) = \sup\{\psi(y) - \psi(x) : \psi \in C^\infty(M;\mathbb{R}) \text{ with } D(\psi) \leq 1\}.$$

Then there is a $K_M < \infty$ with the property that, for each $\delta \in (0,1]$,

$$p(t,x,y) \leq \frac{K_N e^t}{(\delta t)^{\frac{N}{2}}} \exp\left(\frac{6\|B\|^2_u t}{\delta} - \frac{D_M(x,y)^2}{2(1+5\delta)t}\right),$$

where $\|B\|_u = \sup_{x \in M} |B(x)|_{T_x M}$.

PROOF: Assume that $u \in C^{1,2}\left([0,\infty) \times M;(0,\infty)\right)$ is a solution to an equation of the form

$$\partial_t u(t,x) = \tfrac{1}{2}\Delta u + Bu - \operatorname{div}(u\check{B}) + cu,$$

where B and $\check{B}$ are smooth vector fields on M and c is a smooth function on M. Given $r \geq 1$, set $v_r(t) = \|u(t)\|_{L^r(\lambda_M;\mathbb{R})}$. Then, proceeding in exactly the same way as we did at the beginning of § 4.4.1, we find that

$$v^{2r-1}_{2r}\dot{v}_{2r} \leq -\frac{r-1}{2r^2}\|\nabla u^r\|^2_{L^2(\lambda_M;TM)} + r\left(c_r, u^{2r}\right)_{L^2(\lambda_M;\mathbb{R})},$$

where

$$c_r(x) = \frac{c(x)}{r} + \left|B_r(x)\right|^2_{T_x M} \quad \text{with } B_r = \frac{B}{r} + \left(2 - \tfrac{1}{r}\right)\check{B}.$$

In particular, this means that

(*) $$v_2(t) \le e^{t\|c_1\|_\mathrm{u}} \|\varphi\|_\mathrm{u}.$$

When $r \ge 2$, we add and subtract $\frac{r-1}{r^2}\|u^r\|^2_{L^2(M;\mathbb{R})}$ to the right hand side and apply (6.4.2) to thereby obtain

$$\dot v_{2r} \le -\frac{r-1}{2C_M r^2} \frac{v_{2r}^{1+\frac{4r}{N}}}{v_r^{\frac{4r}{N}}} + r\big(r^{-2} + \|c_r\|_\mathrm{u}\big)v_{2r}.$$

Thus, by repeating the argument given in the proof of Lemma 4.1.7, we arrive at

(**)
$$w_{2r}(t) \le \left(\frac{2NC_M r^2}{\delta}\right)^{\frac{N}{4r}} \exp\left(\frac{t\delta}{r}\big(\|c_r\|_\mathrm{u} + r^{-2}\big)\right) w_r(t)$$
$$\text{when } w_r(t) \equiv \sup_{\tau \in [0,t]} \tau^{\frac{N}{4}(1-\frac{2}{r})} v_r(\tau)$$

for each $\delta \in (0,1]$.

Now let $\psi \in C^\infty(M;\mathbb{R})$ be given, and, for $\varphi \in C^\infty\big(M;[0,\infty)\big) \setminus \{0\}$, set

$$u_\varphi^\psi(t,x) = \mathbf{Q}_t^\psi \varphi(x) \equiv e^{-\psi(x)} \int p(t,x,y) e^{\psi(y)} \varphi(y)\, \lambda_M(dy).$$

Then

$$\partial_t u_\varphi^\psi = \tfrac{1}{2}\Delta u_\varphi^\psi + B^\psi u_\varphi^\psi - \mathrm{div}\big(u^\psi \check B^\psi\big) + c^\psi u_\varphi^\psi,$$

where

$$B^\psi = B + \tfrac{1}{2}\nabla\psi, \quad \check B^\psi = -\tfrac{1}{2}\nabla\psi, \text{ and } c^\psi = B\psi + \tfrac{1}{2}|\nabla\psi|^2_{TM}.$$

Combining (*) with (**) and arguing as we did at the corresponding point in §4.4.1, we conclude that there is a $K_M < \infty$ such that

$$\|\mathbf{Q}_t^\psi\|_{2\to\infty} \le \frac{K_M e^{\frac{t(1+6\|B\|_\mathrm{u})}{\delta}}}{(\delta t)^{\frac{N}{4}}} e^{\frac{t(1+5\delta)D_M(\psi)^2}{2}}.$$

Similarly, if

$$\check u_\varphi^\psi(y) = (\mathbf{Q}_t^\psi)^\top \varphi(y) = e^{-\psi(y)} \int p(t,y,x) e^{\psi(x)}\, \lambda_M(dx),$$

then

$$\partial_t \check u_\varphi^\psi = \tfrac{1}{2}\Delta \check u_\varphi^\psi - \check B^\psi \check u_\varphi^\psi + \mathrm{div}\big(B^\psi \check u_\varphi^\psi\big) + c^\psi \check u_\varphi^\psi,$$

and so we can conclude that $\|\mathbf{Q}_t^\psi\|_{1\to 2}$ satisfies an estimate of the same sort as $\|\mathbf{Q}_t^\psi\|_{2\to\infty}$. Finally, after combining these two, using $\mathbf{Q}_t = \mathbf{Q}_{\frac{t}{2}} \circ \mathbf{Q}_{\frac{t}{2}}$, and making a small adjustment in K_M, we get the asserted conclusion. $\quad\square$

What remains to be proved is that $D_M(x,y) = d_M(x,y)$, and, just as in §4.1.2, the proof that $D_M(x,y) \leq d_M(x,y)$ is trivial. However, the proof of the opposite inequality requires some thought. Namely, one would like to prove that $d_M(x,y) \leq D_M(x,y)$ by taking $\psi(y) = d_M(x,y)$. However, as was the case earlier, this choice of ψ is not smooth and therefore ineligible. The problem therefore is to show again that one can approximate $y \rightsquigarrow d_M(x,y)$ by a sequence $\{\psi_n : n \geq 1\}$ of smooth functions in such a way that $\psi_n \longrightarrow d_M(x, \cdot)$ uniformly and $\overline{\lim}_{n\to\infty} \|\nabla\psi_n\|_u \leq 1$.

LEMMA 6.4.4.   *Given $\psi \in C(M;\mathbb{R})$ satisfying $|\psi(y) - \psi(x)| \leq d_M(x,y)$ for all $x,y \in M$, there exists a sequence $\{\psi_n : n \geq 1\} \subseteq C^\infty(M;\mathbb{R})$ with the properties that $\psi_n \longrightarrow f$ uniformly on M and $\overline{\lim}_{n\to\infty} \|\nabla\psi_n\|_u \leq 1$.*

PROOF: The most direct (see Remark 6.4.5 below for another approach) proof of this assertion relies on basic facts about Riemannian geometry, the first of which is that there exists an $R > 0$, the injectivity radius (cf. §2 in Chapter V of [9]), with the property that $(x,y) \rightsquigarrow d_M(x,y)^2$ is a smooth function as long as $d_M(x,y) < R$. Now choose $\rho \in C_c^\infty\big((-R^2, R^2); [0,\infty)\big)$ so that $\omega_{N-1} \int \rho(\tau^2)\tau^{N-1}\, d\tau = 1$, where ω_{N-1} denotes the area of the unit sphere $\mathbb{S}^{N-1}$ in $\mathbb{R}^N$, and, for $\epsilon \in (0,1]$, set

$$\psi_\epsilon(x) = \int \rho_\epsilon\big(d_M(x,y)^2\big)\psi(y)\,\lambda_M(dy) \quad \text{where } \rho_\epsilon(\tau) \equiv \epsilon^{-N}\rho(\epsilon^{-2}\tau).$$

Clearly, $\psi_\epsilon \in C^\infty(M;\mathbb{R})$ for each $\epsilon \in (0,1]$.

In order to show that $\psi_\epsilon \longrightarrow f$ uniformly and that $\overline{\lim}_{\epsilon\searrow 0} \|\psi_\epsilon\|_u \leq 1$, for each $x \in M$ choose an orthogonal transformation $E(x) : \mathbb{R}^N \longrightarrow T_xM$. Then there is a function $w : M \times B_{\mathbb{R}^N}(0, R) \longrightarrow (0,\infty)$ with the property that

$$\psi_\epsilon(x) = \int \rho_\epsilon(|\xi|^2)\psi\big(\exp_x(E(x)\xi)\big)w(x,\xi)\,d\xi,$$

where $\exp_x : T_xM \longrightarrow M$ is the exponential map based at x. Indeed, $\xi \in B_{\mathbb{R}^N}(0, R) \longmapsto \exp_x(E(x)\xi) \in M$ is a diffeomorphism onto $B_M(x, R)$, and $w(x, \cdot)$ is the Jacobian associated with this choice of coordinates. In particular, because (cf. §8 in Chapter XII of [9]) these coordinates are normal at x, $|w(x,\xi) - 1| \leq C|\xi|^2$ for $|\xi| \leq \frac{R}{2}$, where the choice of finite C can be made independent of $x \in M$. Therefore

$$|\psi_\epsilon(x) - \psi(x)| \leq \int \rho_\epsilon(|\xi|^2)\big|\psi\big(\exp_x(E(x)\xi) - \psi(x)\big|\,d\xi$$

$$+ \|\psi\|_u \int \rho_\epsilon(|\xi|^2)|w(x,\xi) - 1|\,d\xi$$

$$\leq \epsilon R \int \rho(|\xi|^2)|\xi|\,d\xi + C\|\psi\|_u \epsilon^2 \int \rho(|\xi|^2)|\xi|^2\,d\xi.$$

Finally, to control the gradient of ψ_ϵ, let x be given. Then, because $B_M(x, R)$ admits a coordinate system, one can make a choice of $y \rightsquigarrow E(y)$ which is smooth on $B_M(x, R)$, in which case $w(y, \xi)$ is a smooth function of $(y, \xi) \in B_M(x, R) \times B_{\mathbb{R}^N}(0, R)$ and there exists a $C < \infty$ such that

$$d_M \left(\exp_{y_1} \big(E(y_1)\xi \big), \exp_{y_2} \big(E(y_2)\xi \big) \right) \leq (1 + C|\xi|) d_M(y_1, y_2)$$

and $|\nabla_y w(y, \xi)|_{T_y M} \leq C|\xi|^2$ for $\xi \in B_{\mathbb{R}^N}\left(0, \frac{R}{2}\right)$ and $y, y_1, y_2 \in B_M\left(x, \frac{R}{2}\right)$. Hence, if $y \in B_M\left(x, \frac{R}{2}\right)$ and $\Pi : [0, \infty) \longrightarrow M$ is a geodesic with $\Pi(0) = y$, then for sufficiently small $t > 0$ and $\epsilon \in \left(0, \frac{1}{2}\right]$,

$$\left| \psi_\epsilon \big(\Pi(t) \big) - \psi_\epsilon(y) \right| \leq (1 + C'\epsilon)t,$$

which means that $|\nabla \psi_\epsilon(y)|_{T_y M} \leq (1 + C'\epsilon)$. $\square$

Given the result in Lemma 6.4.4, one sees that $D_M(x, y) \geq \psi(y) - \psi(x)$ for any ψ with Lipschitz constant less than or equal to 1 relative to $d_M(x, y)$. In particular, one can take $\psi(y) = d_M(x, y)$ and thereby conclude that $D_M(x, y) \geq d_M(x, y)$.

Finally, by combining this with the result in Lemma 6.4.3, we get the upper bound in Theorem 6.4.1 for $t \in (0, 1]$, and to get it for $t > 1$ one can use the Chapman–Kolmogorov equation in the same way as we have done previously. Thus, Theorem 6.4.1 has now been proved.

REMARK 6.4.5. Another way in which one can prove Lemma 6.4.4 is to take $\psi_t = \mathbf{H}_t \psi$ and use the regularity results which we know about $h(t, x, y)$. Indeed, we know that, for each $t > 0$, ψ_t is smooth and that $\psi_t \longrightarrow \psi$ uniformly on M as $t \searrow 0$. Thus, the only question is whether $\overline{\lim}_{t \searrow 0} \|\nabla \psi_t\|_u \leq 1$.

There are many ways to prove this. Perhaps the most elegant is to use a coupling argument based on Bochner's identity. Namely (cf. Theorem 10.37 in [54]), if κ is a lower bound for the Ricci curvature on M, then, for each $x_1, x_2 \in M$, one can construct a coupling $\mathbb{P}_{x,y}$ on $C\big([0, \infty), M \times M\big)$ with the properties that the $\mathbb{P}_{x_1,x_2}$-distribution of $\omega \rightsquigarrow \omega_i$ is $\mathbb{P}_{x_i}$ for $i \in \{1, 2\}$ and

$$d_M\big(\omega_1(t), \omega_2(t)\big) \leq e^{-\kappa t} d_M(x_1, x_2) \quad \mathbb{P}_{x_1,x_2}\text{-almost surely.}$$

Hence,

$$\left| \mathbf{H}_t \psi(x_2) - \mathbf{H}_t \psi(x_1) \right| \leq e^{-\kappa t} |\psi(x_2) - \psi(x_1)| \leq e^{-\kappa t} d_M(x_1, x_2),$$

from which it is an easy step to $|\nabla \mathbf{H}_t \psi(x)|_{T_x M} \leq e^{-\kappa t}$.

An alternative, more pedestrian, approach to the same conclusion is to work in a coordinate patch and apply the results, particularly Lemma 2.2.5, in Chapter 2.

REMARK 6.4.6. It should be apparent that, given a Riemannian structure on M, we could have applied the reasoning in this section to any operator of the form $L = \frac{1}{2}\mathrm{div}(a\nabla) + aB$, where $x \in M \longmapsto a(x) \in (T_xM)^* \times (T_xM)^*$ is symmetric and uniformly positive definite. If we had, we would have obtained results analogous to those which we proved in Chapter 4. In particular, the constants appearing in those results would have enjoyed the same independence from the regularity of a and B as the constants did there.

6.5 Long Time Behavior

The result in Theorem 6.4.1 does not give much information when t is large, it merely says that $p(t, x, y)$ is bounded above and below by positive constants uniformly on $[1, \infty) \times M \times M$. In this section we will show that there is an $f \in C^\infty(M; (0, \infty))$ and a $\lambda > 0$ with the property that

$$(6.5.1) \qquad \sup_{(x,y)\in M^2} \|p(t, x, \,\cdot\,) - f(y)\|_{\mathrm{u}} \leq e^{-\lambda t} \quad \text{for } t \in [1, \infty).$$

Moreover, f is the unique solution to

$$(6.5.2) \qquad \tfrac{1}{2}\Delta f - \mathrm{div}(fB) = 0 \text{ with } \int f(y)\,\lambda_M(dy) = 1.$$

6.5.1. Doeblin's Theorem: The basic result which underlies (6.5.1) goes back to Doeblin.

In the following statement, $Q(t, x, \,\cdot\,)$ is a transition probability function on M and $\{\mathbf{Q}_t : t \geq 0\}$ is the associated semigroup of operators: $\mathbf{Q}_t\varphi(x) = \int \varphi(y)Q(t, x, dy)$ if φ is a bounded measurable function on M and $\rho\mathbf{Q}_t(dy) = \int Q(t, x, dy)\,\rho(dx)$ if ρ is a finite, signed measure on M. Clearly, $\rho\mathbf{Q}_t(M) = \rho(M)$, $\|\mathbf{Q}_t\varphi\|_{\mathrm{u}} \leq \|\varphi\|_{\mathrm{u}}$ and $\|\rho\mathbf{Q}_t\|_{\mathrm{var}} \leq \|\rho\|_{\mathrm{var}}$, where $\|\cdot\|_{\mathrm{var}}$ is the total variation norm.

LEMMA 6.5.3. *Suppose that* $Q(t, x, \,\cdot\,)$ *is a transition probability function on* M *with the property that* $Q(1, x, dy) \geq (1 - \beta)\nu(dy)$ *for some* $\beta \in (0, 1)$ *and probability measure* ν. *Then there is a unique probability measure* μ *such that*

$$\|Q(t, x, \,\cdot\,) - \mu\|_{\mathrm{var}} \leq 2\beta^{[t]}, \quad t \geq 0,$$

where $[t]$ *is the integer part of* t. *In particular,* μ *is* $\{\mathbf{Q}_t : t \geq 0\}$-*invariant in the sense that* $\mu = \mu\mathbf{Q}_t$ *for all* $t \geq 0$, *and so* $\mu \geq (1 - \beta)\nu$.

PROOF: Let ρ be a finite, signed measure with $\rho(M) = 0$, and let $|\rho|$ be the associated variation measure. That is, if $\rho = \rho_+ - \rho_-$, where ρ_+ and ρ_- are the non-negative, mutually singular measures, then $|\rho| = \rho_+ + \rho_-$. Then, for any bounded, measurable $\varphi : M \longrightarrow \mathbb{R}$,

$$\langle \varphi, \rho\mathbf{Q}_1 \rangle = \int \langle \varphi, Q(1, x) \rangle\,\rho(dx) = \int \langle \varphi, Q(1, x) - (1 - \beta)\nu \rangle\,\rho(dx),$$

and so

$$\left|\langle \varphi, \rho \mathbf{Q}_1 \rangle\right| \le \int \langle |\varphi|, Q(1,x) - (1-\beta)\nu \rangle \, |\rho|(dx) \le \beta \|\varphi\|_{\mathrm{u}} \|\rho\|_{\mathrm{var}},$$

since $Q(1,x) - (1-\beta)\nu \ge 0$ and $Q(1,x,M) - (1-\beta)\nu(M) = \beta$. Hence, we have shown that

(*) $\qquad\qquad \|\rho \mathbf{Q}_1\|_{\mathrm{var}} \le \beta \|\rho\|_{\mathrm{var}} \quad \text{if } \rho(M) = 0.$

From (*), it is clear that, for any $t > 0$,

$$\|\rho \mathbf{Q}_t\|_{\mathrm{var}} = \|(\rho \mathbf{Q}_{t-[t]}) \mathbf{Q}_1^{[t]}\|_{\mathrm{var}} \le \beta^{[t]} \|\rho \mathbf{Q}_{t-[t]}\|_{\mathrm{var}} \le \beta^{[t]} \|\rho\|_{\mathrm{var}}$$

if $\rho(M) = 0$. In particular, if $s, t \ge 0$, $x \in M$, and $\rho = Q(t,x) - \delta_x$, then $\|Q(s+t,x) - Q(s,x)\|_{\mathrm{var}} = \|\rho \mathbf{Q}_s\|_{\mathrm{var}} \le 2\beta^{[s]}$. Thus, by the Cauchy criterion for convergence in the variation metric, there is a probability measure μ to which $Q(t,x)$ converges in variation as $t \to \infty$. In particular, for any $s \ge 0$,

$$\|\mu \mathbf{Q}_s - \mu\|_{\mathrm{var}} = \lim_{t\to\infty} \|\mu \mathbf{Q}_s - Q(s+t,x)\|_{\mathrm{var}} = \lim_{t\to\infty} \|(\mu - Q(t,x))\mathbf{Q}_s\|_{\mathrm{var}}$$
$$\le \lim_{t\to\infty} \|\mu - Q(t,x)\|_{\mathrm{var}} = 0,$$

and so $\mu = \mu \mathbf{Q}_s$. Hence, for any $t > 0$ and $y \in M$,

$$\|\mu - Q(t,y)\|_{\mathrm{var}} = \|\mu \mathbf{Q}_t - Q(t,y)\|_{\mathrm{var}} = \|(\mu - \delta_y)\mathbf{Q}_t\|_{\mathrm{var}} \le 2\beta^{[t]}.$$

Finally, if μ' is a second probability measure satisfying $\mu' = \mu' \mathbf{Q}_t$ for all $t \ge 0$, then $\|\mu - \mu'\|_{\mathrm{var}} = \|(\mu - \mu')\mathbf{Q}_t\|_{\mathrm{var}} \le 2\beta^{[t]}$ for all $t \ge 0$, and so $\mu = \mu'$. $\square$

6.5.2. Ergodic Property: As an essentially immediate consequence of Lemma 6.5.3 and the lower bound in Theorem 6.4.1, we get the following result about the long time behavior of $p(t,x,y)$.

THEOREM 6.5.4.   *There is a unique $f \in C(M;\mathbb{R})$ such that $\langle f, \lambda_M \rangle = 1$ and*

$$f(y) = \int p(t,x,y) f(x) \, \lambda_M(dx) \quad \text{for all } (t,y) \in (0,\infty) \times M.$$

Moreover,

$$\sup\{|p(1+t,x,y) - f(y)| : (x,y) \in M^2\} \le A\beta^{[t]},$$

where

$$\beta = 1 - \lambda_M(M)^{-1} \inf\{p(1,x,y) : (x,y) \in M^2\} < 1$$

and

$$A = 2\sup\{p(1,x,y) : (x,y) \in M^2\} < \infty.$$

In particular, $f > 0$. Finally, f is the unique solution $h \in C^\infty(M;\mathbb{R})$ to $L^\top h = 0$ with $\langle h, \lambda_M \rangle = 1$.

PROOF: By Theorem 6.4.1, $\beta < 1$.

Set $\overline{\lambda_M} = \frac{\lambda_M}{V}$, where $V \equiv \lambda_M(M)$ is the volume of M. Then $P(1, x, \cdot) \geq (1 - \beta)\overline{\lambda_M}$ and so, by Lemma 6.5.3, there exists a μ such that $\mu = \mu \mathbf{P}_t$ for all $t \geq 0$. In fact, for all $x \in M$, $\|P(t, x, \cdot) - \mu\|_{\mathrm{var}} \leq 2\beta^{[t]}$.

Now set $f(y) = \int p(1, x, y)\, \mu(dx)$. Clearly, $\langle f, \lambda_M \rangle = 1$, and, by Theorem 6.3.3, $f \in C^\infty\big(M; (0, \infty)\big)$. In addition, for any $\varphi \in C(M; \mathbb{R})$,

$$\big(\varphi, f\big)_{L^2(\lambda_M;\mathbb{R})} = \langle \mathbf{P}_1 \varphi, \mu \rangle = \langle \varphi, \mu \mathbf{P}_1 \rangle = \langle \varphi, \mu \rangle,$$

and so $\mu(dy) = f(y)\, \lambda_M(dy)$. Hence, $f = \int p(t, x, \cdot) f(x)\, \lambda_M(dx)$ follows immediately from $\mu = \mu \mathbf{P}_t$. To check the asserted convergence result, simply note that

$$\big| p(1 + t, x, y) - f(y) \big| = \left| \int p(1, z, y)\big(p(t, x, z) - f(z)\big)\, \lambda_M(dz) \right|$$

$$\leq \frac{A}{2} \|p(t, x, \cdot) - f\|_{L^1(\lambda_M;\mathbb{R})} = \frac{A}{2} \|P(t, x) - \mu\|_{\mathrm{var}} \leq A\beta^{[t]}.$$

Finally, for any $\varphi \in C^\infty(M; \mathbb{R})$,

$$\big(\varphi, L^\top f\big)_{L^2(M;\mathbb{R})} = \big(L\varphi, f\big)_{L^2(M;\mathbb{R})} = \langle L\varphi, \mu \rangle = \frac{d}{dt} \langle \mathbf{P}_t \varphi, \mu \rangle \Big|_{t=0}$$

$$= \frac{d}{dt} \langle \varphi, \mu \mathbf{P}_t \rangle \Big|_{t=0} = 0,$$

and so $L^\top f = 0$. Conversely, if $h \in C^\infty(M; \mathbb{R})$ satisfies $L^\top h = 0$, then

$$\frac{d}{dt} \int h(x)p(t, x, y)\, \lambda_M(dx) = \int h(x)[Lp(t, \cdot, y)](x)\, \lambda_M(dx)$$

$$= \int L^\top h(x)p(t, x, y)\, \lambda_M(dx) = 0.$$

Since $h(y) = \lim_{t \searrow 0} \int h(x)p(t, x, y)\, \lambda_M(dx)$, when $\langle h, \lambda_M \rangle = 1$ this proves that

$$h(y) = \lim_{t \to \infty} \int h(x)p(t, x, y)\, \lambda_M(dx) = \int h(x)f(y)\, \lambda_M(dx) = f(y). \quad \square$$

6.5.3. A Poincaré Inequality: By applying Theorem 6.5.4 in the case when $L = \frac{1}{2}\Delta$, one gets the *Poincaré inequality*

$$(6.5.5) \qquad 2\alpha \big\| \varphi - \langle \varphi, \overline{\lambda_M} \rangle \big\|^2_{L^2(\lambda_M;\mathbb{R})} \leq \|\nabla\varphi\|^2_{L^2(\lambda_M;TM)},$$

where again $\overline{\lambda_M} = \frac{\lambda_M}{V}$ with $V = \lambda_M(M)$, and $\alpha = -\log \beta$, β being the quantity in Theorem 6.5.4 when $L = \frac{1}{2}\Delta$.

To see how this is done, first observe that, because $h(t, x, y) = h(t, y, x)$, $\langle \mathbf{H}_t\varphi, \overline{\lambda_M}\rangle = \langle \varphi, \overline{\lambda_M}\rangle$. Hence, $\overline{\lambda_M}$ is the unique $\mu \in \mathbf{M}_1(M)$ satisfying $\mu = \mu\mathbf{H}_t$. Equivalently, the f in Theorem 6.5.4 is the constant function $\frac{1}{V}$, and so

$$\sup_{(x,y)\in M^2} \left| h(1 + t, x, y) - V^{-1} \right| \le e^\alpha A e^{-\alpha t}.$$

In particular, if $(\varphi, \mathbf{1})_{L^2(\lambda_M;\mathbb{R})} = 0$, then

$$\|\mathbf{H}_{1+t}\varphi\|_{L^2(\lambda_M;\mathbb{R})} \le e^\alpha A V e^{-\alpha t} \|\varphi\|_{L^2(\lambda_M;\mathbb{R})}.$$

Equivalently, if $\tilde{\mathbf{H}}_t$ is defined by $\tilde{\mathbf{H}}_t\varphi = \mathbf{H}_t\varphi - \langle \varphi, \overline{\lambda_M}\rangle$, then

(*) $$\|\tilde{\mathbf{H}}_t\|_{\mathrm{op}} \le e^{2\alpha} A V e^{-\alpha t} \quad \text{for } t \ge 1.$$

In order to get (6.5.5) from the above, we must first show that the estimate in (*) can be improved to the estimate[6]

(**) $$\|\tilde{\mathbf{H}}_t\|_{\mathrm{op}} \le e^{-\alpha t} \quad \text{for } t > 0.$$

To this end, check that $\tilde{\mathbf{H}}_t$ is self-adjoint and satisfies $\tilde{\mathbf{H}}_{s+t} = \tilde{\mathbf{H}}_s \circ \tilde{\mathbf{H}}_t$, and conclude that

$$\|\tilde{\mathbf{H}}_{2t}\|_{\mathrm{op}} = \sup\left\{ (\varphi, \tilde{\mathbf{H}}_{2t}\varphi)_{L^2(\lambda_M;\mathbb{R})} : \|\varphi\|_{L^2(\lambda_M;\mathbb{R})} \le 1 \right\}$$
$$= \sup\left\{ (\|\tilde{\mathbf{H}}_t\varphi\|_{L^2(\lambda_M;\mathbb{R})}^2 : \|\varphi\|_{L^2(\lambda_M;\mathbb{R})}^2 \le 1 \right\} = \|\tilde{\mathbf{H}}_t\|_{\mathrm{op}}^2,$$

since

$$\|T\|_{\mathrm{op}} = \sup\left\{ (\varphi, T\varphi)_{L^2(\lambda_M;\mathbb{R})} : \|\varphi\|_{L^2(\lambda_M;\mathbb{R})} \le 1 \right\}$$

for a non-negative definite, self-adjoint operator T on $L^2(\lambda_M;\mathbb{R})$. Hence, for any $n \ge 1$, $\|\tilde{\mathbf{H}}_{2^n t}\|_{\mathrm{op}} = \|\tilde{\mathbf{H}}_t\|_{\mathrm{op}}^{2^n}$, and so, by (*), $\|\tilde{\mathbf{H}}_t\|_{\mathrm{op}} \le \left(e^{2\alpha} A V \right)^{2^{-n}} e^{-\alpha t}$ for $n \ge 1$ such that $2^n t \ge 1$. Clearly (**) follows from this.

Given (**), the rest is easy. Namely, if $\varphi \in C^\infty(M;\mathbb{R})$ with $\langle \varphi, \lambda_M\rangle = 0$, then $\mathbf{H}_t\varphi = \tilde{\mathbf{H}}_t\varphi$ and therefore

$$\int_0^t \left\| \nabla \mathbf{H}_\tau\varphi \right\|_{L^2(\lambda_m;TM)}^2 d\tau = -\int_0^t \left(\mathbf{H}_\tau\varphi, \Delta\mathbf{H}_\tau\varphi \right)_{L^2(\lambda_M;\mathbb{R})} d\tau$$
$$= \|\varphi\|_{L^2(\lambda_M;\mathbb{R})}^2 - \|\mathbf{H}_t\varphi\|_{L^2(\lambda_M;\mathbb{R})}^2 \ge \left(1 - e^{-2\alpha t} \right) \|\varphi\|_{L^2(\lambda_M;\mathbb{R})}^2.$$

Hence, after dividing through by t and letting $t \searrow 0$, we get $\|\nabla\varphi\|_{L^2(\lambda_M;TM)}^2 \ge 2\alpha\|\varphi\|_{L^2(\lambda_M;\mathbb{R})}^2$.

---

[6] For those who know the Spectral Theorem, this conclusion is more or less obvious.

6.6 Historical Notes and Commentary

There are many ways to study parabolic and elliptic equations on a manifold. Most geometric analysts prefer an approach in which Sobolev theory plays the central role, although S.T. Yau and R. Hamilton have made spectacular contributions based on the minimum principle.

The first systematic treatment of diffusions on manifolds was given by K. Itô [25], using stochastic differential equations. Although it works, Itô's approach suffers from a basic problem: The square root of the diffusion matrix is not intrinsically defined, and so, each time that the diffusion enters a new coordinate chart he had to make a new choice of square root. Itô masterfully handled this difficulty by taking advantage of the rotation invariance of Brownian motion, but the result lacks æsthetic appeal and is less than compelling from a geometric standpoint. Following, and, in a sense, reversing a key observation by J. Eells and D. Elworthy [13], P. Malliavin [37] showed that life becomes infinitely simpler if one carries out the construction on the orthogonal frame bundle instead of the manifold itself. Loosely speaking, on the orthogonal frame bundle, there is a canonical way to take the square root. For further details, see [54].

Because it characterizes the process directly in terms of the operator L, which is intrinsically defined on the manifold, the martingale problem formulation provides a simple and natural way to patch local results together on a manifold. The first person to take advantage of this observation was R. Azencott [5], who used essentially the same line of reasoning as the one given in § 6.1. The techniques used in §§ 6.2 and 6.3 are more or less obvious once one knows those in Chapter 5. As for § 6.4, the upper bound in § 6.4.2 is due to E.B. Davies [10], but I do not know another place in which the lower bound in § 6.4.1 appears. Nonetheless, the last part of Theorem 6.4.1 can and has been proved elsewhere by large deviations methods.

CHAPTER 7

Subelliptic Estimates
and Hörmander's Theorem

Up until now I have assiduously avoided the use of many of the modern analytic techniques which have become essential tools for experts working in partial differential equations. In particular, nearly all my reasoning has been based on the minimum principle, and I have made no use so far of either Sobolev spaces or the theory of pseudodifferential operators. In this concluding chapter, I will attempt to correct this omission by first giving a brief review of the basic theories of Sobolev spaces and pseudodifferential operators and then applying them to derive significant extensions of the sort of hypoellipticity results proved in § 3.4.

Because the approach taken in this chapter is such a dramatic departure from what has come before, it may be helpful to explain the origins of the analysis which follows and of the goals toward which it is directed. For this purpose, consider the Laplace operator Δ for $\mathbb{R}^N$, and ask yourself what you can say about u on the basis of information about Δu. In particular, what can you say about $\partial_i \partial_j u$? One of the most bedeviling facts with which one has to contend is that (except in one dimension) u need not be twice continuous differentiable just because $\Delta u \in C(\mathbb{R}^N; \mathbb{C})$ in the sense of Schwartz distributions. On the other hand, if one replaces continuity by integrability, this uncomfortable fact disappears. To be precise, if $u \in L^2(\mathbb{R}^N; \mathbb{C})$ and, in the sense of distributions, $\Delta u \in L^2(\mathbb{R}^N; \mathbb{C})$, then all second order derivatives of u are also in $L^2(\mathbb{R}^N; \mathbb{C})$ and, in fact, simple Fourier analysis shows that $\|\partial_i \partial_j u\|_{L^2(\mathbb{R}^N;\mathbb{C})} \le \|\Delta u\|_{L^2(\mathbb{R}^N;\mathbb{C})}$. Thus, as this trivial example indicates, there is good reason to believe that one should use integral, as opposed to pointwise, estimates to measure smoothness. Although this observation goes back to the nineteenth century, it came of age in the first part of the twentieth century with the work of H. Weyl and, in a more profound and systematic form, of S. Sobolev.

Weyl and Sobolev's ideas were developed further and sharpened in the second half of the twentieth century by L. Schwartz, A.P. Calderón, L. Nirenberg, L. Hörmander, C. Fefferman, and a host of others. In applications to the sort of equations with which we have been dealing, an important goal of this line of research was the extension to other operators of the sort of inequality just discussed for the Laplacian. To understand how such estimates, which became known as *elliptic a priori estimates*, relate to the sort of results in § 3.4, think about using the one above for the Laplacian to prove that if $u \in L^2(\mathbb{R}^N; \mathbb{C})$ and $\Delta u = f$, where f and all its derivatives are in $L^2(\mathbb{R}^N; \mathbb{C})$, then $u \in C^\infty(\mathbb{R}^N; \mathbb{C})$. There are two ingredients in the proof. The first is the trivial observations that, for any $m \geq 1$, $\Delta^m u = \Delta^{m-1} f \in L^2(\mathbb{R}^N; \mathbb{C})$. The second ingredient is the rather trivial form of the Sobolev Embedding Theorem, which says that if $\Delta^m u \in L^2(\mathbb{R}^N; \mathbb{C})$, then $u \in C^\infty(\mathbb{R}^N; \mathbb{C})$. Indeed,

$$(2\pi)^N \|\partial^\alpha u\|_{\mathrm{u}} \leq \|\widehat{\partial^\alpha u}\|_{L^1(\mathbb{R}^N;\mathbb{C})} = \int |\xi^\alpha| |\hat{u}(\xi)| \, d\xi$$

$$\leq \left(\int \frac{|\xi|^{2\|\alpha\|}}{1 + |\xi|^{2(\|\alpha\|+N)}} \, d\xi \right)^{\frac{1}{2}} \left(\int \left(1 + |\xi|^{2(\|\alpha\|+N)} \right) |\hat{u}(\xi)|^2 \, d\xi \right)^{\frac{1}{2}}$$

$$\leq C(\alpha, N) \left(\|u\|_{L^2(\mathbb{R}^N;\mathbb{C})} + \|\Delta^{\|\alpha\|+N} u\|_{L^2(\mathbb{R}^N;\mathbb{C})} \right).$$

An examination of the preceding argument reveals two places where patches must be applied if one intends to prove the result in § 3.4.1. First, one needs to localize it so that it applies to a u about which one knows only that Δu is smooth in an open set. Secondly, one has to learn how to handle operators whose coefficients are variable. It turns out that these two are closely related and that both are really problems coming from commutation. It is to deal with these and related problems that we will introduce the class of pseudodifferential operators.

7.1 Elementary Facts about Sobolev Spaces

We will need some background material from functional analysis. Readers who are unfamiliar with these matters should consult any one of the many thorough treatments for more details. See, for example, [1].

7.1.1. Tempered Distributions: We begin by recalling a few familiar facts from Schwartz distribution theory.

Let $\mathscr{S}(\mathbb{R}^N; \mathbb{C})$ be Schwartz's test function space of smooth functions $\varphi : \mathbb{R}^N \longrightarrow \mathbb{C}$ which, together with all their derivatives, are rapidly decreasing.[1] $\mathscr{S}(\mathbb{R}^N; \mathbb{C})$ becomes a Fréchet space under the metric

(7.1.1)
$$d(\varphi, \psi) \equiv \sum_{n=0}^{\infty} \frac{1}{2^n} \frac{\|\varphi - \psi\|_{(n)}}{1 + \|\varphi - \psi\|_{(n)}}, \quad \text{where}$$

$$\|\varphi\|_{(n)} = \left(\sum_{\|\alpha\| \leq n} \int_{\mathbb{R}^N} (1 + |x|^2)^{\frac{n}{2}} |\partial^\alpha \varphi(x)|^2 \, dx \right)^{\frac{1}{2}},$$

---

[1] A function is said to be rapidly decreasing if it tends to 0 faster than $(1 + |x|^2)^{-m}$ for all $m \geq 1$.

and the associated topological dual space $\mathscr{S}'(\mathbb{R}^N; \mathbb{C})$ is the space of *tempered distributions*. We give $\mathscr{S}'(\mathbb{R}^N; \mathbb{C})$ the weak* topology. That is, the topology such that, for each $u \in \mathscr{S}'(\mathbb{R}^N; \mathbb{C})$, finite intersections of sets of the form $\{w \in \mathscr{S}'(\mathbb{R}^N; \mathbb{C}) : |\langle \varphi, w \rangle - \langle u, \varphi \rangle| < \epsilon\}$ are a neighborhood basis at u. Notice that if $f : \mathbb{R}^N \longrightarrow \mathbb{C}$ is Borel measurable and *tempered* in the sense that $(1 + |x|^2)^{-L} f$ is integrable for some $L \geq 0$, then $\varphi \rightsquigarrow \int f(x)\varphi(x) \, dx$ determines a tempered distribution, which we will again denote by f.

There are lots of operations which can be performed on $\mathscr{S}'(\mathbb{R}^N; \mathbb{C})$. In particular, if $f \in C^\infty(\mathbb{R}^N; \mathbb{C})$ and all derivatives of f are tempered, then we can define a continuous map $u \in \mathscr{S}'(\mathbb{R}^N; \mathbb{C}) \longmapsto fu \in \mathscr{S}'(\mathbb{R}^N; \mathbb{C})$ by $\langle \varphi, fu \rangle = \langle f\varphi, u \rangle$. Also, for each $\alpha \in \mathbb{N}^\alpha$, $u \in \mathscr{S}'(\mathbb{R}^N; \mathbb{C}) \longmapsto \partial^\alpha u \in \mathscr{S}'(\mathbb{R}^N; \mathbb{C})$ given by $\langle \varphi, \partial^\alpha u \rangle = (-1)^{\|\alpha\|} \langle \partial^\alpha \varphi, u \rangle$ is continuous.

A great virtue of $\mathscr{S}(\mathbb{R}^N; \mathbb{C})$ is that, as distinguished from $C_c^\infty(\mathbb{R}^N; \mathbb{C})$, the Fourier transform map $\varphi \rightsquigarrow \hat{\varphi}$ is a homeomorphism of $\mathscr{S}(\mathbb{R}^N; \mathbb{C})$ onto itself. Hence, for any $u \in \mathscr{S}'(\mathbb{R}^N; \mathbb{C})$, we can define its Fourier transform $\hat{u} \in \mathscr{S}'(\mathbb{R}^N; \mathbb{C})$ by $\hat{u}(\varphi) = \langle \hat{\varphi}, u \rangle$, in which case $u \rightsquigarrow \hat{u}$ is continuous. Similarly, if $\check{\varphi}(\xi) = \hat{\varphi}(-\xi)$ for $\varphi \in \mathscr{S}(\mathbb{R}^N; \mathbb{C})$, then the map $u \in \mathscr{S}'(\mathbb{R}^N; \mathbb{C}) \longmapsto \check{u} \in \mathscr{S}'(\mathbb{R}^N; \mathbb{C})$ given by $\langle \varphi, \check{u} \rangle = \langle \check{\varphi}, u \rangle$ is continuous, and the Fourier inversion formula says that $\hat{u}^\vee = (2\pi)^N u$. Once one knows how to define the Fourier transform of tempered distributions, one knows how to define the operation of convolution by an element of $\mathscr{S}(\mathbb{R}^N; \mathbb{C})$ on $\mathscr{S}'(\mathbb{R}^N; \mathbb{C})$. Namely, if $\rho \in \mathscr{S}(\mathbb{R}^N; \mathbb{C})$, then $\rho \star u$ is defined to be the element of $\mathscr{S}'(\mathbb{R}^N; \mathbb{C})$ whose Fourier transform is $\hat{\rho}\hat{u}$.

It should be clear that if u is a tempered distribution which is given by a function f for which any one of these operations is classically defined, then the operation takes u into the tempered distribution determined by the function obtained from f by acting on it with the classical operation. That is, the definitions of these operations on $\mathscr{S}'(\mathbb{R}^N; \mathbb{C})$ extend their classical antecedents.

7.1.2. The Sobolev Spaces: Given $s \in \mathbb{R}$, define $\|\varphi\|_s$ for $\varphi \in \mathscr{S}(\mathbb{R}^N; \mathbb{C})$ by

$$\|\varphi\|_s = \left(\frac{1}{(2\pi)^N} \int_{\mathbb{R}^N} (1 + |\xi|^2)^s |\hat{\varphi}(\xi)|^2 \, d\xi \right)^{\frac{1}{2}}.$$

We define the *Sobolev space* $H_s(\mathbb{R}^N; \mathbb{C})$ to be the space of $u \in \mathscr{S}'(\mathbb{R}^N; \mathbb{C})$ for which

$$\|u\|_s \equiv \sup\{|\langle \varphi, u \rangle| : \|\varphi\|_{-s} \leq 1\} < \infty.$$

Observe that for any $m \in \mathbb{N}$ and $s > \frac{m+N}{2}$, elementary Fourier analysis shows that, for some $C_m < \infty$,

$$(7.1.2) \quad H_s(\mathbb{R}^N; \mathbb{C}) \subseteq C_b^m(\mathbb{R}^N; \mathbb{C}) \text{ and } \|u\|_u^{(m)} \equiv \sum_{\|\alpha\| \leq m} \|\partial^\alpha u\|_u \leq C_m \|u\|_s.$$

It should be evident that each $H_s(\mathbb{R}^N; \mathbb{C})$ is closed in $\mathscr{S}'(\mathbb{R}^N; \mathbb{C})$, the spaces $H_s(\mathbb{R}^N; \mathbb{C})$ decrease as s increases, and the space $H_\infty(\mathbb{R}^N; \mathbb{C}) \equiv$

$\bigcap_{s\in\mathbb{R}} H_s(\mathbb{R}^N;\mathbb{C})$ consists of those $\varphi \in C^\infty(\mathbb{R}^N;\mathbb{C}) \cap L^2(\mathbb{R}^N;\mathbb{C})$ all of whose derivatives are also in $L^2(\mathbb{R}^N;\mathbb{C})$. In addition, $H_0(\mathbb{R}^N;\mathbb{C})$ can be identified with $L^2(\mathbb{R}^N;\mathbb{C})$ and $\|u\|_0 = \|u\|_{L^2(\mathbb{R}^N;\mathbb{C})}$, and so we will use the space $H_0(\mathbb{R}^N;\mathbb{C})$, the norm $\|\cdot\|_0$, and the inner product $(\cdot,\cdot)_0$ interchangeably with, respectively, $L^2(\mathbb{R}^N;\mathbb{C})$, $\|\cdot\|_{L^2(\mathbb{R}^N;\mathbb{C})}$, and $(\cdot,\cdot)_{L^2(\mathbb{R}^N;\mathbb{C})}$. To better understand the space $H_s(\mathbb{R}^N;\mathbb{C})$ for general s, define the *Bessel operator* $B^s = (I-\Delta)^{-\frac{s}{2}}$ on $\mathscr{S}'(\mathbb{R}^N;\mathbb{C})$ by $\widehat{B^s u} = (1+|\xi|^2)^{-\frac{s}{2}}\hat{u}$. It is then an easy matter to check that $H_s(\mathbb{R}^N;\mathbb{C}) = \{B^s f : f \in L^2(\mathbb{R}^N;\mathbb{C})\}$ and that $\|u\|_s = \|f\|_{L^2(\mathbb{R}^N;\mathbb{C})}$ if $u = B^s f$. Hence, each $H_s(\mathbb{R}^N;\mathbb{C})$ is a separable Hilbert space with inner product $(u,v)_s \equiv (B^{-s}v, B^{-s}u)_{L^2(\mathbb{R}^N;\mathbb{C})}$. Alternatively, one can think of $H_s(\mathbb{R}^N;\mathbb{C})$ as the Hilbert space obtained by completing $\mathscr{S}(\mathbb{R}^N;\mathbb{C})$ with respect to the norm $\|\cdot\|_s$. Finally, it should be clear that $H_{-\infty}(\mathbb{R}^N;\mathbb{C}) \equiv \bigcup_{s\in\mathbb{R}} H_s(\mathbb{R}^N;\mathbb{C})$ is the space of $u \in \mathscr{S}'(\mathbb{R}^N;\mathbb{C})$ such that

$$|\langle\varphi,u\rangle| \leq C \sum_{\|\alpha\|\leq n} \|\partial^\alpha\varphi\|_{L^2(\mathbb{R}^N;\mathbb{C})}$$

for some $C < \infty$ and $n \geq 0$. In particular, any Schwartz distribution with compact support is an element of $H_{-\infty}(\mathbb{R}^N;\mathbb{C})$.

As a Hilbert space, the dual space $H_s(\mathbb{R}^N;\mathbb{C})^*$ of $H_s(\mathbb{R}^N;\mathbb{C})$ is naturally identifiable with itself. On the other hand, there is another natural identification of $H_s(\mathbb{R}^N;\mathbb{C})^*$, one which is preferable when one is thinking about $H_s(\mathbb{R}^N;\mathbb{C})$ as a subspace of $\mathscr{S}'(\mathbb{R}^N;\mathbb{C})$. Namely, given $s \in \mathbb{R}$ and $v \in H_{-s}(\mathbb{R}^N;\mathbb{C})$, $\varphi \in \mathscr{S}(\mathbb{R}^N;\mathbb{C}) \longmapsto \langle\varphi,v\rangle \in \mathbb{C}$ admits a unique continuous extension as a linear functional on $H_s(\mathbb{R}^N;\mathbb{C})$ with norm $\|v\|_{-s}$. Conversely, if $\Lambda \in H_s(\mathbb{R}^N;\mathbb{C})^*$, then $|\Lambda\varphi| \leq \|\Lambda\|_{H_s(\mathbb{R}^N;\mathbb{C})^*}\|\varphi\|_s$, and so Λ can be identified with the element u of $H_{-s}(\mathbb{R}^N;\mathbb{C})$ determined by $\langle\varphi,u\rangle = \Lambda\varphi$, in which case $\|u\|_{-s} = \|\Lambda\|_{H_s(\mathbb{R}^N;\mathbb{C})^*}$. For this reason, we are justified in writing $\langle u,v\rangle$ for the action of the linear functional determined by $v \in H_{-s}(\mathbb{R}^N;\mathbb{C})$ on $u \in H_s(\mathbb{R}^N;\mathbb{C})$. In this connection, notice that another expression for $\langle u,v\rangle$ is $(B^{-s}u, \overline{B^s v})_{L^2(\mathbb{R}^N;\mathbb{C})}$.

7.2 Pseudodifferential Operators

Pseudodifferential operators will facilitate the following sort of reasoning. Let $L = \sum_{i,j=1}^N a(x)_{ij}\partial_{x_i}\partial_{x_j}$, and assume that $a(x)$ is symmetric and uniformly positive definite. Given $f \in \mathscr{S}(\mathbb{R}^N;\mathbb{C})$, set

$$u(x) = \frac{1}{(2\pi)^N} \int_{\mathbb{R}^N} \left(1 + \sum_{i,j=1}^N a(x)_{ij}(x)\xi_i\xi_j\right)^{-1} \hat{f}(\xi)\,d\xi.$$

If a is constant, then $(I-L)u = f$. However, if a is not constant, then $(I-L)u \neq f$. Nonetheless, even when a is not constant, u ought to be an "approximate" solution to $(I-L)u = f$ and should be a good candidate as the first step in a perturbation scheme which converges to a true solution. To make the preceding "ought" and "should" mathematically meaningful, one has to determine the precise sense in which u is an approximate solution, and the machine which will allow us do so is the theory of pseudodifferential operators.

The first step toward our goal is to give a precise meaning to the order of an operator. Namely, a linear map $L : \mathscr{S}(\mathbb{R}^N; \mathbb{C}) \longrightarrow H_\infty(\mathbb{R}^N; \mathbb{C})$ is said to be of *order* $\sigma \in \mathbb{R}$ if, for each $s \in \mathbb{R}$, there exists a $C_s < \infty$ such that $\|L\varphi\|_s \leq C_s \|\varphi\|_{s+\sigma}$. Further, the *true order* of L is the infimum over those $\sigma \in \mathbb{R}$ of which L is of order σ, and we will say that L is of *order* $-\infty$ if L is of order σ for every $\sigma \in \mathbb{R}$. To check that these definitions conform to ones intuition, show that ∂^α is of true order $\|\alpha\|$ and that convolution by a function from $\mathscr{S}(\mathbb{R}^N; \mathbb{C})$ is of order $-\infty$.

Obviously, if L is an operator of order σ, then it admits a unique extension as an operator on $H_{-\infty}(\mathbb{R}^N; \mathbb{C})$ in such a way that, for each $s \in \mathbb{R}$ and with the same constant C_s, $\|Lu\|_s \leq C_s \|u\|_{s+\sigma}$ whenever $u \in H_{s+\sigma}(\mathbb{R}^N; \mathbb{C})$. Next, think of L as a densely defined operator on $L^2(\mathbb{R}^N; \mathbb{C})$, and let $L^\top$ denote its *transpose*.[2] That is, if $\psi \in L^2(\mathbb{R}^N; \mathbb{C})$ and there is a $C < \infty$ such that

$$\left| \int_{\mathbb{R}^N} \psi(x) L\varphi(x) \, dx \right| \leq C \|\varphi\|_{L^2(\mathbb{R}^N; \mathbb{C})} \quad \text{for all } \varphi \in \mathscr{S}(\mathbb{R}^N; \mathbb{C}),$$

then ψ is in the domain of $L^\top$ and $L^\top \psi$ is the unique $f \in L^2(\mathbb{R}^N; \mathbb{C})$ such that

$$\int f(x) \varphi(x) \, dx = \int \psi(x) L\varphi(x) \, dx \quad \text{for all } \varphi \in \mathscr{S}(\mathbb{R}^N; \mathbb{C}).$$

We want to show that $\mathscr{S}(\mathbb{R}^N; \mathbb{C})$ is contained in the domain of $L^\top$ and that $L^\top : \mathscr{S}(\mathbb{R}^N; \mathbb{C}) \longrightarrow H_\infty(\mathbb{R}^N; \mathbb{C})$ has order σ with the constants $\{C_{-s-\sigma} : s \in \mathbb{R}\}$. To this end, let $\varphi, \varphi \in \mathscr{S}(\mathbb{R}^N; \mathbb{C})$ be given, and note that

$$\left| \int \psi(x) L\varphi(x) \, dx \right| \leq \|\psi\|_{s+\sigma} \|L\varphi\|_{-s-\sigma} \leq C_{-s-\sigma} \|\psi\|_{s+\sigma} \|\varphi\|_{-s}.$$

Applying this with $s = 0$, we see that ψ is in the domain of $L^\top$. In addition, for any s, it says that

$$\left| \int \varphi(x) L^\top \psi(x) \, dx \right| = \left| \int \psi(x) L\varphi(x) \, dx \right| \leq C_{-s-\sigma} \|\psi\|_{s+\sigma} \|\varphi\|_{-s},$$

and therefore that $\|L^\top \psi\|_s \leq C_{-s-\sigma} \|\psi\|_{s+\sigma}$.

---

[2] The reason why we deal with transposes instead of adjoints is that $\langle \varphi, u \rangle$ generalizes the non-Hermitian inner product on $L^2(\mathbb{R}^N; \mathbb{C})$, not the unusual Hermitian one. Thus, the domain of $L^\top$ is same as that of L^*, the adjoint of L as an operator on $L^2(\mathbb{R}^N; \mathbb{C})$, and $L^\top \varphi = \overline{L^* \bar{\varphi}}$.

7.2.1. Symbols: We introduce the notation $\overset{\circ}{\mathbb{R}}^N = \mathbb{R}^N \setminus \{0\}$ and $D^\alpha = (\sqrt{-1})^{\|\alpha\|}\partial^\alpha$. Given a measurable function $a : \mathbb{R}^N \times \overset{\circ}{\mathbb{R}}^N \longrightarrow \mathbb{C}$ which satisfies $|a(x,\xi)| \leq C(1+|\xi|^2)^s$ for some $C < \infty$ and $s \in \mathbb{R}$, define the operator $a(x,D)$ on $\mathscr{S}(\mathbb{R}^N;\mathbb{C})$ so that

$$\left(a(x,D)\varphi\right)^\wedge(\xi) = \int_{\mathbb{R}^N} e^{\sqrt{-1}\,(x,\xi)_{\mathbb{R}^N}} a(x,\xi)\varphi(x)\,dx.$$

The function a is called the *symbol* of the operator $a(x,D)$. It is important to recognize that if $a(x,\xi) = f(x)\xi^\alpha$, then $a(x,D)\varphi = D^\alpha(f\varphi)$, not $fD^\alpha\varphi$.

We now want to describe a class of symbols which will be the basic building blocks out of which we will construct our pseudodifferential operators. We will say that $a : \mathbb{R}^N \times \overset{\circ}{\mathbb{R}}^N \longrightarrow \mathbb{C}$ is a *symbol of order* 0 if a is smooth, $\xi \rightsquigarrow a(x,\xi)$ is homogeneous of degree 0 for each $x \in \mathbb{R}^N$, and $a(x,\xi) = a(\infty,\xi) + a'(x,\xi)$, where $a(\infty,\cdot)$ is a smooth function on $\overset{\circ}{\mathbb{R}}^N$ which is homogeneous of order 0 and

$$(7.2.1) \quad \sup_{\substack{x \in \mathbb{R}^N \\ \xi \in \mathbb{S}^{N-1}}} (1+|x|^2)^m |\partial_x^\alpha \partial_\xi^\beta a'(x,\xi)| < \infty \quad \text{for all } m \in \mathbb{N} \text{ and } \alpha, \beta \in \mathbb{N}^N.$$

Because $\widehat{fg} = (2\pi)^{-N}\hat{f} \star \hat{g}$, we can describe the action of $a(x,D)$ as

$$(7.2.2) \quad \begin{aligned} &\left(a(x,D)\varphi\right)^\wedge(\xi) = a(\infty,\xi)\hat{\varphi}(\xi) + \int_{\mathbb{R}^N} \tilde{a}(\xi-\eta,\xi)\hat{\varphi}(\eta)\,d\eta \quad \text{where} \\ &\tilde{a}(\eta,\xi) = \frac{1}{(2\pi)^N}\widehat{a'(\,\cdot\,,\xi)}^x(\eta) = \frac{1}{(2\pi)^N}\int_{\mathbb{R}^N} e^{\sqrt{-1}\,(\eta,x)_{\mathbb{R}^N}} a'(x,\xi)\,dx. \end{aligned}$$

It is important to notice that $\tilde{a}$ is also a function which is smooth on $\mathbb{R}^N \times \overset{\circ}{\mathbb{R}}^N$, homogeneous of degree 0 for each $\eta \in \mathbb{R}^N$, and satisfies

$$(7.2.3) \quad \sup_{\substack{\eta \in \mathbb{R}^N \\ \xi \in \mathbb{S}^{N-1}}} (1+|\eta|^2)^m |\partial_\eta^\alpha \partial_\xi^\beta \tilde{a}(\eta,\xi)| < \infty \quad \text{for all } m \in \mathbb{N} \text{ and } \alpha, \beta \in \mathbb{N}^N.$$

Although $a(x,D)\varphi$ does not map $\mathscr{S}(\mathbb{R}^N;\mathbb{C})$ into itself, we will show that it maps $\mathscr{S}(\mathbb{R}^N;\mathbb{C})$ into $H_\infty(\mathbb{R}^N;\mathbb{C})$ and that it has order 0. In order to do so, we will use the following simple lemmas.

7.2.4 LEMMA. *If $K : \mathbb{R}^N \times \mathbb{R}^N \longrightarrow \mathbb{C}$ is a measurable function with*

$$A \equiv \left(\sup_{\xi \in \mathbb{R}^N} \int |K(\xi,\eta)|\,d\eta\right) \vee \left(\sup_{\eta \in \mathbb{R}^N} \int |K(\xi,\eta)|\,d\xi\right) < \infty,$$

and if $K\varphi(\xi) = \int K(\xi,\eta)\varphi(\eta)\,d\eta$ for $\varphi \in \mathscr{S}(\mathbb{R}^N;\mathbb{C})$, then $\|K\varphi\|_0 \leq A\|\varphi\|_0$ for all $\varphi \in \mathscr{S}(\mathbb{R}^N;\mathbb{C})$. Hence, K has a unique extension as a bounded operator on $H_0(\mathbb{R}^N;\mathbb{C})$ into itself.

Proof: By Schwarz's inequality,

$$|K\varphi(\xi)| \le \left(\int |K(\xi,\eta)| \, d\eta \right)^{\frac{1}{2}} \left(\int |K(\xi,\eta)||\varphi(\eta)|^2 \, d\eta \right)^{\frac{1}{2}}$$

$$\le A^{\frac{1}{2}} \left(\int |K(\xi,\eta)||\varphi(\eta)|^2 \, d\eta \right)^{\frac{1}{2}},$$

and

$$\left\| \left(\int |K(\,\cdot\,,\eta)||\varphi(\eta)|^2 \, d\eta \right)^{\frac{1}{2}} \right\|_0^2 = \iint |K(\xi,\eta)||\varphi(\eta)|^2 \, d\xi d\eta$$

$$= \int |\varphi(\eta)|^2 \left(\int |K(\xi,\eta)| \, d\xi \right) d\eta \le A\|\varphi\|_0^2. \quad \square$$

7.2.5 Lemma.   For $\xi, \eta \in \mathbb{R}^N$ and $t \in \mathbb{R}$,

$$\left(\frac{1 + |\xi|^2}{1 + |\eta|^2} \right)^t \le 2^{|t|} \left(1 + |\eta - \xi|^2 \right)^{|t|}.$$

Proof: There is nothing to do when $t = 0$. Furthermore, by reversing the roles of ξ and η, the case when $t < 0$ can be reduced to the case when $t > 0$. Finally, to handle $t > 0$, use

$$1 + |\xi|^2 \le 1 + \left(|\xi - \eta| + |\eta| \right)^2 \le 1 + 2|\xi - \eta|^2 + 2|\eta|^2$$

$$\le 2\left(1 + |\xi - \eta|^2 + |\eta|^2 \right) \le 2\left(1 + |\xi - \eta|^2 \right)\left(1 + |\eta|^2 \right). \quad \square$$

7.2.6 Theorem.   If $a : \mathbb{R}^N \times \overset{\circ}{\mathbb{R}}{}^N \longrightarrow \mathbb{C}$ is a measurable function and $a(x,\xi) = a(\infty,\xi) + a'(x,\xi)$, where $\sup_{\xi \in \overset{\circ}{\mathbb{R}}{}^N} (1 + |\xi|^2)^{-\frac{\sigma}{2}} |a(\infty,\xi)| < \infty$ for some $\sigma \in \mathbb{R}$ and, for each ξ, $x \rightsquigarrow a'(x,\xi)$ is a smooth function which satisfies

$$\sup_{\xi \in \overset{\circ}{\mathbb{R}}{}^N} (1 + |\xi|^2)^{-\frac{\sigma}{2}} \sum_{\|\alpha\| \le m} \|\partial_x^\alpha a'(\,\cdot\,,\xi)\|_{L^1(\mathbb{R}^N;\mathbb{C})} < \infty$$

for all $m \in \mathbb{N}$, then $a(x,D)$ maps $\mathscr{S}(\mathbb{R}^N;\mathbb{C})$ into $H_\infty(\mathbb{R}^N;\mathbb{C})$ and has order σ. In particular, if $a(x,D)$ is a symbol of order 0, then $a(x,D)$ is of order 0. Also, if $a(\infty,\xi)$ and, for all $m \in \mathbb{N}$, $\sum_{\|\alpha\| \le m} |\partial_x^\alpha a(\,\cdot\,,\xi)\|_{L^1(\mathbb{R}^N;\mathbb{C})}$ are bounded and rapidly decreasing in ξ, then $a(x,D)$ has order $-\infty$.

Proof: Let a be as in the first assertion. Then, by the same reasoning that led to (7.2.2) and (7.2.3), the Fourier transform of $a(x,D)\varphi$ equals

$$a(\infty,\xi)\hat{\varphi}(\xi) + \int \tilde{a}(\xi - \eta,\xi)\hat{\varphi}(\eta) \, d\eta,$$

where $\tilde{a}(\eta,\xi) \equiv (2\pi)^{-N} \widehat{a'(\,\cdot\,,\xi)}^{\,x}(\eta)$ satisfies

$$(*) \qquad \sup_{(\eta,\xi)\in\mathbb{R}^N\times\overset{\circ}{\mathbb{R}}{}^N} (1+|\xi|^2)^{-\frac{\sigma}{2}}(1+|\eta|^2)^{-m}|\tilde{a}(\eta,\xi)| < \infty$$

for every $m \in \mathbb{N}$. Hence

$$(2\pi)^{\frac{N}{2}}\|a(x,D)\varphi\|_s = \left\|(1+|\xi|^2)^{\frac{s}{2}}\big(a(x,D)\varphi\big)^{\wedge}\right\|_0$$
$$\leq \left\|(1+|\xi|^2)^{\frac{s}{2}}a(\infty,\,\cdot\,)\hat{\varphi}\right\|_0 + \left\|K\big(B^{-s-\sigma}\varphi\big)^{\wedge}\right\|_0,$$

where K is the integral operator whose kernel is

$$K(\xi,\eta) = \left(\frac{1+|\xi|^2}{1+|\eta|^2}\right)^{\frac{s+\sigma}{2}}(1+|\xi|^2)^{-\frac{\sigma}{2}}\tilde{a}(\xi-\eta,\xi).$$

Since $a(\infty,\xi)$ is bounded by a constant times $(1+|\xi|^2)^{\frac{\sigma}{2}}$, the first term requires no further comment. At the same time, by Lemma 7.2.5 and $(*)$, for each $m \geq 0$ there is a $C_m < \infty$ such that

$$|K(\xi,\eta)| \leq C_m \big(1+|\xi-\eta|^2\big)^{|\frac{s+\sigma}{2}|-m}.$$

Thus, by choosing $m > \frac{|s+\sigma|+N}{2}$ and applying Lemma 7.2.4, we see that there is an $A < \infty$ such that $\|K(B^{-s-\sigma}\varphi)^{\wedge}\|_0 \leq A\|(B^{-s-\sigma}\varphi)^{\wedge}\|_0 = (2\pi)^{\frac{N}{2}}A\|\varphi\|_{s+\sigma}$. $\square$

7.2.2. Homogeneous Pseudodifferential Operators: Thus far we have not discussed the composition of operators of the form $a(x,D)$. In part this is because we like to think of $a(x,D)$ as operators on $\mathscr{S}(\mathbb{R}^N;\mathbb{C})$, and, in general, with the definition we have given, $a(x,D)$ will not always take $\mathscr{S}(\mathbb{R}^N;\mathbb{C})$ back into itself. Further, we have discussed operators only of order 0. In this subsection we will modify our operators to produce operators of all orders which map $\mathscr{S}(\mathbb{R}^N;\mathbb{C})$ into itself.

Choose and fix a $\zeta \in C^{\infty}(\mathbb{R}^N;[0,1])$ which vanishes on $\overline{B(0,\frac{1}{2})}$ and equals 1 off of $B(0,\frac{3}{4})$, and, for $\sigma \in \mathbb{R}$, set $\zeta^{\sigma}(\xi) = |\xi|^{\sigma}\zeta(\xi)$. Given a symbol a of order 0, define the operator P_a^{σ} by

$$(7.2.7) \qquad P_a^{\sigma} = \zeta^{\sigma}(D)a(x,D).$$

It is important to notice that if η is a second function of the same sort, then, because $\zeta^{\sigma}(\xi) - \eta^{\sigma}(\xi)$ vanishes off of $B(0,1)$, $P_a^{\sigma} - \eta^{\sigma}(D)a(x,D)$ maps $\mathscr{S}(\mathbb{R}^N;\mathbb{C})$ into $H_{\infty}(\mathbb{R}^N;\mathbb{C})$ and has order $-\infty$. In particular, as long as one is interested in properties modulo order $-\infty$, the particular choice of ζ is irrelevant. Also, Theorem 7.2.6 shows that, for each $s \in \mathbb{R}$,

$$\|P_a^{\sigma}\varphi\|_s \leq \|a(x,D)\phi\|_{s+\sigma} \leq C_s\|\varphi\|_{s+\sigma}$$

for some $C_s < \infty$. Hence, P_a^{σ} takes $\mathscr{S}(\mathbb{R}^N;\mathbb{C})$ into $H_{\infty}(\mathbb{R}^N;\mathbb{C})$ and has order σ.

Given $\sigma \in \mathbb{R}$ and a symbol a of order 0, the operator P_a^{σ} is called the *homogeneous pseudodifferential operator of order σ with symbol a*. The great advantage of P_a^{σ} over $|D|^{\sigma}a(x,D)$ is the content of the next theorem.

7.2.8 THEOREM.   *Given $\sigma \in \mathbb{R}$ and a symbol a of order 0, the operator P_a^σ maps $\mathscr{S}(\mathbb{R}^N; \mathbb{C})$ continuously into itself and has order σ.*

PROOF: All that we have to show is that P_a^σ maps $\mathscr{S}(\mathbb{R}^N; \mathbb{C})$ continuously into itself. That is, we must show that for each $\alpha \in \mathbb{N}^N$ and $s \geq 0$ there is a $C < \infty$ and $n \geq 0$ such that (cf. (7.1.1))

$$(*)\qquad \left(\int (1 + |\xi|^2)^s \left| D^\alpha \widehat{P_a^\sigma \varphi}(\xi) \right|^2 d\xi \right)^{\frac{1}{2}} \leq C \|\varphi\|_{(n)}.$$

For this purpose, use (7.2.2) to write

$$\widehat{P_a^\sigma \varphi}(\xi) = \zeta^\sigma(\xi) a(\infty, \xi) \hat{\varphi}(\xi) + \zeta^\sigma(\xi) \int \tilde{a}(\xi - \eta, \xi) \hat{\varphi}(\eta) \, d\eta.$$

Thus $D^\alpha \widehat{P_a^\sigma \varphi}$ is a finite linear combination of terms of the form

$$I(\xi) = \left(D^{\alpha^1} \zeta^\sigma(\xi) \right) \left(b_{\alpha^2}(\infty, \xi) \right) \left(D^{\alpha^3} \hat{\varphi}(\xi) \right),$$

where $b_\beta(\infty, \xi) = D_\xi^\beta a(\infty, \xi)$ and $\alpha^1 + \alpha^2 + \alpha^3 = \alpha$, or

$$J(\xi) = \left(D^{\alpha^1} \zeta^\sigma(\xi) \right) \left(\int b_{\alpha^2, \alpha^3}(\xi - \eta, \xi) D^{\alpha^4} \hat{\varphi}(\eta) \, d\eta \right),$$

where $b_{\beta, \gamma}(\eta, \xi) = D_\eta^\beta \partial_\xi^\gamma \tilde{a}$ and $\alpha^1 + \alpha^2 + \alpha^3 + \alpha^4 = \alpha$. Set $a_\beta(\infty, \xi) = |\xi|^{\|\beta\|} b_\beta(\infty, \xi)$ and $a'_{\beta, \gamma}(x, \xi) = (-x)^\beta |\xi|^{\|\gamma\|} D_\xi^\gamma a'(x, \xi)$. The key observation is that, because $a(\infty, \cdot)$ and $a'(x, \cdot)$ are homogeneous of order 0, $a_\beta(\infty, \xi)$ and $a'_{\beta, \gamma}(x, \xi)$ are both symbols of order 0. In addition, if $f_{\beta, \gamma}(\xi) = |\xi|^{-\|\gamma\|} D^\beta \zeta^\sigma(\xi)$ and $\varphi_\beta(x) = (-x)^\beta \varphi(x)$, then I and J are the Fourier transforms of, respectively,

$$f_{\alpha^1, \alpha^2}(D) a_{\alpha^2}(\infty, D) \varphi_{\alpha^3} \quad \text{and} \quad f_{\alpha^1, \alpha^3}(D) a'_{\alpha^2, \alpha^3}(x, D) \varphi_{\alpha^4}.$$

Hence, since $f_{\beta, \gamma}(D) : \mathscr{S}(\mathbb{R}^N; \mathbb{C}) \longrightarrow H_\infty(\mathbb{R}^N; \mathbb{C})$ has order $\sigma - \|\gamma\| \leq \sigma$ and $a_{\alpha_2}(\infty, D)$ has order 0,

$$\left(\frac{1}{(2\pi)^N} \int (1 + |\xi|^2)^s |I(\xi)|^2 d\xi \right)^{\frac{1}{2}} = \left\| f_{\alpha^1, \alpha^2}(D) a_{\alpha^2}(\infty, D) \varphi_{\alpha^3} \right\|_s,$$

which is dominated by a constant times $\|\varphi_{\alpha^3}\|_{s+\sigma}$. Similarly,

$$\left(\int (1 + |\xi|^2)^s |J(\xi)|^2 d\xi \right)^{\frac{1}{2}} \leq C \|\varphi_{\alpha^4}\|_{s+\sigma}.$$

Thus, we have now proved that $(*)$ holds for some $C < \infty$ when $n \geq \sigma + \|\alpha\|$. $\square$

REMARK 7.2.9. Although it will not play a role in our applications, at some point one should ask why an operator like P_a^σ should have acquired the name "pseudodifferential." A primary reason is that P_a^σ is *quasilocal*. That is, if φ vanishes in an open neighborhood of a point x, then $P^\sigma \varphi(x)$ can be bounded in terms of $\|\varphi\|_0$. In fact, for all $s \in \mathbb{R}$ and $\alpha \in \mathbb{N}^N$, $\partial^\alpha P_a^\sigma \varphi$ at x can be bounded in terms of $\|\varphi\|_s$. Thus, if one uses the Schwartz kernel representation of P_a^σ, then the kernel will be smooth away from the diagonal.

To prove quasilocality, we will show that for each $r > 0$, $\alpha \in \mathbb{N}^N$, and $s \in \mathbb{R}$, there is a constant $A(r, \alpha, s) < \infty$, depending only on σ and a in addition to r, α, and s, such that

$$(*) \qquad \left|\partial^\alpha P_a^\sigma \varphi(x)\right| \le A(r, \alpha, s)\|\varphi\|_s \quad \text{if } \varphi = 0 \text{ on } B(x, r).$$

For this purpose, first use (7.2.2) to write $(2\pi)^N \partial^\alpha P_a^\sigma \varphi(x)$ as

$$\int e_x(-\xi)b(\xi)\hat{\varphi}(\xi)\,d\xi + \iint e_x(-\xi)c(\xi - \eta, \xi)\hat{\varphi}(\eta)\,d\xi d\eta,$$

where $e_x(\xi) \equiv e^{\sqrt{-1}(x,,\xi)_{\mathbb{R}^N}}$, $b(\xi) = \zeta^\sigma(\xi)(-\sqrt{-1}\xi)^\alpha a(\infty, \xi)$, and $\tilde{c}(\eta, \xi) = \zeta^\sigma(\xi)(-\sqrt{-1}\xi)^\alpha \tilde{a}(\eta, \xi)$. Given a $\varphi \in \mathscr{S}(\mathbb{R}^N; \mathbb{C})$ which vanishes in $B(0, r)$, define $\varphi_n(y) = (-|y|^2)^{-n}\varphi(y)$ for $n > \frac{N}{2}$. Because $\varphi = 0$ in $B(0, r)$, we can write $\varphi_n = \psi_n \varphi$ where $\psi_n \in C_b^\infty(\mathbb{R}^N; \mathbb{R})$ and $\psi_n(y) = (-1)^n |y|^{-2n}$ for $|y| \ge r$. Hence $\varphi_n \in \mathscr{S}(\mathbb{R}^N; \mathbb{C})$ and, by Theorem 7.2.6 applied to the symbol $\psi_n(x)$, we know that, for each s, there is a $C(n, r, s) < \infty$ such that $\|\varphi_n\|_s \le C(n, r, s)\|\varphi\|_s$. In addition, $\Delta^n \hat{\varphi}_n = \hat{\varphi}$.

Now, let x be given, suppose that $\varphi \in \mathscr{S}(\mathbb{R}^N; \mathbb{C})$ vanishes on $B(x, r)$, and set $\varphi_{x,n} = (\tau_x \varphi)_n$, where $\tau_x : \mathscr{S}(\mathbb{R}^N; \mathbb{C}) \longrightarrow \mathscr{S}(\mathbb{R}^N; \mathbb{C})$ is defined so that $\tau_x \varphi(y) = \varphi(x + y)$. Then $\|\varphi_{x,n}\|_s \le C(n, r, s)\|\tau_x \varphi\|_s = C(n, r, s)\|\varphi\|_s$ and $\Delta^n \widehat{\varphi_{x,n}}(\xi) = \widehat{\tau_x \varphi}(\xi) = e_x(-\xi)\hat{\varphi}(\xi)$. Thus, we can now write

$$\int e_x(-\xi)b(\xi)\hat{\varphi}(\xi)\,d\xi = \int b(\xi)\Delta^n \widehat{\varphi_{x,n}}(\xi)\,d\xi = \int \Delta^n b(\xi)\widehat{\varphi_{x,n}}(\xi)\,d\xi.$$

In addition, because $a(\infty, \xi)$ is homogeneous of order 0, it is easy to check that there exists a $K(n, \sigma, \alpha) < \infty$ such that

$$|\Delta^n b(\xi)| \le K(n, \sigma, \alpha)(1 + |\xi|^2)^{\frac{\sigma + \|\alpha\|}{2} - n},$$

which means that, for any $s \in \mathbb{R}$,

$$\left|\int e_x(-\xi)b(\xi)\hat{\varphi}(\xi)\,d\xi\right| \le K(n, \sigma, \alpha) \int (1 + |\xi|^2)^{\frac{\sigma + \|\alpha\|}{2} - n}\left|\widehat{\varphi_{x,n}}(\xi)\right|\,d\xi$$

$$\le \left(\int (1 + |\xi|^2)^{\sigma + \|\alpha\| - s - 2n}\,d\xi\right)^{\frac{1}{2}} (2\pi)^{\frac{N}{2}}\|\varphi_{x,n}\|_s.$$

By taking $n > \frac{N}{2} \vee \left(\frac{\sigma + \|\alpha\| - s}{2} + \frac{N}{4} \right)$ and combining this with $\|\varphi_{x,n}\|_\sigma \leq C(n,r,s)\|\varphi\|_s$, we get the desired sort of estimate on this term.

The reasoning for the second term is similar. By making a change of coordinates, one sees that

$$\iint e_x(-\xi)\tilde{c}(\xi - \eta, \xi)\hat{\varphi}(\eta) \, d\xi d\eta = \iint e_x(-\xi - \eta)\tilde{c}(\xi, \xi + \eta)\hat{\varphi}(\eta) \, d\xi d\eta$$

$$= \iint e_x(-\xi)\tilde{c}(\xi, \xi + \eta)\widehat{\tau_x\varphi}(\eta) \, d\xi d\eta.$$

Again, replace $\widehat{\tau_x\varphi}$ by $\Delta^n \widehat{\varphi_{x,n}}$, use integration by parts in η to move Δ^n to $\tilde{c}$, and apply (7.2.3) to see that, for each $m \in \mathbb{N}$, there is a $K(m,n,\sigma,\alpha) < \infty$ such that the preceding is dominated by $K(m,n,\sigma,\alpha)$ times

$$(2\pi)^{-\frac{N}{2}} \iint (1 + |\xi|^2)^{-m}(1 + |\xi + \eta|^2)^{\frac{\sigma + \|\alpha\|}{2} - n}\left|\widehat{\varphi_{x,n}}\right| d\xi d\eta$$

$$\leq \|\varphi_{x,n}\|_s \int (1 + |\xi|^2)^{-m} \left(\int (1 + |\xi + \eta|^2)^{\sigma + \|\alpha\| - 2n}(1 + |\eta|^2)^{-s} \, d\eta \right)^{\frac{1}{2}} d\xi.$$

Thus, all that remains is to show that the iterated integral on the right is finite when m and n are sufficiently large. When $s \geq 0$, this is obvious. When $s \leq 0$, apply Lemma 7.2.5 to dominate it by

$$2^{-\frac{s}{2}} \int (1 + |\xi|^2)^{m - \frac{s}{2}} \left(\int (1 + |\xi + \eta|^2)^{\sigma + \|\alpha\| - s - 2n} \, d\eta \right)^{\frac{1}{2}} d\xi.$$

An important property revealed by the preceding line of reasoning is the role played by the homogeneity of $a(x,\xi)$ in ξ. In particular, it is homogeneity which allows us to say that differentiation of a with respect to ξ increasing the rate at which a decays as $|\xi| \to \infty$.

7.2.3. Composition of Pseudodifferential Operators: Theorem 7.2.8 makes it comfortable to compose homogeneous pseudodifferential operators. Furthermore, given symbols a and b of order 0 and σ, $\tau \in \mathbb{R}$, it is clear that $P_a^\sigma \circ P_b^\tau$ maps $\mathscr{S}(\mathbb{R}^N; \mathbb{C})$ into itself and has order $\sigma + \tau$. In fact, one can hope that $P_a^\sigma \circ P_b^\tau$ is closely related to $P_{ab}^{\sigma + \tau}$, and the goal of this section is to prove that it is. In this direction, we will prove the following theorem.

7.2.10 THEOREM. *Let a and b be symbols of order 0. Given σ, $\tau \in \mathbb{R}$, set $a^\sigma(x,\xi) = |\xi|^\sigma a(x,\xi)$ and $b^\tau(x,\xi) = |\xi|^\tau b(x,\xi)$, and define*

$$c_m^{\sigma,\tau}(x,\xi) = (-1)^m |\xi|^{m - \sigma - \tau} \sum_{\|\alpha\| = m} \frac{\left(D_x^\alpha a^\sigma(x,\xi)\right)\left(\partial_\xi^\alpha b^\tau(x,\xi)\right)}{\alpha!}$$

for $m \in \mathbb{N}$. Then $c_m^{\sigma,\tau}$ is a symbol of order 0 for each $m \in \mathbb{N}$. Moreover, for each $n \geq 1$, the difference

$$P_a^\sigma \circ P_b^\tau - \sum_{m=0}^{n-1} P_{c_m^{\sigma,\tau}}^{\sigma + \tau - m}$$

maps $\mathscr{S}(\mathbb{R}^N;\mathbb{C})$ *into itself and has order* $\sigma + \tau - n$. *In particular, the commutator* $[P_a^\sigma, P_b^\tau]$ *of* P_a^σ *and* P_b^τ *has order* $\sigma + \tau - 1$.

It is an easy matter to check that the $c_m^{\sigma,\tau}$'s are symbols of order 0. In addition, since $c_0^{\sigma,\tau}(x,\xi) = a(x,\xi)b(x,\xi)$, it will follow from the first assertion that $P_a^\sigma \circ P_b^\tau - P_{ab}^{\sigma+\tau}$ and $P_b^\tau \circ P_a^\sigma - P_{ab}^{\sigma+\tau}$ have order $\sigma + \tau - 1$ and therefore that $[P_a^\sigma, P_b^\tau]$ also has order $\sigma + \tau - 1$. Thus, the only part of Theorem 7.2.10 which requires comment is the assertion about the order of the difference, and the first step is to observe that, because $\big(I - \zeta(D)\big)P_{c_0^{\sigma,\tau}}^{\sigma+\tau}$ has order $-\infty$, it suffices to prove the result when $P_{c_0^{\sigma,\tau}}^{\sigma+\tau}$ is replaced by $\zeta(D)P_{c_0^{\sigma,\tau}}^{\sigma+\tau}$. Next, set $f(x,\xi) = \zeta^\sigma(\xi)a(x,\xi)$, $g(x,\xi) = \zeta^\tau(\xi)b(s,\xi)$, and decompose $f(x,\xi)$ and $g(x,\xi)$ into $f(\infty,\xi) + f'(x,\xi)$ and $g(\infty,\xi) + g'(x,\xi)$, corresponding to the decompositions of a and b. Then the Fourier transform of $P_a^\sigma \circ P_b^\tau \varphi$ equals

$$f(\infty,\xi)g(\infty,\xi)\hat{\varphi}(\xi) + f(\infty,\xi)\int \tilde{g}(\xi - \eta, \xi)\hat{\varphi}(\eta)\,d\eta$$

$$+ \int \tilde{f}(\xi - \eta, \xi)g(\infty,\eta)\hat{\varphi}(\eta)\,d\eta + \int \left(\int \tilde{f}(\xi - \eta, \xi)\tilde{g}(\mu - \eta, \mu)\,d\mu\right)\hat{\varphi}(\mu)d\eta$$

and that of $\zeta(D)P_{c_0^{\sigma,\tau}}^{\sigma+\tau}\varphi$ equals

$$f(\infty,\xi)g(\infty,\xi)\hat{\varphi}(\xi) + f(\infty,\xi)\int \tilde{g}(\xi - \eta, \xi)\hat{\varphi}(\eta)\,d\eta$$

$$+ \int \tilde{f}(\xi - \eta, \xi)g(\infty,\xi)\hat{\varphi}(\eta)\,d\eta + \int \left(\int \tilde{f}(\xi - \mu, \xi)\tilde{g}(\mu - \eta, \xi)\,d\mu\right)\hat{\varphi}(\mu)d\eta,$$

where $\tilde{f}$ and $\tilde{g}$ are defined from f' and g' in the same way as $\tilde{a}$ was defined from a' in (7.2.2). Hence, the Fourier transform of $\big(P_a^\sigma \circ P_b^\tau - \zeta(D)P_{c_0^{\sigma,\tau}}^{\sigma+\tau}\big)\varphi$ equals the sum of

$$\int \tilde{f}(\xi - \eta, \xi)\big(g(\infty,\eta) - g(\infty,\xi)\big)\hat{\varphi}(\eta)\,d\eta$$

and

$$\int \left(\int \tilde{f}(\xi - \mu, \xi)\big(\tilde{g}(\mu - \eta, \mu) - \tilde{g}(\mu - \eta, \xi)\big)\,d\mu\right)\hat{\varphi}(\eta)\,d\eta.$$

Next expand $g(\infty, \cdot)$ and $\tilde{g}(\mu - \eta, \cdot)$ in Taylor's series to obtain

$$\tilde{f}(\xi - \eta, \xi)\big(g(\infty,\eta) - g(\infty,\xi)\big)$$

$$= \sum_{1 \le \|\alpha\| \le n-1} \frac{(\eta - \xi)^\alpha}{\alpha!}\tilde{f}(\xi - \eta, \xi)\partial_\xi^\alpha g(\infty,\xi) + \tilde{f}(\xi - \eta, \xi)R_n(\xi, \eta)$$

$$= \sum_{1 \le \|\alpha\| \le n-1} \frac{(-1)^{\|\alpha\|}}{\alpha!}\widetilde{D_x^\alpha f}(\xi - \eta, \xi)\partial_\xi^\alpha g(\infty,\xi) + \tilde{f}(\xi - \eta, \xi)R_n(\xi, \eta)$$

and, similarly,

$$\int \tilde{f}(\xi - \mu, \xi)\big(\tilde{g}(\mu - \eta, \mu) - \tilde{g}(\mu - \eta, \xi)\big)\, d\mu$$

$$= \sum_{1 \leq \|\alpha\| \leq n-1} \frac{(-1)^{\|\alpha\|}}{\alpha!} \int \widetilde{D_x^\alpha f}(\xi - \mu, \xi)\widetilde{\partial_\xi^\alpha g}(\mu - \eta, \xi)\, d\mu$$

$$+ \int \tilde{f}(\xi - \mu, \xi)\tilde{R}_n(\mu - \eta; \xi, \eta)\, d\mu.$$

Because

$$\int \widetilde{D_x^\alpha f}(\xi - \eta, \xi)\partial_\xi^\alpha g(\infty, \xi)\hat{\varphi}(\eta)\, d\eta$$

$$+ \int \left(\int \widetilde{D_x^\alpha f}(\xi - \mu, \xi)\widetilde{\partial_\xi^\alpha g}(\mu - \xi, \xi)\, d\mu \right) \hat{\varphi}(\eta)\, d\eta$$

is the Fourier transform of $(D_x^\alpha f \partial_\xi^\alpha g)(x, D)\varphi$, we now see that

$$\big(P_a^\sigma \circ P_b^\tau - \zeta(D)P_{c_0^{\sigma,\tau}}^{\sigma+\tau}\big)\varphi$$

$$= \sum_{1 \leq \|\alpha\| \leq n-1} \frac{(-1)^{\|\alpha\|}}{\alpha!}(D_x^\alpha f \partial_\xi^\alpha g)(x, D)\varphi + E_n\varphi + \tilde{E}_n\varphi,$$

where

$$\widehat{E_n\varphi}(\xi) = \int \tilde{f}(\xi - \eta, \xi)R_n(\xi, \eta)\hat{\varphi}(\eta)\, d\eta$$

and

$$\widehat{\tilde{E}_n\varphi}(\xi) = \int \left(\int \tilde{f}(\xi - \mu, \xi)\tilde{R}_n(\mu - \eta; \xi, \eta)\, d\mu \right) \hat{\varphi}(\eta)\, d\eta.$$

The next step is to notice that because, for each α,

$$(D_x^\alpha f \partial_\xi^\alpha g)(x, \xi) - \zeta^{\sigma+\tau-\|\alpha\|}(\xi)D_x^\alpha a^\sigma(x, \xi)\partial_\xi^\alpha b^\tau(x, \xi)$$

vanishes when $|\xi| > 1$, the last part of Theorem 7.2.6 says that the difference

$$\sum_{1 \leq \|\alpha\| \leq n-1} \frac{(-1)^{\|\alpha\|}}{\alpha!}(D_x^\alpha f \partial_\xi^\alpha g)(x, D) - \sum_{m=1}^{n-1} P_{c_m^{\sigma,\tau}}^{\sigma+\tau-m}$$

is an operator of order $-\infty$. Hence, what remains is to prove that the operators E_n and $\tilde{E}_n$ are of order $\sigma + \tau - n$. Equivalently, for each $s \in \mathbb{R}$, we must show that the operators K and $\tilde{K}$ whose kernels are

$$K(\xi, \eta) = \frac{(1 + |\xi|^2)^{\frac{s}{2}}}{(1 + |\eta|^2)^{\frac{s+\sigma+\tau-n}{2}}} \tilde{f}(\xi - \eta, \xi)R_n(\xi, \eta)$$

and

$$\tilde{K}(\xi, \eta) = \frac{(1 + |\xi|^2)^{\frac{s}{2}}}{(1 + |\eta|^2)^{\frac{s+\sigma+\tau-n}{2}}} \int \tilde{f}(\xi - \mu, \xi)\tilde{R}_n(\mu - \eta; \xi, \eta)\, d\mu$$

are bounded on $L^2(\mathbb{R}^N; \mathbb{C})$ into itself. In order to check these, we want to apply Lemma 7.2.4, and for this we will use the following.

7.2.11 LEMMA.   Let $h : \mathring{\mathbb{R}}^N \longrightarrow \mathbb{C}$ be a smooth function which is homogeneous of degree λ, and define $f = \zeta h$. Then, for each $n \geq 1$, there exists a $C < \infty$, which depends only on n, λ, and $\max_{\|\alpha\| \leq n} \sup_{\xi \in \mathbb{S}^{N-1}} |\partial^\alpha h(\xi)|$, such that

$$
r_n(\xi, \eta) \equiv \left| f(\eta) - f(\xi) - \sum_{\|\alpha\| \leq n-1} \frac{(\eta - \xi)^\alpha}{\alpha!} \partial^\alpha f(\xi) \right|
$$
$$
\leq C |\eta - \xi|^n \left((1 + |\xi|^2)^{\frac{\lambda - n}{2}} + (1 + |\eta|^2)^{\frac{\lambda - n}{2}} \right).
$$

PROOF: We work by cases. Throughout, C will denote a finite constant with the asserted dependence, but it can change from expression to expression.

(1) $|\xi| \vee |\eta| \leq 3$: In this case, simply use the fact that f is smooth and the standard remainder estimate in Taylor's theorem.

(2) $|\xi| \leq 2$ & $|\eta| \geq 3$ or $|\eta| \leq 2$ & $|\xi| \geq 3$: By symmetry, it suffices to treat the first of these, in which case $1 \leq \frac{1}{3}|\eta| \leq |\eta - \xi| \leq \frac{5}{3}|\eta|$, $1 + |\xi|^2 \leq 5$, and $f(\eta) = h(\eta)$. Thus,

$$
r_n(\xi, \eta) \leq |h(\eta)| + \sum_{\|\alpha\| \leq n-1} \frac{|\eta - \xi|^{\|\alpha\|}}{\alpha!} |\partial^\alpha f(\xi)|
$$
$$
\leq C \left[|\eta|^\lambda + \sum_{\|\alpha\| \leq n-1} \left(\frac{5|\eta|}{3} \right)^{\|\alpha\|} \right] \leq C (|\eta|^\lambda + |\eta|^n).
$$

At the same time,

$$
|\eta - \xi|^n \left((1 + |\xi|^2)^{\frac{\lambda - n}{2}} + (1 + |\eta|^2)^{\frac{\lambda - n}{2}} \right)
$$
$$
\geq \frac{1}{3^n 5^{\frac{|\lambda - n|}{2}}} |\eta|^n \left(1 + (1 + |\eta|^2)^{\frac{\lambda - n}{2}} \right) \geq \frac{1}{3^n 5^{\frac{|\lambda - n|}{2}}} (|\eta|^n + |\eta|^\lambda).
$$

(3) $|\eta| \wedge |\xi| \geq 2$ & $(\xi, \eta)_{\mathbb{R}^N} \geq 0$: In this case, $|\xi + t(\eta - \xi)| \geq 1$ for $t \in [0, 1]$ and therefore the remainder term in Taylor's Theorem is the sum of terms of the form

$$
\frac{(\eta - \xi)^\alpha}{\alpha!} \int_0^1 (1 - t)^{n-1} \partial^\alpha h\big(\xi + t(\eta - \xi)\big) \, dt,
$$

where $\|\alpha\| = n$. Note that $\left| \partial^\alpha h\big(\xi + t(\eta - \xi)\big) \right| \leq C |\xi + t(\eta - \xi)|^{\lambda - n}$. If $\lambda \geq n$, we have

$$
|\xi + t(\eta - \xi)|^{\lambda - n} \leq \big(|\xi| \vee |\eta| \big)^{\lambda - n} \leq (1 + |\xi|^2)^{\frac{\lambda - n}{2}} + (1 + |\eta|^2)^{\frac{\lambda - n}{2}}.
$$

If $\lambda < n$, we use

$$|\xi + t(\eta - \xi)|^2 \geq (1-t)^2 |\xi|^2 + t^2 |\eta|^2 \geq \frac{|\xi|^2 \wedge |\eta|^2}{4}$$

to see that

$$|\xi + t(\eta - \xi)|^{\lambda - n} \leq 2^{n-\lambda}\big(|\xi| \wedge |\eta|\big)^{\lambda - n} \leq 2^{n-\lambda}\big(|\xi|^{\lambda - n} + |\eta|^{\lambda - n}\big).$$

Thus, since $\rho \geq 2 \implies \frac{\rho^2}{1+\rho^2} \geq \frac{1}{2}$, we have

$$|\xi + t(\eta - \xi)|^{\lambda - n} \leq 4^{n-\lambda}\Big((1 + |\xi|^2)^{\frac{\lambda - n}{2}} + (1 + |\eta|^2)^{\frac{\lambda - n}{2}}\Big).$$

(4) $|\eta| \wedge |\xi| \geq 2$ & $(\xi, \eta)_{\mathbb{R}^N} \leq 0$: In this case, $|\eta - \xi|^2 \geq |\eta|^2 + |\xi|^2 \geq \big(|\xi| \vee |\eta|\big)^2$, $f(\xi) = h(\xi)$, and $f(\eta) = h(\eta)$. Thus

$$r_n(\xi, \eta) \leq |h(\eta)| + \sum_{\|\alpha\| \leq n-1} \frac{|\eta - \xi|^{\|\alpha\|}}{\alpha!} |\partial^\alpha h(\xi)|$$

$$\leq C\left(|\eta|^\lambda + \sum_{m=0}^{n-1} |\xi|^{\lambda - m} |\eta - \xi|^m\right)$$

$$= C\left[|\eta|^\lambda + \sum_{m=0}^{n-1} \left(\frac{|\xi|}{|\eta - \xi|}\right)^{n-m} |\xi|^{\lambda - n} |\eta - \xi|^n\right] \leq C\big(|\eta|^\lambda + |\xi|^{\lambda - n} |\eta - \xi|^n\big).$$

At the same time, since $|\xi| \wedge |\eta| \geq 2$,

$$|\eta - \xi|^n \Big((1 + |\xi|^2)^{\frac{\lambda - n}{2}} + (1 + |\eta|^2)^{\frac{\lambda - n}{2}}\Big)$$

$$\geq |\eta - \xi|^n (1 + |\xi|^2)^{\frac{\lambda - n}{2}} + |\eta|^n (1 + |\eta|^2)^{\frac{\lambda - n}{2}}$$

$$\geq 2^{-\frac{|\lambda - n|}{2}}\big(|\xi|^{\lambda - n} |\eta - \xi|^n + |\eta|^\lambda\big). \quad \square$$

Return to the consideration of the kernels $K(\xi, \eta)$ and $\tilde{K}(\xi, \eta)$. By combining (7.2.3) with Lemma 7.2.11 and the fact that $\zeta^\sigma(\xi) \leq 2^{|\sigma|}(1 + |\xi|^2)^{\frac{\sigma}{2}}$, one sees that, for each $m \in \mathbb{N}$, there is a $C_m < \infty$ such that

$$|\tilde{f}(\xi - \eta, \xi) R_n(\xi, \eta)| \leq C_m (1 + |\xi|^2)^{\frac{\sigma}{2}} (1 + |\eta - \xi|^2)^{-m} |\xi - \eta|^n$$
$$\times \Big((1 + |\xi|^2)^{\frac{\tau - n}{2}} + (1 + |\eta|^2)^{\frac{\tau - n}{2}}\Big).$$

Hence, if $\chi \equiv s + \sigma + \tau - n$, then, by Lemma 7.2.5, $|K(\xi, \eta)|$ is dominated by

$$C_m \left(\frac{1 + |\xi|^2}{1 + |\eta|^2}\right)^{\frac{\chi}{2}} (1 + |\eta - \xi|^2)^{\frac{n}{2} - m} \leq 2^{\frac{|\chi|}{2}} C_m \big(1 + |\eta - \xi|^2\big)^{\frac{|\chi| + n}{2} - m}$$

plus

$$C_m \left(\frac{1 + |\xi|^2}{1 + |\eta|^2} \right)^{\frac{s+\sigma}{2}} \left(1 + |\eta - \xi|^2 \right)^{\frac{n}{2} - m} \leq 2^{\frac{|s+\sigma|}{2}} C_m \left(1 + |\eta - \xi|^2 \right)^{\frac{|s+\sigma|+n}{2} - m}.$$

Thus, when m is sufficiently large, Lemma 7.2.4 applies to K.

The proof that $\tilde{K}$ is bounded is similar. Proceeding as above, one sees that, for $m \in \mathbb{N}$, $|\tilde{K}(\xi, \eta)|$ is dominated by a constant times

$$\frac{(1 + |\xi|^2)^{\frac{s+\sigma}{2}}}{(1 + |\eta|^2)^{\frac{\chi}{2}}} |\xi - \eta|^n (1 + |\xi|^2)^{\frac{\tau - n}{2}} \int (1 + |\xi - \mu|^2)^{-m} (1 + |\eta - \mu|^2)^{-m} d\mu$$

plus

$$\frac{(1 + |\xi|^2)^{\frac{s+\sigma}{2}}}{(1 + |\eta|^2)^{\frac{\chi}{2}}} |\xi - \eta|^n (1 + |\eta|^2)^{\frac{\tau - n}{2}} \int (1 + |\xi - \mu|^2)^{-m} (1 + |\eta - \mu|^2)^{-m} d\mu.$$

Next note that either $|\mu - \xi|$ or $|\mu - \eta|$ must be at least $\frac{|\xi - \eta|}{2}$, and conclude that, for $m > \frac{N}{2}$,

$$\int (1 + |\xi - \mu|^2)^{-m} (1 + |\eta - \mu|^2)^{-m} d\mu \leq C_m (1 + |\xi - \eta|^2)^{-m}.$$

Combining this with the preceding, one finds that $|\tilde{K}(\xi, \eta)|$ is dominated by a constant times

$$\left[\frac{(1 + |\xi|^2)^{\frac{\chi}{2}}}{(1 + |\eta|^2)^{\frac{\chi}{2}}} + \frac{(1 + |\xi|^2)^{\frac{s+\sigma}{2}}}{(1 + |\eta|^2)^{\frac{s+\sigma}{2}}} \right] (1 + |\xi - \eta|^2)^{\frac{n}{2} - m},$$

and so, by Lemma 7.2.5,

$$|\tilde{K}(\xi, \eta)| \leq C_m \left[(1 + |\xi - \eta|^2)^{\frac{\chi+n}{2} - m} + (1 + |\xi - \eta|^2)^{\frac{s+\sigma+n}{2} - m} \right].$$

Thus, by taking m sufficiently large, we see that Lemma 7.2.4 applies and says that $\tilde{K}$ is bounded on $L^2(\mathbb{R}^N; \mathbb{C})$.

7.2.4. General Pseudodifferential Operators: The result in Theorem 7.2.10 makes it clear that the composition of P_a^σ with P_b^τ will seldom be a homogeneous pseudodifferential operator. Nonetheless, that same theorem shows that $P_a^\sigma \circ P_b^\tau$ will, in the sense of order, be arbitrarily well approximated by the sum of homogeneous pseudodifferential operators of decreasing order. With this in mind, we introduce the class $\mathscr{P}$ of operators $P : \mathscr{S}(\mathbb{R}^N; \mathbb{C}) \longrightarrow \mathscr{S}(\mathbb{R}^N; \mathbb{C})$ with the property that there exists a strictly decreasing sequence $\{\sigma_m : m \in \mathbb{N}\} \subseteq \mathbb{R}$ and a sequence $\{a_m : m \in \mathbb{N}\}$ of symbols of order 0 such that $\sigma_m \searrow -\infty$ and

$$(7.2.12) \qquad P - \sum_{m=0}^{n} P_{a_m}^{\sigma_m} \quad \text{has order } \sigma_{n+1} \text{ for each } n \in \mathbb{N}.$$

An element $P \in \mathscr{P}$ is called a *pseudodifferential operator*. Given $\sigma_m \searrow -\infty$ and a sequence $\{a_m : m \in \mathbb{N}\}$ of symbols of order 0, we will write $P \sim \sum_{m=0}^{\infty} P_{a_m}^{\sigma_m}$ if (7.2.12) holds, in which case we will say that $\sum_{m=0}^{\infty} P_{a_m}^{\sigma_m}$ is an *asymptotic series* for P.

It is should be obvious that $\mathscr{P}$ is a vector space over $\mathbb{C}$. In addition, it follows from Theorem 7.2.10 that it is closed under composition. Thus it forms an algebra over $\mathbb{C}$ when composition plays the role of multiplication.

There are a great many non-trivial bookkeeping matters which ought to be addressed at this point. The most important of these deal with the relationship between a pseudodifferential and its asymptotic series. For instance, it is clear that two pseudodifferential operators admit the same asymptotic series if and only if they differ by an operator of order $-\infty$, and for this reason it is often useful to replace each element $P \in \mathscr{P}$ by the equivalence class of operators which differ from P by an operator of order $-\infty$. On the other hand, it is not immediately clear that a pseudodifferential operator admits only one asymptotic series or that each asymptotic series has an operator for which it is the asymptotic series. Both these are true, but their proofs require some work, and so it is fortunate that, for the applications that we will be making, neither of them is necessary. Thus, instead of proving them, we will simply give a couple of important examples of pseudodifferential operators which are not homogeneous.

Partial Differential Operators: Suppose that $L = \sum_{\|\alpha\| \le m} c_\alpha D^\alpha$ is a partial differential operator. If each of the coefficients c_α can be decomposed into the sum of a constant plus an element of $\mathscr{S}(\mathbb{R}^N; \mathbb{C})$, then L is a pseudodifferential operator. To see this, first note that $L\varphi = \sum_{\|\alpha\| \le m} D^\alpha(c'_\alpha \varphi)$, where the coefficients $\{c'_\alpha : \|\alpha\| \le m\}$ are linear combinations of the c_α's and their derivatives. Thus, it suffices to check that if $c(x) = c(\infty) + c'(x)$, where $c(\infty) \in \mathbb{C}$ and $c' \in \mathscr{S}(\mathbb{R}^N; \mathbb{C})$, then the operator P given by $P\varphi = D^\alpha(c\varphi)$ is a pseudodifferential operator for each $\alpha \in \mathbb{N}^N$. But if $a(x, \xi) = |\xi|^{-\|\alpha\|} c(x) \xi^\alpha$, then $P = |D|^{\|\alpha\|} a(x, D)$ and so P equals $P_a^{\|\alpha\|}$ plus $(1 - \zeta(D))|D|^{\|\alpha\|} a(x, D)$, which, by Theorem 7.2.6, is an operator of order $-\infty$. Notice that, in this case, the asymptotic series is finite.

Bessel Operators: The Bessel operators $B^s = (I - \Delta)^{-\frac{s}{2}}$, $s \in \mathbb{R}$, are pseudodifferential operators. Indeed, by Theorem 7.2.6, $B^s - \zeta(D)B^s$ has order $-\infty$. Moreover,

$$(1 + |\xi|^2)^{-\frac{s}{2}} = |\xi|^{-s} \sum_{m=0}^{n} \binom{-\frac{s}{2}}{m} |\xi|^{-2m} + |\xi|^{-s-2(n+1)} R_n(s, \xi),$$

where $R_n(s, \xi)$ is bounded for $n > -\frac{s}{2}$. Thus,

$$B^s \sim \sum_{m=0}^{\infty} \binom{-\frac{s}{2}}{m} \zeta^{-\frac{s}{2}-2m}(D).$$

7.3 Hypoellipticity Revisited

In this section we will use the theory in § 7.2 to give a far-reaching generalization of the result proved in § 3.4.1. Namely, given a differential operator

$$(7.3.1) \qquad\qquad L = \sum_{\|\alpha\| \leq m} c_\alpha D^\alpha,$$

with $m \geq 1$ and smooth, complex valued coefficients, we will say that L is *elliptic* on an open set $W \subseteq \mathbb{R}^N$ if the function

$$(x, \xi) \in W \times \mathbb{S}^{N-1} \longmapsto \sum_{\|\alpha\| = m} c_\alpha(x) \xi^\alpha$$

never vanishes, and we will show that ellipticity on W implies hypoellipticity on W.

Obviously, if an L given by (1.1.8) has smooth, real valued coefficients and if the matrix $a(x)$ is symmetric and strictly positive definite for each $x \in W$, then L is elliptic on W. On the other hand, the associated heat operator $L + \partial_t$ is not elliptic on $\mathbb{R} \times W$. Thus, although the result here generalizes the one in § 3.4.1, it does not cover the one in § 3.4.2. Nonetheless, it covers situations which the results in § 3.4 cannot touch. For example, although it contains only first order derivatives, the Cauchy–Riemann operator $\partial_x - \sqrt{-1}\, \partial_y$ is elliptic.

7.3.1. Elliptic Estimates: In this subsection we will prove the following *elliptic estimate*.

7.3.2 THEOREM. *Suppose that L is given by $(7.3.1)$ and that L is elliptic on W, and let ψ_1, $\psi_2 \in C_c^\infty\big(W; [0, 1]\big)$ with $\psi_2 = 1$ on a open neighborhood of* $\mathrm{supp}(\psi_1)$*. If u is a distribution for which $\psi_2 u$ and $\psi_2 L u$ are both in $H_s(\mathbb{R}^N; \mathbb{C})$ for some $s \in \mathbb{R}$, then $\psi_1 u \in H_{s+m}(\mathbb{R}^N; \mathbb{C})$. In fact, there exists a $C_s < \infty$ such that*

$$\|\psi_1 u\|_{s+m} \leq C_s \big(\|\psi_2 L u\|_s + \|\psi_2 u\|_s \big).$$

In particular, if each c_α can be written as a constant plus an element of $\mathscr{S}(\mathbb{R}^N; \mathbb{C})$, then L has true order m.

First observe that, given the earlier assertions, the final assertion is more or less obvious. Indeed, under the stated condition on the c_α's, the operation $c_\alpha D^\alpha$ has order $\|\alpha\|$, and so L has order m. On the other hand, if L had order σ for some $\sigma < m$, then we would have

$$\|\psi_1 u\|_m \leq C_0 \big(\|\psi_2 L u\|_0 + \|\psi_2 u\|_0 \big) \leq C \|\psi_2 u\|_{\sigma+}, \quad u \in L^2(\mathbb{R}^N; \mathbb{C}),$$

which is impossible.

Turning to the first part of the theorem, we begin with the observation that is suffices to show that

(*)
$$\psi_2 u,\ \psi_2 Lu \in H_s(\mathbb{R}^N; \mathbb{C}) \implies \psi_1 u \in H_{s+1}(\mathbb{R}^N; \mathbb{C})\ \text{and}$$
$$\|\psi_1 u\|_{s+1} \leq A_s(\psi_1, \psi_2)\big(\|\psi_2 Lu\|_{s+1-m} + \|\psi_2 u\|_s\big),$$

for some $A_s(\psi_1, \psi_2) < \infty$. Indeed, assume that (*) holds, and suppose $\psi_2 u,\ \psi_2 Lu \in H_s(\mathbb{R}^N; \mathbb{C})$. We will use induction on $1 \leq \ell \leq m$ to show that, for each $s \in \mathbb{R}$,

(**)
$$\|\psi_1 u\|_{s+\ell} \leq B_\ell(\psi_1, \psi_2)\big(\|\psi_2 Lu\|_{s+\ell-m} + \|\psi_2 u\|_s\big).$$

There is nothing to do when $\ell = 1$. Thus, assume (**) holds for some $1 \leq \ell < m$, and choose $\psi \in C^\infty\big(W; [0, 1]\big)$ so that $\psi_2 = 1$ on an open neighborhood of $\mathrm{supp}(\psi)$ and $\psi = 1$ on an open neighborhood of $\mathrm{supp}(\psi_1)$. Then

$$\|\psi_1 u\|_{s+\ell+1} \leq A_{s+\ell}(\psi_1, \psi)\big(\|\psi Lu\|_{s+\ell+1-m} + \|\psi u\|_{s+\ell}\big),$$
$$\|\psi u\|_{s+\ell} \leq B_\ell(\psi, \psi_2)\big(\|\psi_2 Lu\|_{s+\ell-m} + \|\psi_2 u\|_s\big),$$

and so $\|\psi_1 u\|_{s+\ell+1}$ is dominated by

$$A_{s+\ell+1}(\psi_1, \psi)\Big(\|\psi Lu\|_{s+\ell+1-m} + B_\ell(\psi, \psi_2)\big(\|\psi_2 Lu\|_{s+\ell-m} + \|\psi_2 u\|_s\big)\Big)$$
$$\leq B_{\ell+1}(\psi_1, \psi_2)\big(\|\psi_2 Lu\|_{s+\ell+1-m}\|\psi_2 u\|_s\big)$$

for some $B_{\ell+1}(\psi_1, \psi_2) < \infty$.

To prove (*), choose a symbol $a(x, \xi)$ of order 0 with the properties that $|a(x, \xi)| \geq \epsilon > 0$ for some $\epsilon > 0$ and all $(x, \xi) \in \mathbb{R}^N \times \mathbb{S}^{N-1}$,

$$a(x, \xi) = |\xi|^{-m} \sum_{\|\alpha\|=m} c_\alpha(x)\xi^\alpha$$

for x in a bounded, open neighborhood of $\mathrm{supp}(\psi_2)$, and $a(x, \xi) = 1$ for x outside of a bounded set. Now set $u_1 = \psi_1 u$, $u_2 = \psi_2 u$, and $f = \psi_1 Lu$, and assume that $u_2,\ f \in H_s(\mathbb{R}^N; \mathbb{C})$. Then, since L is local, $Lu_1 = f + L'u_2$, where $L' = [L, \psi_1]$ is the commutator of L and multiplication by ψ_1. By either direct computation or Theorem 7.2.10, we know that L' has order $m - 1$. At the same time, since $L^\top = |D|^m a(x, D) + \sum_{\|\alpha\| \leq m-1} D^\alpha c_\alpha$, for a given $\varphi \in \mathscr{S}(\mathbb{R}^N; \mathbb{C})$, we have

$$\langle \varphi, Lu_1 \rangle = \langle \psi_2 L^\top \varphi, u_1 \rangle$$
$$= \langle |D|^m a(x, D)\varphi, u_1 \rangle + \Big\langle \sum_{\|\alpha\| \leq m-1} D^\alpha \psi_2 c_\alpha \varphi, u_1 \Big\rangle$$
$$= \langle P_a^m \varphi, u_1 \rangle + \langle K\varphi, u_2 \rangle,$$

where (remember that $|D|^m a(x,D) - P_a^m$ has order $-\infty$) $K : \mathscr{S}(\mathbb{R}^N;\mathbb{C}) \longrightarrow H_\infty(\mathbb{R}^N;\mathbb{C})$ has order $m-1$. After combining this with $Lu_1 = f + L'u_2$, we see that

$$\langle P_a^m \varphi, u_1 \rangle = \langle \varphi, f \rangle + \langle K'\varphi, u_2 \rangle,$$

where $K' : \mathscr{S}(\mathbb{R}^N;\mathbb{C}) \longrightarrow H_\infty(\mathbb{R}^N;\mathbb{C})$ again has order $m-1$. Now set $b(x,\xi) = \frac{1}{a(x,\xi)}$, and note that b is also a symbol of order 0. In addition, by Theorem 7.2.10, $H \equiv P_a^m \circ P_b^{-m} - I$ maps $\mathscr{S}(\mathbb{R}^N;\mathbb{C})$ into itself and has order -1. Hence, after replacing φ by $P_b^{-m}\varphi$ in the proceeding, we conclude that

$$\langle \varphi, u_1 \rangle = \langle P_b^{-m}\varphi, f \rangle + \langle H'\varphi, u_2 \rangle,$$

where H' maps $\mathscr{S}(\mathbb{R}^N;\mathbb{C})$ into $H_\infty(\mathbb{R}^N;\mathbb{C})$ and has order -1. In particular, because

$$\left| \langle P_b^{-m}\varphi, f \rangle \right| \leq \|P_b^{-m}\varphi\|_{-s-1+m} \|f\|_{s+1-m} \leq C\|f\|_{s+1-m} \|\varphi\|_{-s-1}$$

while

$$\left| \langle H'\varphi, u_2 \rangle \right| \leq \|H'\varphi\|_{-s} \|u_2\|_s \leq C\|u_2\|_s \|\varphi\|_{-s-1},$$

this means that there exists an $A_s(\psi_1, \psi_2)$ such that

$$\left| \langle \varphi, u_1 \rangle \right| \leq A_s(\psi_1, \psi_2) \big(\|f\|_{s+1-m} + \|u_2\|_s \big) \|\varphi\|_{-s-1},$$

from which (*) follows immediately.

7.3.2. Hypoellipticity: Given Theorem 7.3.2, the following theorem is more or less obvious.

7.3.3 THEOREM. *Let L and W be as in Theorem 7.3.2. If u is a distribution and $Lu \in C^\infty(W;\mathbb{C})$, then $u \in C^\infty(W;\mathbb{C})$.*

PROOF: Given an open U with $\bar{U} \subset\subset W$, choose ψ_1, $\psi_2 \in C^\infty\big(W;[0,1]\big)$ so that $\psi_1 = 1$ on an open neighborhood of $\bar{U}$ and $\psi_2 = 1$ on an open neighborhood of supp(ψ_1). Then $\psi_2 u \in H_s(\mathbb{R}^N;\mathbb{C})$ for some $s \in \mathbb{R}$. Now repeat the argument given to pass from (*) to (**) in the preceding subsection to show that $\psi_1 u \in H_\infty(\mathbb{R}^N;\mathbb{C})$. Finally, apply (7.1.2) to conclude that $\psi_1 u \in C_c^\infty(\mathbb{R}^N;\mathbb{C})$. $\square$

7.4 Hörmander's Theorem

As we saw in §3.4.2, an operator need not be elliptic in order to be hypoelliptic. Indeed, the result there says that if L is given by (1.1.8) with smooth coefficients, and if the diffusion matrix a is strictly positive definite, then the parabolic operator $L+\partial_t$ will be hypoelliptic. Further, Kolmogorov gave a beautiful example which shows that $L + \partial_t$ can be hypoelliptic even though L is not elliptic. Namely, he considered the operator $L = \frac{1}{2}\partial_{x_1}^2 + x_1\partial_{x_2}$. This L is certainly not elliptic: The diffusion matrix is everywhere of rank 1. Nonetheless, the associated heat equation $\partial_t u = Lu$ admits a smooth fundamental solution $p(t, x, y)$. In fact, direct calculation can be used to check that

$$p(t, x, y) = \frac{\sqrt{3}}{\pi t^2} \exp\left(-\tfrac{1}{2}\big(y - m(t, x), C(t)^{-1}(y - m(t, x))\big)_{\mathbb{R}^N}\right),$$

where

$$m(t, x) = \begin{pmatrix} x_1 \\ x_2 + t x_1 \end{pmatrix} \quad \text{and} \quad C(t) = \begin{pmatrix} t & \frac{t^2}{2} \\ \frac{t^2}{2} & \frac{t^3}{3} \end{pmatrix}.$$

In particular, proceeding as in § 3.4.2, one can use this $p(t, x, y)$ to verify that $L + \partial_t$ is hypoelliptic.

A more enlightening way to see that $p(t, x, y)$ is a fundamental solution to $\partial_t u = Lu$ is to use the diffusion associated with L. That is, let $t \rightsquigarrow B(t)$ be a standard $\mathbb{R}$-valued Brownian motion, and set

$$X(t, x) = \begin{pmatrix} X_1(t, x) \\ X_2(t, x) \end{pmatrix} = m(t, x) + \begin{pmatrix} B(t) \\ \int_0^t B(\tau)\, d\tau \end{pmatrix}.$$

Then it is an easy matter to check that the distribution of $X(\,\cdot\,, x)$ solves the martingale problem for L starting from x. Furthermore, for each $t > 0$, $X(t, x)$ is a Gaussian random variable with mean $m(t, x)$ and covariance $C(t)$. Hence, the distribution of $X(t, x)$ is the probability measure on $\mathbb{R}^2$ with density

$$\big((2\pi)^2 \det C(t)\big)^{-\frac{1}{2}} \exp\left(-\frac{1}{2}\big(y - m(t, x), C(t)^{-1}(y - m(t, x))\big)_{\mathbb{R}^N}\right),$$

which is another expression for $p(t, x, y)$. The reason why the probabilistic approach is more revealing is that it highlights the reason why, in spite of the degeneracy of the diffusion matrix, the transition probability function admits a density. Specifically, although the $\mathbb{R}^2$-valued path $t \rightsquigarrow X(t, x, B)$ is built out of single $\mathbb{R}$-valued Brownian motion, $B(t)$ and $\int_0^t B(\tau)\, d\tau$ are sufficiently uncorrelated that their joint distribution admits a smooth density.

In this section we will prove a striking result, due to L. Hörmander, which puts Kolmogorov's example into context. To state Hörmander's Theorem, let $\{V_0, \ldots, V_M\}$ be a family of smooth (real) vector fields on $\mathbb{R}^N$, think of the V_k's as directional derivatives, and consider the operator

$$(7.4.1) \qquad\qquad L = V_0 + \frac{1}{2} \sum_{k=1}^M V_k^2,$$

where $V_k^2 = V_k \circ V_k$. Notice that this operator can be written in the form in (1.1.8). Indeed, suppose that $V_k = \sum_{i=1}^N \sigma_{i,k} \partial_{x_i}$, take σ to be the $N \times M$ matrix whose kth column is

$$\begin{pmatrix} \sigma_{1,k} \\ \vdots \\ \sigma_{N,k} \end{pmatrix},$$

and set $b_i = \sigma_{i,0} + \frac{1}{2} \sum_{k=0}^{M} V_k \sigma_{i,k}$. Then an equivalent expression for L is (1.1.8) with $a = \sigma\sigma^\top$ and b. Conversely, if L is given by (1.1.8) and σ is a smooth $N \times M$ square root of a, then L can be written in the form in (7.4.1) when

$$(7.4.2) \qquad V_k = \begin{cases} \sum_{i=1}^{N} \sigma_{i,k} \partial_{x_i} & \text{if } 1 \le k \le M \\ \sum_{i=1}^{N} \left(b_i - \frac{1}{2} \sum_{k=1}^{M} (V_k \sigma_{i,k}) \right) \partial_{x_i} & \text{if } k = 0. \end{cases}$$

In order to state Hörmander's result, we need to introduce the notion of the *Lie algebra* $\mathrm{Lie}(V_0, \ldots, V_M)$ generated by $\{V_0, \ldots, V_M\}$. Namely, $\mathrm{Lie}(V_0, \ldots, V_M)$ is the smallest vector space of smooth vector fields which contains $\{V_0, \ldots, V_M\}$ and is closed under the Lie product given by the commutator $[V, V'] = V \circ V' - V' \circ V$. Equivalently, $\mathrm{Lie}(V_0, \ldots, V_M)$ is the vector space of vector fields spanned by the V_k's together with their multiple commutators

$$\left[V_{k_1}, [V_{k_2}, [\cdots, V_{k_r}] \cdots] \right].$$

Finally, $\mathrm{Lie}_x(V_0, \ldots, V_k)$ is the subspace of the tangent space $T_x \mathbb{R}^N$ obtained by evaluating the elements of $\mathrm{Lie}(V_0, \ldots, V_M)$ at x. Of course, because $\mathbb{R}^N$ is Euclidean, $T_x \mathbb{R}^N$ can be identified with $\mathbb{R}^N$, in which case $\mathrm{Lie}_x(V_0, \ldots, V_M)$ is identified with the column vectors whose components are the coefficients of the elements of $\mathrm{Lie}_x(V_0, \ldots, V_M)$.

7.4.3 THEOREM. (**Hörmander**) *Suppose that L is given by (7.4.1), and let $c \in C^\infty(\mathbb{R}^N; \mathbb{R})$. If $W \subseteq \mathbb{R}^N$ is an open set on which $\mathrm{Lie}_x(V_0, \ldots, V_M) = T_x \mathbb{R}^N$ for each $x \in W$, then $L + c$ is hypoelliptic on W.*

The condition $\mathrm{Lie}_x(V_0, \ldots, V_M) = T_x M$ is called *Hörmander's condition.*

It should be clear that Hörmander's theorem covers both the results about parabolic and elliptic operators in § 3.4 as well as Kolmogorov's example. Indeed, if L has the form in (1.1.8) with smooth coefficients, and if a is non-degenerate, then (cf. Lemma 2.3.1) we can take $\sigma = a^{\frac{1}{2}}$ in the preceding discussion, in which case the coefficients of V_k, $1 \le k \le N$, are the kth column of σ, and therefore $\{V_1, \ldots, V_N\}$ already spans $T_x \mathbb{R}^N$ at each x. Hence, Hörmander's Theorem guarantees that L is hypoelliptic. To see that $L + \partial_t$ is also hypoelliptic, think of the V_k's as vector fields on $\mathbb{R} \times \mathbb{R}^N$ with vanishing coefficient in the first coordinate, and consider the family $\{V_0 + \partial_t, V_1, \ldots, V_N\}$. Because, at each (t, x), $\{V_0 + \partial_t, V_1, \ldots, V_N\}$ spans $T_x(\mathbb{R} \times \mathbb{R}^N)$, once again Hörmander's theorem applies. Finally, to handle Kolmogorov's example, take $V_0 = x_1 \partial_{x_2} + \partial_t$ and $V_1 = \partial_{x_1}$ on $\mathbb{R} \times \mathbb{R}^2$. Then $[V_1, V_0] = \partial_{x_2}$, and so $\{V_0, V_1, [V_0, V_1]\}$ spans $T_x(\mathbb{R} \times \mathbb{R}^2)$ at each point (t, x).

Our proof of Hörmander's will follow the basic strategy used to prove Theorem 7.3.3, although the details are quite different. The similarity is that we will prove it by developing an analog of the elliptic estimate in Theorem 7.3.2. Specifically, we will show that there is a $\delta > 0$ such that

$$(7.4.4) \qquad \|\psi_1 u\|_{s+\delta} \leq C_s \big(\|\psi_2 (L+c)u\|_s + \|\psi_2 u\|_s\big).$$

In the case when L is elliptic on W (i.e., the vector fields $\{V_1, \ldots, V_M\}$ span $T_x \mathbb{R}^N$ at each $x \in W$), Theorem 7.3.2 says that we can take $\delta = 2$. However, the δ in (7.4.4) will, in general, be much smaller than 2 and will depend on the number commutators required to achieve a spanning set. Because δ is smaller than the one in the elliptic result, the estimate in (7.4.4) is called a *subelliptic estimate*.

7.4.1. A Preliminary Reduction: In this subsection we will show that (7.4.4) for $s = 0$ and $\psi_1 = \psi_2 \equiv 1$ implies (7.4.4) for all $s \in \mathbb{R}$. It is extremely important to keep in mind that, as distinguished from earlier considerations in this chapter, *everything here, functions, vector fields, and distributions, is real valued.* Also, we will be assuming that the coefficients of the vector fields $\{V_0, \ldots, V_M\}$ as well as the function c are elements of $\mathscr{S}(\mathbb{R}^N; \mathbb{R})$.

We begin with what looks like a big step toward (7.4.4) with $s = 0$. However, as we will see, it is really only a very small step.

7.4.5 Lemma. *There is a $C < \infty$ such that*

$$\sum_{k=1}^{M} \|V_k \varphi\|_0^2 \leq -4\big(L\varphi, \varphi\big)_0 + C\|\varphi\|_0^2, \quad \varphi \in H_2(\mathbb{R}^N; \mathbb{R}).$$

Proof: Clearly, it suffices to work with $\varphi \in \mathscr{S}(\mathbb{R}^N; \mathbb{R})$.

For each k there is an $\eta_k \in \mathscr{S}(\mathbb{R}^N; \mathbb{R})$ such that $V_k^* = V_k^\top = -V_k + \eta_k$. Thus,

$$\|V_k \varphi\|_0^2 = \big(V_k^* V_k \varphi, \varphi\big)_0 = -\big(V_k^2 \varphi, \varphi\big)_0 + \big(V_k \varphi, \eta_k \varphi\big)_0$$
$$\leq -\big(V_k^2 \varphi, \varphi\big)_0 + \tfrac{1}{2}\big(\|V_k \varphi\|_0^2 + \|\eta_k \varphi\|_0^2\big),$$

and so

$$\sum_{k=1}^{M} \|V_k \varphi\|_0^2 \leq -4\big(L\varphi, \varphi\big)_0 + \sum_{k=1}^{M} \|\eta_k \varphi\|_0^2 + 4\big(V_0 \varphi, \varphi\big)_0.$$

At the same time, because everything is $\mathbb{R}$-valued,

$$\big(V_0 \varphi, \varphi\big)_0 = -\big(V_0 \varphi, \varphi\big)_0 + \big(\eta_0 \varphi, \varphi\big)_0,$$

and so $(V_0 \varphi, \varphi)_0 = \tfrac{1}{2}(\eta_0 \varphi, \varphi)_0$. $\quad\square$

7.4.6 LEMMA. Let $T : H_{-\infty}(\mathbb{R}^N; \mathbb{R}) \longrightarrow H_\infty(\mathbb{R}^N; \mathbb{R})$ be an operator of order $-\infty$ and $\psi \in C_c^\infty(\mathbb{R}^N; \mathbb{R})$. Then, for all $\mathbb{R}$-valued distributions u,

$$
\sum_{k=1}^{M} \|V_k T\psi u\|_0^2 \le C\bigg(\|T\psi Lu\|_0^2 + \|T\psi u\|_0^2
$$

$$
+ \sum_{k=0}^{M} \big\|[V_k, T\psi]u\big\|_0^2 + \sum_{k=1}^{M} \big\| \big[V_k[V_k, T\psi]\big]u \big\|_0^2 \bigg),
$$

where $C < \infty$ can be taken independent of both T and ψ.

PROOF: By Lemma 7.4.5,

$$
\sum_{k=1}^{M} \|V_k T\psi u\|_0^2 \le -4\big(LT\psi u, T\psi u\big)_0 + C\|T\psi u\|_0^2.
$$

Also,

$$
\big|\big(LT\psi u, T\psi u\big)_0\big| \le \big|\big(T\psi Lu, T\psi u\big)_0\big| + \big|\big([L, T\psi]u, T\psi u\big)_0\big|
$$

$$
\le \tfrac{1}{2}\big(\|T\psi Lu\|_0^2 + \|T\psi u\|_0^2\big) + \big|\big([L, T\psi]u, T\psi u\big)_0\big|.
$$

Thus, $\sum_{k=1}^{M} \|V_k T\psi u\|_0^2$ is dominated by $4\big|\big([L, T\psi]u, T\psi u\big)_0\big|$ plus terms of the sort allowed in the desired estimate. To go further, note that

$$
[V_k^2, T\psi] = V_k[V_k, T\psi] + [V_k, T\psi]V_k = 2V_k[V_k, T\psi] - \big[V_k, [V_k, T\psi]\big],
$$

and therefore

$$
(7.4.7) \qquad [L, T\psi] = \sum_{k=1}^{M} V_k[V_k, T\psi] - \frac{1}{2}\sum_{k=1}^{M} \big[V_k, [V_k, T\psi]\big] + [V_0, T\psi].
$$

Now the only remaining problem comes from $\big(V_k[V_k, T\psi]u, T\psi u\big)_0$. But, using the notation introduced in the proof of the preceding lemma, we have

$$
\big|\big(V_k[V_k, T\psi]u, T\psi u\big)_0\big| = \big|\big([V_k, T\psi]u, \eta_k T\psi u\big)_0 - \big([V_k, T\psi]u, V_k T\psi u\big)_0\big|
$$

$$
\le \tfrac{1}{8}\|V_k T\psi u\|_0^2 + \tfrac{5}{2}\big\|[V_k, T\psi]u\big\|_0^2 + \tfrac{1}{2}\|\eta_k T\psi u\|_0^2.
$$

Hence, we have shown that $\sum_{k=1}^{M} \|V_k T\psi u\|_0^2$ is dominated by half itself plus terms of the allowed sort. $\square$

The next two lemmas deal with the commutator of a pseudodifferential operator with a mollifier. Here, $\rho \in C_c^\infty(\mathbb{R}^N; \mathbb{R})$, $\rho_\epsilon(x) = \epsilon^{-N}\rho(\epsilon^{-1}x)$, and, for $m \in \mathbb{N}$, $R_{\rho,\epsilon}^m$ is the convolution operator given by $R_{\rho,\epsilon}^m \varphi = \epsilon^m \rho_\epsilon \star \varphi$.

7.4.8 LEMMA.   Given $\rho \in C_c^\infty(\mathbb{R}^N; \mathbb{R})$ and $m \in \mathbb{N}$, there is a $C_m < \infty$ such that $\sup_{\epsilon \in (0,1]} \|R_{m,\epsilon}^\rho u\|_s \leq C_m \|u\|_{s-m}$ for all $s \in \mathbb{R}$ and $u \in H_{s-m}(\mathbb{R}^N; \mathbb{R})$. Moreover, if $c \in \mathscr{S}(\mathbb{R}^N; \mathbb{R})$, then $[c, R_{m,\epsilon}^\rho]$ can be decomposed into the sum $U_{(\rho,c),\epsilon}^m + V_{(\rho,c),\epsilon}^m + W_{(\rho,c),\epsilon}^m$, where

$$U_{(\rho,c),\epsilon}^m \varphi = \sum_{\|\alpha\|=1} R_{x^\alpha \rho,\epsilon}^{m+1}(\partial^\alpha c)\varphi,$$

$$V_{(\rho,c),\epsilon}^m \varphi = \sum_{\|\alpha\|=2} \frac{1}{\alpha!} R_{x^\alpha \rho,\epsilon}^{m+2}(\partial^\alpha c)\varphi,$$

and, for each $s \in \mathbb{R}$, there is a $C_s < \infty$ such that

$$\sup_{\epsilon \in (0,1]} \|W_{(\rho,c),\epsilon}^m u\|_s \leq C_s \|u\|_{s-m-3} \quad \text{for } u \in H_s(\mathbb{R}^N; \mathbb{R}).$$

Finally, for each $\epsilon \in (0,1]$, $U_{(\rho,c),\epsilon}^m$, $V_{(\rho,c),\epsilon}^m$, and $W_{(\rho,c),\epsilon}^m$ are of order $-\infty$ on $H_{-\infty}(\mathbb{R}^N; \mathbb{R})$ into $H_\infty(\mathbb{R}^N; \mathbb{R})$.

PROOF: Clearly, it suffices to work with $u = \varphi \in H_\infty(\mathbb{R}^N; \mathbb{C})$. Also, for each $\epsilon \in (0,1]$, $[c, R_\epsilon^m]$, $U_{(\rho,c),\epsilon}^m$, and $V_{(\rho,c),\epsilon}^m$ are of order $-\infty$ on $H_{-\infty}(\mathbb{R}^N; \mathbb{R})$ into $H_\infty(\mathbb{R}^N; \mathbb{R})$, and therefore the same is true of $W_{(\rho,c),\epsilon}^m$.

Because $\widehat{R_{\rho,\epsilon}^m \varphi}(\xi) = \epsilon^m \hat{\rho}(\epsilon\xi)\hat{\varphi}(\xi)$ and $\hat{\rho}$ is rapidly decreasing,

$$\|R_{\rho,\epsilon}^m \varphi\|_s^2 = \frac{\epsilon^{2m}}{(2\pi)^N} \int (1+|\xi|^2)^s |\hat{\rho}(\epsilon\xi)|^2 |\hat{\varphi}(\xi)|^2 \, d\xi$$

$$\leq \frac{C_m^2}{(2\pi)^N} \int (1+|\xi|^2)^{s-m} |\hat{\varphi}(\xi)|^2 \, d\xi = C_m^2 \|\varphi\|_{s-m}^2,$$

where

$$C_m = \sup_{\epsilon \in (0,1]} \sup_{\xi \in \mathbb{R}^N} \left(\epsilon^2(1+|\xi|^2)\right)^{\frac{m}{2}} |\hat{\rho}(\epsilon\xi)| \leq \sup_{\xi \in \mathbb{R}^N} (1+|\xi|^2)^{\frac{m}{2}} |\hat{\rho}(\xi)| < \infty.$$

Turning to the second part of the lemma, note that the Fourier transforms of $cR_{\rho,\epsilon}^m \varphi$ and $R_{\rho,\epsilon}^m(c\varphi)$ are, respectively,

$$\frac{\epsilon^m}{(2\pi)^N} \int \hat{c}(\xi - \eta)\hat{\rho}(\epsilon\eta)\hat{\varphi}(\eta) \, d\eta$$

and

$$\frac{\epsilon^m}{(2\pi)^N} \int \hat{c}(\xi - \eta)\hat{\rho}(\epsilon\xi)\hat{\varphi}(\eta) \, d\eta.$$

Thus, the Fourier transform of $[c, R_{\rho,\epsilon}^m]\varphi$ equals $(2\pi)^{-N}$ times

$$\epsilon^m \int \hat{c}(\xi - \eta)\big(\hat{\rho}(\epsilon\eta) - \hat{\rho}(\epsilon\xi)\big)\hat{\varphi}(\eta)\,d\eta$$

$$= \epsilon^m \sum_{1 \le \|\alpha\| \le 2} \frac{\epsilon^{\|\alpha\|}}{\alpha!} \int (\eta - \xi)^\alpha \hat{c}(\xi - \eta)(\partial^\alpha \hat{\rho})(\epsilon\xi)\hat{\varphi}(\eta)\,d\eta$$

$$+ \epsilon^{m+3} \sum_{\|\alpha\|=3} \frac{1}{\alpha!} \int H_\epsilon^\alpha(\xi, \eta)\hat{\varphi}(\eta)\,d\eta,$$

where each $H_\epsilon^\alpha(\xi, \eta)$ is a constant times

$$(*) \qquad\qquad (\eta - \xi)^\alpha \hat{c}(\xi - \eta)(\partial^\alpha \rho)\big(\epsilon\theta^\alpha(\xi, \eta)\big)$$

with $\theta^\alpha(\xi, \eta)$ a point on the line segment connecting ξ to η. Because

$$(\eta - \xi)^\alpha \hat{c}(\xi - \eta)(\partial^\alpha \hat{\rho})(\xi) = \widehat{\partial^\alpha c}(\xi - \eta)\widehat{x^\alpha \rho}(\epsilon\xi),$$

it is easy to identify the sums over $\{\alpha : \|\alpha\| = 1\}$ and $\{\alpha : \|\alpha\| = 2\}$ as $U_{(\rho,c),\epsilon}^m$ and $V_{(\rho,c),\epsilon}^m$, respectively. Hence, all that remains is to show that, for each $s \in \mathbb{R}$ and α with $\|\alpha\| = 3$, the operator with kernel

$$K_\epsilon^\alpha(\xi, \eta) = \epsilon^{m+3} \frac{(1 + |\xi|^2)^{\frac{s}{2}}}{(1 + |\eta|^2)^{\frac{s-m-3}{2}}} H_\epsilon^\alpha(\xi, \eta)$$

is bounded on $L^2(\mathbb{R}^N; \mathbb{C})$ independent of $\epsilon \in (0, 1]$. But, because both $\widehat{\partial^\alpha c}$ and $\widehat{x^\alpha \rho}$ are rapidly decreasing, we know that, for each $n \in \mathbb{N}$, there is a $C_n < \infty$ such that

$$|K_\epsilon^\alpha(\xi, \eta)| \le C_n \epsilon^{m+3} \frac{(1 + |\xi|^2)^{\frac{s}{2}}}{(1 + |\eta|^2)^{\frac{s-m-3}{2}}} \frac{(1 + |\xi - \eta|^2)^{-n-\frac{s}{2}}}{(1 + \epsilon^2 |\xi|^2 \wedge |\eta|^2)^{\frac{m+3}{2}}}$$

$$\le C_n 2^{\frac{|s|}{2}} \frac{(1 + |\xi - \eta|^2)^{-n}(1 + |\eta|^2)^{\frac{m+3}{2}}}{(1 + |\xi|^2 \wedge |\eta|^2)^{\frac{m+3}{2}}}$$

$$\le C_n 2^{\frac{|s|}{2}} (1 + |\xi - \eta|^2)^{-n} \left[1 + \frac{(1 + |\eta|^2)^{\frac{m+3}{2}}}{(1 + |\xi|^2)^{\frac{m+3}{2}}} \right]$$

$$\le 2C_n 2^{\frac{|s|+m+3}{2}+1}(1 + |\xi - \eta|^2)^{-n+\frac{m+3}{2}},$$

where we have made two applications of Lemma 7.2.5. Thus, by taking $n > \frac{m+3+N}{2}$, we see that Lemma 7.2.4 applies. $\square$

To facilitate the notation in the following, we will use $\mathscr{P}'$ to denote the class of pseudodifferential operators which admit an asymptotic series $\sum_{m=0}^\infty P_{a_m}^{\sigma_m}$ with the property that $\{\sigma_m - \sigma_0 : m \ge 1\} \subseteq \mathbb{Z}^+$. Using Theorems 7.2.10, check that $\mathscr{P}'$ is closed under composition and commutation.

We will use P^σ to denote a generic element of $\mathscr{P}'$ whose asymptotic series begins with $\sigma_0 = \sigma$ and will use R^m to denote a generic operator of the form $R_{\rho,\epsilon}^m$.

Finally, ∂ will be used to denote any of the operators ∂^α with $\|\alpha\| = 1$.

7.4.9 Lemma.   Let c and ψ be elements of $C_c^\infty(\mathbb{R}^N;\mathbb{R})$, and use ψ' to denote a generic element of $C_c^\infty(\mathbb{R}^N;\mathbb{R})$ whose support is contained in $\operatorname{supp}(\psi)$. Then the operator $[c\partial, P^\sigma R_\epsilon^m \psi]$ can be written as a finite, linear combination of operators of the form $P^\sigma R_\epsilon^m \psi'$ and $P^\sigma R_\epsilon^{m+1}\psi'$ plus an operator $E_\epsilon^{\sigma-m-2}\psi'$, where $E_\epsilon^{\sigma-m-2} : H_{-\infty}(\mathbb{R}^N;\mathbb{R}) \longrightarrow H_\infty(\mathbb{R}^N;\mathbb{R})$ has order $-\infty$ for each $\epsilon \in (0,1]$, and, for each $s \in \mathbb{R}$, admits a $C_s < \infty$ such that

$$\sup_{\epsilon\in(0,1]} \|E_\epsilon^{\sigma-m-2}u\|_s \le C_s\|u\|_{s+\sigma-m-2}, \quad u \in H_{s+\sigma-m-2}(\mathbb{R}^N;\mathbb{R}).$$

Proof: In the calculations which follow, a "$\sqrt{}$" will be placed after terms which are either of the form $P^\sigma R_\epsilon^m \psi'$ or $P^\sigma R_\epsilon^{m+1}\psi'$.

Using $[\partial, R_\epsilon^m] = 0$ and applying the last part of Theorem 7.2.10 where necessary, one can check that

$$[c\partial, P^\sigma R_\epsilon^m \psi] = c\partial P^\sigma R_\epsilon^m \psi - P^\sigma R_\epsilon^m \psi c\partial$$
$$= c\partial P^\sigma R_\epsilon^m \psi - P^\sigma R_\epsilon^m \partial c\psi + P^\sigma R_\epsilon^m \big(\partial(c\psi)\big)\sqrt{},$$

$$c\partial P^\sigma R_\epsilon^m \psi - P^\sigma R_\epsilon^m \partial c\psi = P^\sigma c\partial R_\epsilon^m \psi - P^\sigma \partial R_\epsilon^m c\psi + [c\partial, P^\sigma]R_\epsilon^m \psi\sqrt{},$$

and

$$P^\sigma c\partial R_\epsilon^m \psi - P^\sigma \partial R_\epsilon^m c\psi = P^\sigma c\partial R_\epsilon^m \psi - P^\sigma \partial c R_\epsilon^m \psi + P^\sigma \partial[c, R_\epsilon^m]\psi$$
$$= -P^\sigma \big(\partial c\big)R_\epsilon^m \psi\sqrt{} + P^\sigma \partial[c, R_\epsilon^m]\psi.$$

Finally, by Lemma 7.4.8, $P^\sigma \partial[c, R_\epsilon^m]\psi$ is a finite sum of terms of the form $P^\sigma \partial R_\epsilon^{m+1}c\psi$ and $P^\sigma \partial R_\epsilon^{m+2}c\psi$ plus a term $P^\sigma \partial W_\epsilon^m \psi$, where W_ϵ^m satisfies the estimate at the end of that lemma. Since $\partial R_\epsilon^{m+1}$ and $\partial R_\epsilon^{m+2}$ are, respectively, of the forms R_ϵ^m and R_ϵ^{m+1}, while

$$\|(P^\sigma \partial W_\epsilon^m \psi)u\|_s \le C_s\|\partial W_\epsilon^m(\psi u)\|_{s+\sigma} \le C_s'\|W_\epsilon^m(\psi u)\|_{s+\sigma+1}$$
$$\le C_s''\|\psi u\|_{s+\sigma+m-2} \le C_s'''\|u\|_{s+\sigma+m-2},$$

this completes the proof.   □

7.4.10 Lemma.   Let P^σ be an operator of the sort in the preceding, $\psi_1 \in C_c^\infty(\mathbb{R}^N;\mathbb{R})$, and vector fields X, Y, and Z with coefficients from $\mathscr{S}(\mathbb{R}^N;\mathbb{R})$ be given. If $\psi_2 \in C_c^\infty(\mathbb{R}^N;\mathbb{R})$ satisfies $\psi_2 = 1$ on an open neighborhood of $\operatorname{supp}(\psi_1)$, then

$$[X, P^\sigma R_{\rho,\epsilon}^0 \psi_1] = T_{1,\epsilon}^\sigma \psi_2, \quad [Y, [X, P^\sigma R_{\rho,\epsilon}^0 \psi_1]] = T_{2,\epsilon}^\sigma \psi_2,$$
$$\text{and } \Big[Z, [Y, [X, P^\sigma R_{\rho,\epsilon}^0 \psi_1]]\Big] = T_{3,\epsilon}^\sigma \psi_2,$$

where the $T_{\ell,\epsilon}^\sigma$'s are of order $-\infty$ from $H_{-\infty}(\mathbb{R}^N;\mathbb{R}$ into $H_\infty(\mathbb{R}^N;\mathbb{R})$ for each $\epsilon \in (0,1]$ and, for each $s \in \mathbb{R}$, there exists a $C_s < \infty$ such that

$$\max_{1\le\ell\le3} \sup_{\epsilon\in(0,1]} \|T_{\ell,\epsilon}^\sigma u\|_s \le C_s\|u\|_{s+\sigma}, \quad u \in H_{s+\sigma}(\mathbb{R}^N;\mathbb{R}).$$

PROOF: By Lemma 7.4.9, $[X, P^\sigma R^0_{\rho,\epsilon} \psi_1]$ is a finite linear combination of terms having the form $P^\sigma R^0_\epsilon \psi'_1$ and $P^\sigma R^1_\epsilon \psi'_1$ plus a term $E^{\sigma-2}_\epsilon \psi'_1$, where $\psi'_1 \in C^\infty_c(\mathbb{R}^N; \mathbb{R})$ with $\mathrm{supp}(\psi'_1) \subseteq \mathrm{supp}(\psi_1)$ and $E^{\sigma-2}_\epsilon : H_{-\infty}(\mathbb{R}^N; \mathbb{R}) \longrightarrow H_\infty(\mathbb{R}^N; \mathbb{R})$ has order $-\infty$ for each $\epsilon \in (0,1]$ and satisfies $\sup_{\epsilon \in (0,1]} \|E^{\sigma-2}_\epsilon \varphi\|_s \leq C_s \|\varphi\|_{s+\sigma-2}$. Thus, we have shown that $[X, P^\sigma R^0_{\rho,\epsilon} \psi_1]$ has the desired form.

To handle the higher order commutators, one repeats the preceding argument to conclude that

$$[Y, [X, P^\sigma R^0_{\rho,\epsilon} \psi]] \quad \text{and} \quad \left[Z, \left[Y, [X, P^\sigma R^0_{\rho,\epsilon} \psi]\right]\right]$$

are finite linear combinations of, respectively, operators

$$P^\sigma R^0_\epsilon \psi'_1, \ P^\sigma R^1_\epsilon \psi'_1, \ P^\sigma R^2_\epsilon \psi'_1, \ E^{\sigma-2}_\epsilon \psi'_1, \ E^{\sigma-1}_\epsilon \psi'_1, \ \text{and} \ [Y, E^{\sigma-2}_\epsilon \psi'_1]$$

and

$$P^\sigma R^0_\epsilon \psi'_1, \ P^\sigma R^1_\epsilon \psi'_1, \ P^\sigma R^2_\epsilon \psi'_1, \ P^\sigma R^3_\epsilon \psi'_1, E^{\sigma-2}_\epsilon \psi'_1, \ E^{\sigma-1}_\epsilon \psi'_1, \ E^\sigma_\epsilon \psi'_1,$$
$$[Z, E^{\sigma-2}_\epsilon \psi'_1], \ [Z, E^{\sigma-1}_\epsilon \psi'_1], \ \text{and} \ \left[Z, [Y, E^{\sigma-2}_\epsilon \psi'_1]\right].$$

Finally, simply observe that

$$[Y, E^{\sigma-m}_\epsilon \psi'_1] = T^{\sigma-m+1}_\epsilon \psi_2 \ \text{and} \ \left[Z, [Y, E^{\sigma-m}_\epsilon \psi'_1]\right] = T^{\sigma-m+2}_\epsilon \psi_2,$$

where, for each $\epsilon \in (0,1]$, T^τ_ϵ has order $-\infty$ and, for each $s \in \mathbb{R}$, there is a $C_s < \infty$ for which $\sup_{\epsilon \in (0,1]} \|T^\tau_\epsilon \varphi\|_s \leq C_s \|\varphi\|_{s+\tau}$. $\square$

We have, at long last, arrived at our goal in this subsection.

7.4.11 THEOREM. *Let U be an open subset of $\mathbb{R}^N$ and assume that there exist a $\delta \in (0,1]$ and $C < \infty$ such that*

$$(7.4.12) \qquad \|\varphi\|_\delta \leq C(\|L\varphi\|_0 + \|\varphi\|_0) \quad \text{for all } \varphi \in C^\infty_c(U; \mathbb{R}).$$

Then, for any $c \in \mathscr{S}(\mathbb{R}^N; \mathbb{R})$ and $\psi_1, \psi_2 \in C^\infty_c(U; \mathbb{R})$ with $\psi_2 = 1$ on an open neighborhood of $\mathrm{supp}(\psi_1)$, there exists for each $s \in \mathbb{R}$ a $C_s < \infty$ such that

$$\|\psi_1 u\|_{s+\delta} \leq C_s(\|\psi_2(L+c)u\|_s + \|\psi_2 u\|_s)$$

whenever u is an $\mathbb{R}$-valued distribution for which $\psi_2 u, \psi_2 L u \in H_s(\mathbb{R}^N; \mathbb{C})$.

PROOF: Once again, we use C to denote a finite constant which can change from expression to expression but is independent of ϵ. Also, it should be clear that the result for $c = 0$ implies the result for general $c \in \mathscr{S}(\mathbb{R}^N; \mathbb{R})$. Thus, we will assume that $c = 0$.

Let r be half the distance between $\mathrm{supp}(\psi_2)$ and the complement of U, choose $\rho \in C^\infty_c(B(0,r); [0,\infty))$ with total integral 1, and set $R_\epsilon =$

$R_{\rho,\epsilon}^0$ for $\epsilon \in (0,1]$. Finally, choose $\eta \in C_c^\infty(U;\mathbb{R})$ so that $\eta(x) = 1$ when $|x - \text{supp}(\psi_2)| \leq r$.

To get started, note that, by (7.4.12) and the fact that $[\eta, B^{-s}]$ has order $s - 1$,

$$\|R_\epsilon \psi_1 u\|_{s+\delta} = \|B^{-s}\eta R_\epsilon \psi_1 u\|_\delta \leq \|\eta B^{-s} R_\epsilon \psi_1 u\|_\delta + \|[\eta, B^{-s}] R_\epsilon \psi_1 u\|_\delta$$
$$\leq C\big(\|L\eta B^{-s} R_\epsilon \psi_1 u\|_0 + \|\eta B^{-s} R_\epsilon \psi_1 u\|_0 + \|R_\epsilon \psi_1 u\|_{s-1+\delta}\big)$$
$$\leq C\big(\|L\eta B^{-s} R_\epsilon \psi_1 u\|_0 + \|\psi_2 u\|_s\big).$$

Thus, it remains to estimate $\|L\eta B^{-s} R_\epsilon \psi_1 u\|_0$. To this end, set $P^s = \eta B^{-s}$, and note that, by the discussion about Bessel operators in § 7.4.2, $P^s \in \mathscr{P}'$ and $\sigma_0 = s$ in its asymptotic expansion. Clearly

$$\|L\eta B^{-s} R_\epsilon \psi_1 u\|_0 \leq \|P^s R_\epsilon \psi_1 L u\|_0 + \|[L, P^s R_\epsilon \psi_1] u\|_0,$$

and, since $\|P^s R_\epsilon \psi_1 L u\|_0 \leq C\|L\psi_2 u\|_s$, we are left with the estimation of $\|[L, P^s R_\epsilon \psi_1] u\|_0$.

Using (7.4.7), we can write

$$[L, P^s R_\epsilon \psi_1] = \sum_{k=1}^M V_k [V_k, P^s R_\epsilon \psi_1] - \frac{1}{2}\sum_{k=1}^M [V_k, [V_k, P^s R_\epsilon \psi_1]] + [V_0, P^s R_\epsilon \psi_1].$$

By Lemma 7.4.10 (with $\sigma = s$), the actions on u of the last two terms on the right have $L^2(\mathbb{R}^N;\mathbb{C})$-norms bounded by $C\|\psi_2 u\|_s$. To handle the first term, we use Lemma 7.4.9 to justify taking $T = [V_k, P^s R_\epsilon \psi_1]$ in Lemma 7.4.6 and thereby conclude that $\|V_k[V_k, P^s R_\epsilon \psi_1] u\|_0$ can be estimated in terms of

$$\|T\psi_1 u\|_0, \ \|T\psi_1 L u\|_0, \ \|[V_k, T\psi_1] u\|_0, \ \text{and} \ \|[V_k, [V_k, T\psi_1]] u\|_0.$$

Clearly the first of these causes no problem, the second is dominated by a constant times $\|L\psi_2 u\|_s$, while, by another application of Lemma 7.4.9, the third and fourth can be dominated by a constant times $\|\psi_2 u\|_s$. □

7.4.2. Completing Hörmander's Theorem: Let $\mathcal{B}$ denote the set of $\beta \in \bigcup_{\ell=1}^\infty \{0,\ldots,M\}^\ell$, and define $\ell(\beta) = \ell$ if $\beta \in \{0,\ldots,M\}^\ell$. Next, define V_β for $\beta \in \mathcal{B}$ so that

$$V_\beta = \begin{cases} V_k & \text{if } \ell(\beta) = 1 \text{ and } \beta_1 = k \\ [V_{\beta_\ell}, V_{(\beta_1,\ldots,\beta_{\ell-1})}] & \text{if } \ell(\beta) = \ell \geq 2. \end{cases}$$

Then $\text{Lie}_x(V_0,\ldots,V_M)$ is the span of $\{V_\beta : \beta \in \mathcal{B}\}$ at x. In particular, $\text{Lie}_x(V_0,\ldots,V_M) = T_x\mathbb{R}^N$ if and only if one can find $\{\beta^i : 1 \leq i \leq N\} \subseteq \mathcal{B}$ such that $\{V_{\beta^i} : 1 \leq i \leq N\}$ at x is a basis for $T_x\mathbb{R}^N$. Now assume that $\text{Lie}_x(V_0,\ldots,V_M) = T_x\mathbb{R}^N$, choose $\{\beta^i : 1 \leq i \leq N\}$ accordingly, and let $A(x)$ be the $N \times N$-matrix whose ith column is the coefficients of V_{β^i}. Then $\det\big(A(x)\big) \neq 0$, and so, by continuity, $\det\big(A(x')\big) \neq 0$ for x' in an open neighborhood of x, and therefore $\{V_{\beta^i} : 1 \leq i \leq N\}$ at x' is a basis for $T_{x'}\mathbb{R}^N$ when x' is that neighborhood. Thus the set of $x \in \mathbb{R}^N$ such $\mathrm{Lie}_x(V_0, \ldots, V_M) = T_x\mathbb{R}^N$ is open. Moreover, by an easy Heine–Borel covering argument, if W is an open subset for which $\mathrm{Lie}_x(V_0, \ldots, V_M) = T_x\mathbb{R}^N$ when $x \in W$, then, for each $K \subset\subset W$, there is an $\ell \geq 1$ such that $\{V_\beta : \ell(\beta) \leq \ell\}$ spans $T_x\mathbb{R}^N$ for all $x \in K$. In fact, using an elementary partition of unity, one can see that, for each α with $\|\alpha\| = 1$, there exist $\{v_{\alpha,\beta} : \ell(\beta) \leq \ell\} \subseteq C_c^\infty(\mathbb{R}^N; \mathbb{R})$ such that

$$(7.4.13) \qquad \partial^\alpha = \sum_{\{\beta : \ell(\beta) \leq \ell\}} v_{\alpha,\beta} V_\beta \quad \text{on } K.$$

Now refer to Theorem 7.4.3. In view of the preceding, we see that for each open U with $\bar{U} \subset\subset W$ there is an $\ell \geq 1$ and $\{v_{\alpha,\beta} : \ell(\beta) \leq \ell\} \subseteq C_c^\infty(\mathbb{R}^N; \mathbb{R})$ for which (7.4.13) holds on $\bar{U}$. Hence, since, for any $\delta > 0$, there exists a $C < \infty$ such that

$$\|\varphi\|_\delta \leq C \left(\|\varphi\|_0 + \sum_{\|\alpha\|=1} \|\partial^\alpha \varphi\|_{\delta-1} \right),$$

we will know that (7.4.12) holds once we show that, for each $\ell \geq 1$, there is a $\delta \in (0,1]$ and a $C < \infty$ such that

$$(7.4.14) \qquad \max_{\{\beta : \ell(\beta) \leq \ell\}} \|V_\beta \varphi\|_{\delta-1} \leq C\big(\|L\varphi\|_0 + \|\varphi\|_0\big)$$
$$\text{for all } \varphi \in C_c^\infty(U, \mathbb{R}).$$

7.4.15 LEMMA. Let $\epsilon \in \big(0, \frac{1}{2}\big]$. If X is a real vector field on $\mathbb{R}^N$ with coefficients in $\mathscr{S}(\mathbb{R}^N; \mathbb{R})$ and $P^{2\epsilon-1} \in \mathscr{P}'$ has an asymptotic expansion with $\sigma_0 = 2\epsilon - 1$, then there exists a $C < \infty$ such that

$$\max_{1 \leq k \leq M} \big|\big(V_k X\varphi, P^{2\epsilon-1}\varphi\big)_0\big| \vee \big|\big(X V_k\varphi, P^{2\epsilon-1}\varphi\big)_0\big|$$
$$\leq C\big(\|X\varphi\|_{2\epsilon-1}^2 + \|L\varphi\|_0^2 + \|\varphi\|_0^2\big) \quad \text{for } \varphi \in \mathscr{S}(\mathbb{R}^N; \mathbb{R}).$$

In particular, there is a $C < \infty$ such that

$$\max_{1 \leq k \leq M} \big\|[V_k, X]\varphi\big\|_{\epsilon-1} \leq C\big(\|X\varphi\|_{2\epsilon-1} + \|L\varphi\|_0 + \|\varphi\|_0\big)$$

for all $\varphi \in \mathscr{S}(\mathbb{R}^N; \mathbb{R})$.

Proof: Let $1 \le k \le M$ be given. Then

$$
\begin{aligned}
\left|\left(V_k X\varphi, P^{2\epsilon-1}\varphi\right)_0\right| &= \left|\left(X\varphi, (V_k)^* P^{2\epsilon-1}\varphi\right)_0\right| \\
&\le \left|\left((P^{2\epsilon-1})^* X\varphi, (V_k)^*\varphi\right)_0\right| + \left|\left(X\varphi, [(V_k)^*, P^{2\epsilon-1}]\varphi\right)_0\right| \\
&\le \tfrac{1}{2}\left(\|(P^{2\epsilon-1})^* X\varphi\|_0^2 + \|(V_k)^*\varphi\|_0^2 + \|[(V_k)^*, P^{2\epsilon-1}]^* X\varphi\|_0^2 + \|\varphi\|_0^2\right) \\
&\le C\left(\|X\varphi\|_{2\epsilon-1}^2 + \|V_k\varphi\|_0^2 + \|\varphi\|_0^2\right) \le C'\left(\|X\varphi\|_{2\epsilon-1}^2 + \|L\varphi\|_0^2 + \|\varphi\|_0^2\right),
\end{aligned}
$$

where we have used the fact that $P^{2\epsilon-1}$ and $[(V_k)^*, P^{2\epsilon-1}]$ both have order $2\epsilon - 1$ and applied Lemma 7.4.5 in the last step. Similarly,

$$
\begin{aligned}
\left|\left(X V_k\varphi, P^{2\epsilon-1}\varphi\right)_0\right| &= \left|\left(V_k\varphi, X^* P^{2\epsilon-1}\varphi\right)_0\right| \\
&\le \left|\left(V_k\varphi, P^{2\epsilon-1} X^*\varphi\right)_0\right| + \left|\left(V_k\varphi, [X^*, P^{2\epsilon-1}]\varphi\right)_0\right| \\
&\le \tfrac{1}{2}\left(2\|V_k\varphi\|_0^2 + \|P^{2\epsilon-1} X^*\varphi\|_0^2 + \|[X^*, P^{2\epsilon-1}]\varphi\|_0^2\right) \\
&\le C\left(\|X\varphi\|_{2\epsilon-1}^2 + \|L\varphi\|_0^2 + \|\varphi\|_0^2 + \|\varphi\|_{2\epsilon-1}^2\right) \\
&\le C'\left(\|X\varphi\|_{2\epsilon-1}^2 + \|L\varphi\|_0^2 + \|\varphi\|_0^2\right),
\end{aligned}
$$

where, in the passage to the last line, we have used $2\epsilon - 1 \le 0$.

Turning to the final assertion, take $P^{2\epsilon-1} = B^{2-2\epsilon}[V_k, X]$. Then

$$
\begin{aligned}
\left\|[V_k, X]\varphi\right\|_{\epsilon-1}^2 &= \left\|B^{1-\epsilon}[V_k, X]\varphi\right\|_0^2 = \left([V_k, X]\varphi, P^{2\epsilon-1}\varphi\right)_0 \\
&= \left(V_k X\varphi, P^{2\epsilon-1}\varphi\right)_0 - \left(X V_k\varphi, P^{2\epsilon-1}\varphi\right)_0,
\end{aligned}
$$

and so the desired conclusion follows from the preceding ones. $\qquad\square$

By combining Lemmas 7.4.5 and 7.4.15 with an obvious induction argument, we see that

$$
\|V_\beta\varphi\|_{2^{1-\ell(\beta)}-1} \le C\left(\|L\varphi\|_0 + \|\varphi\|_0\right)
$$

when $\beta \in \mathcal{B}$ has no coordinate equal to 0. Thus, if, instead of the hypothesis in Theorem 7.4.3, we had made the stronger hypothesis that $\mathrm{Lie}_x(V_1, \ldots, V_M) = T_x\mathbb{R}^N$, then we would be done. Indeed, in that case, (7.4.13) would hold with $v_{\alpha,\beta} = 0$ whenever $\beta_j = 0$ for some $1 \le j \le \ell(\beta)$, and therefore (7.4.14) would hold with $\delta = 2^{1-\ell}$.

To bring V_0 into play, we need the analog of Lemma 7.4.15 for V_0.

7.4.16 LEMMA. Let $\epsilon \in \left(0, \frac{1}{4}\right]$. If X is as in Lemma 7.4.15, then there is a $C < \infty$ such that

$$
\left\|[V_0, X]\varphi\right\|_{\epsilon-1} \le C\left(\|X\varphi\|_{4\epsilon-1} + \|L\varphi\|_0 + \|\varphi\|_0\right) \quad \text{for all } \varphi \in \mathscr{S}(\mathbb{R}^N; \mathbb{R}).
$$

Proof: Just as in the proof of the last part of Lemma 7.4.15, one sees that it suffices to know that

$$
\begin{aligned}
\left|\left(V_0 X\varphi, P^{2\epsilon-1}\varphi\right)_0\right| &\vee \left|\left(X V_0\varphi, P^{2\epsilon-1}\varphi\right)_0\right| \\
&\le C\left(\|X\varphi\|_{4\epsilon-1}^2 + \|L\varphi\|_0^2 + \|\varphi\|_0^2\right),
\end{aligned}
$$

where $P^{2\epsilon-1} = B^{2-2\epsilon}[V_0, X]$.

To estimate $\left|\left(V_0 X\varphi, P^{2\epsilon-1}\varphi\right)_0\right|$, begin by noting that

$$V_0^* = \left(L - \frac{1}{2}\sum_{k=1}^{M} V_k^2\right)^* = L^* - \frac{1}{2}\sum_{k=1}^{M} V_k^2 + \sum_{k=1}^{M} \psi_k V_k + \psi,$$

where $\eta_0, \psi, \psi_1, \ldots, \psi_M \in \mathscr{S}(\mathbb{R}^N; \mathbb{R})$. Hence, $\left|\left(V_0 X\varphi, P^{2\epsilon-1}\varphi\right)_0\right|$ is dominated by

$$\left|\left(X\varphi, LP^{2\epsilon-1}\varphi\right)_0\right| + \frac{1}{2}\sum_{k=1}^{M}\left|\left(V_k^2 X\varphi, P^{2\epsilon-1}\varphi\right)_0\right|$$

$$+ \sum_{k=1}^{M}\left|\left(V_k X\varphi, \psi_k P^{2\epsilon-1}\varphi\right)_0\right| + \left|\left(X\varphi, \psi P^{2\epsilon-1}\varphi\right)_0\right|.$$

Clearly

$$\left|\left(X\varphi, \psi P^{2\epsilon-1}\varphi\right)_0\right| = \left|\left((P^{2\epsilon-1})^* \psi X\varphi, \varphi\right)_0\right| \leq C\left(\|X\varphi\|_{2\epsilon-1}^2 + \|\varphi\|_0^2\right).$$

Also, by Lemma 7.4.15,

$$\sum_{k=1}^{M}\left|\left(V_k X\varphi, \psi_k P^{2\epsilon-1}\varphi\right)_0\right| \leq C\left(\|X\varphi\|_{2\epsilon-1}^2 + \|L\varphi\|_0^2 + \|\varphi\|_0^2\right).$$

Thus, it suffices to estimate $\left|\left(X\varphi, LP^{2\epsilon-1}\varphi\right)_0\right|$ and, for $1 \leq k \leq M$, $\left|\left(V_k^2 X\varphi, P^{2\epsilon-1}\varphi\right)_0\right|$. Note that the first of these is dominated by

$$\left|\left((P^{2\epsilon-1})^* X\varphi, L\varphi\right)_0\right| + \left|\left(X\varphi, [L, P^{2\epsilon-1}]\varphi\right)_0\right|$$

$$\leq C\left(\|X\varphi\|_{2\epsilon-1}^2 + \|L\varphi\|_0^2\right) + \left|\left(X\varphi, [L, P^{2\epsilon-1}]\varphi\right)_0\right|.$$

To handle $\left|\left(X\varphi, [L, P^{2\epsilon-1}]\varphi\right)_0\right|$, we again use (7.4.7) to see that

$$[L, P^{2\epsilon-1}] = \sum_{k=1}^{M} V_k[V_k, P^{2\epsilon-1}] - \frac{1}{2}\sum_{k=1}^{M}[V_k, [V_k, P^{2\epsilon-1}]] + [V_0, P^{2\epsilon-1}]$$

$$= \sum_{k=1}^{M}(V_k)^* P_k^{2\epsilon-1} + P_0^{2\epsilon-1},$$

where the $P_k^{2\epsilon-1}$'s are again elements of $\mathscr{P}'$ whose asymptotic series have $\sigma_0 = 2\epsilon - 1$. Thus, by Lemma 7.4.15,

$$\left|\left(X\varphi, [L, P^{2\epsilon-1}]\varphi\right)_0\right| \leq \sum_{k=1}^{M}\left|\left(V_k X\varphi, P_k^{2\epsilon-1}\varphi\right)_0\right| + \left|\left(X\varphi, P_0^{2\epsilon-1}\varphi\right)_0\right|$$

$$\leq C\left(\|X\varphi\|_{2\epsilon-1}^2 + \|L\varphi\|_0^2 + \|\varphi\|_{2\epsilon-1}^2\right).$$

Turning to $\left|\left(V_k^2 X\varphi, P^{2\epsilon-1}\varphi\right)_0\right|$, begin by dominating it by

$$\left|\left(V_k X\varphi, P^{2\epsilon-1}V_k\varphi\right)_0\right| + \left|\left(V_k X\varphi, \tilde{P}_k^{2\epsilon-1}\varphi\right)_0\right|,$$

where $\tilde{P}_k^{2\epsilon-1} = \eta_k P^{2\epsilon-1} - [V_k, P^{2\epsilon-1}]$ is again an operator of the sort in Lemma 7.4.15. Hence, by that lemma, the second term causes no problem. As for the first term, use

$$\left|\left(V_k X\varphi, P^{2\epsilon-1}V_k\varphi\right)_0\right| \leq \tfrac{1}{2}\left(\|(P^{2\epsilon-1})^* V_k X\varphi\|_0^2 + \|V_k\varphi\|_0^2\right),$$

and apply Lemma 7.4.5 to handle $\|V_k\varphi\|_0^2$. At the same time,

$$\|(P^{2\epsilon-1})^* V_k X\varphi\|_0 \leq \|V_k(P^{2\epsilon-1})^* X\varphi\|_0 + \left\|[(P^{2\epsilon-1})^*, V_k]X\varphi\right\|_0,$$

and the second of these is dominated by a constant times $\|X\varphi\|_{2\epsilon-1}$. To treat $\|V_k(P^{2\epsilon-1})^* X\varphi\|_0$, apply Lemma 7.4.5 to dominate its square by a constant times

$$\left|\left(L(P^{2\epsilon-1})^* X\varphi, (P^{2\epsilon-1})^* X\varphi\right)_0\right| + \|X\varphi\|_{2\epsilon-1}^2.$$

Next, set

$$P^{2\epsilon} = (P^{2\epsilon-1})^* X = [V_0, X]^* B^{2-2\epsilon}X = -[X, V_0]B^{2-2\epsilon}X + \eta B^{2-2\epsilon}X,$$

where $\eta \in \mathscr{S}(\mathbb{R}^N; \mathbb{R})$, observe that $P^{2\epsilon} \in \mathscr{P}'$ with $\sigma_0 = 2\epsilon$, and check that

$$\begin{aligned}
\left|\left(L(P^{2\epsilon-1})^* X\varphi, (P^{2\epsilon-1})^* X\varphi\right)_0\right| &= \left|\left(LP^{2\epsilon}\varphi, (P^{2\epsilon-1})^* X\varphi\right)_0\right| \\
&\leq \left|\left(L\varphi, (P^{2\epsilon-1}P^{2\epsilon})^* X\varphi\right)_0\right| + \left|\left([L, P^{2\epsilon}]\varphi, (P^{2\epsilon-1})^* X\varphi\right)_0\right| \\
&\leq C\left(\|L\varphi\|_0^2 + \|X\varphi\|_{4\epsilon-1}^2\right) + \left|\left([L, P^{2\epsilon}]\varphi, (P^{2\epsilon-1})^* X\varphi\right)_0\right|.
\end{aligned}$$

Now (cf. (7.4.7)) write $[L, P^{2\epsilon}] = \sum_{k=1}^{M} \tilde{P}_k^{2\epsilon} V_k + \tilde{P}_0^{2\epsilon}$, where

$$\tilde{P}_k^{2\epsilon} = \begin{cases} [V_0, P^{2\epsilon}] - \tfrac{1}{2}\sum_{k'=1}^{M}[V_{k'}, [V_{k'}, P^{2\epsilon}]] & \text{if } k = 0 \\ [V_k, P^{2\epsilon}] & \text{if } 1 \leq k \leq M, \end{cases}$$

and note that the $\tilde{P}_k^{2\epsilon}$'s are again in $\mathscr{P}'$ with $\sigma_0 = 2\epsilon$. Thus

$$\begin{aligned}
\left|\left([L, P_k^{2\epsilon}]\varphi, (\tilde{P}_k^{2\epsilon-1})^* X\varphi\right)_0\right| & \\
\leq \sum_{k=1}^{M}&\left|\left(V_k\varphi, (P^{2\epsilon-1}\tilde{P}_k^{2\epsilon})^* X\varphi\right)_0\right| + \left|\left(\varphi, (P^{2\epsilon-1}\tilde{P}_0^{2\epsilon})^* X\varphi\right)_0\right| \\
\leq C&\left(\sum_{k=1}^{M}\|V_k\varphi\|_0^2 + \|X\varphi\|_{4\epsilon-1}^2 + \|\varphi\|_0^2\right) \\
\leq C'&\left(\|L\varphi\|_0^2 + \|X\varphi\|_{4\epsilon-1}^2 + \|\varphi\|_0^2\right),
\end{aligned}$$

where Lemma 7.4.5 was applied at the end.

To estimate $\left|\left(XV_0\varphi, P^{2\epsilon-1}\varphi\right)_0\right|$, we write $V_0 = L - \frac{1}{2}\sum_{k=1}^M V_k^2$ and find that we must estimate

$$\left|\left(XL\varphi, P^{2\epsilon-1}\varphi\right)_0\right| \quad \text{and} \quad \left|\left(XV_k^2\varphi, P^{2\epsilon-1}\varphi\right)_0\right|.$$

Writing $X^* = -X + \eta$, one sees that the first of these is dominated by

$$\left|\left(L\varphi, XP^{2\epsilon-1}\varphi\right)_0\right| + \left|\left(L\varphi, \eta P^{2\epsilon-1}\varphi\right)_0\right|$$
$$\leq \left|\left(L\varphi, P^{2\epsilon-1}X\varphi\right)_0\right| + \left|\left(L\varphi, [X, P^{2\epsilon-1}]\varphi\right)_0\right| + \left|\left(L\varphi, \eta P^{2\epsilon-1}\varphi\right)_0\right|$$
$$\leq \tfrac{3}{2}\|L\varphi\|_0^2 + C\|\varphi\|_{2\epsilon-1}^2 \leq \tfrac{3}{2}\|L\varphi\|_0^2 + C\|\varphi\|_0^2.$$

As for the second,

$$\left|\left(XV_k^2\varphi, P^{2\epsilon-1}\varphi\right)_0\right| = \left|\left(V_k\varphi, (V_k)^* X^* P^{2\epsilon-1}\varphi\right)_0\right|$$
$$\leq \left|\left(V_k\varphi, (V_k)^* P^{2\epsilon-1}X^*\varphi\right)_0\right| + \left|\left(V_k\varphi, (V_k)^*[X^*, P^{2\epsilon-1}]\varphi\right)_0\right|$$
$$\leq C\left(\|V_k\varphi\|_0^2 + \|\varphi\|_0^2 + \|(V_k)^* P^{2\epsilon-1}X^*\varphi\|_0^2\right),$$

since both $P^{2\epsilon-1}$ and its commutator with X^* have order $2\epsilon - 1 \leq 0$. Because $1 \leq k \leq M$, $\|V_k\varphi\|_0^2$ is handled by Lemma 7.4.5. Also,

$$\|(V_k)^* P^{2\epsilon-1}X^*\varphi\|_0 \leq \|V_k P^{2\epsilon-1}X^*\varphi\|_0 + C\left(\|X\varphi\|_{2\epsilon-1} + \|\varphi\|_0\right).$$

Finally, by the procedure used above to estimate $\|V_k(P^{2\epsilon-1})^* X\varphi\|_0$, we can estimate $\|V_k P^{2\epsilon-1}X^*\varphi\|_0$ in terms of $\|L\varphi\|_0$, $\|\varphi\|_0$, and, $\|X^*\varphi\|_{4\epsilon-1}$, which, since $4\epsilon - 1 \leq 0$, is dominated by $\|X\varphi\|_{4\epsilon-1} + C\|\varphi\|_0$. $\square$

Arguing as we did in the paragraph following Lemma 7.4.15, and combining this with the discussion preceding that lemma, we can now show that if $\{V_\beta : \ell(\beta) \leq \ell\}$ spans $T_x\mathbb{R}^N$ at every x in an open set of which $\bar{U}$ is a compact subset, then

$$\|\varphi\|_{4^{1-\ell}} \leq C\left(\|L\varphi\|_0 + \|\varphi\|_0\right), \quad \varphi \in C_c^\infty(U;\mathbb{R}).$$

In particular, after putting this together with Theorem 7.4.11, we have proved the following subellipticity statement.

7.4.17 THEOREM.   If $\mathrm{Lie}_x(V_0, \ldots, V_M) = T_x\mathbb{R}^N$ for each $x \in W$ and $c \in C^\infty(\mathbb{R}^N;\mathbb{R})$, then for each $\psi_1\,\psi_2 \in C_c^\infty(W;\mathbb{R})$ with $\psi_2 = 1$ on an open neighborhood of $\mathrm{supp}(\psi_1)$ there exists a $\delta > 0$ and a map $s \in \mathbb{R} \longmapsto C_s \in (0,\infty)$ such that (7.4.4) holds for all $s \in \mathbb{R}$ and $u \in H_s(\mathbb{R}^N;\mathbb{C})$.

Once one has Theorem 7.4.17, Theorem 7.4.3 follows easily by the same sort of bootstrap procedure which we used to prove Theorem 7.3.3 on the basis of the elliptic estimate in Theorem 7.3.2.

7.4.3. Some Applications: Having worked so hard to prove Theorem 7.4.3, it seems appropriate to point out a few of its applications.

For applications to probability theory, the following corollary is important. In its proof, it will be convenient to have introduced the Lie module

$$\widetilde{\mathrm{Lie}}(V_0, \ldots, V_M) = \left\{ \sum_{m=1}^{n} \psi_m V_{\beta^m} : n \geq 1, \ \beta^1, \ldots, \beta^n \in \mathcal{B}, \right.$$

$$\left. \text{and } \psi_1, \ldots, \psi_n \in C^\infty(\mathbb{R}^N; \mathbb{R}) \right\}$$

over $C^\infty(\mathbb{R}^N; \mathbb{R})$ generated by $\{V_0, \ldots, V_M\}$. It is elementary to check that $\widetilde{\mathrm{Lie}}(V_0, \ldots, V_M)$ is again a Lie algebra of vector fields and that, for each $x \in \mathbb{R}^N$, the subspace $\widetilde{\mathrm{Lie}}_x(V_0, \ldots, V_M)$ of $T_x\mathbb{R}^N$ obtained by evaluating the elements of $\widetilde{\mathrm{Lie}}(V_0, \ldots, V_M)$ at x coincides with $\mathrm{Lie}_x(V_0, \ldots, V_M)$.

7.4.18 THEOREM. *Let $V_0, \ldots, V_M$ be smooth, real vector fields on $\mathbb{R}^N$, and define L as in (7.4.1). If $\mathrm{Lie}_x(V_0, \ldots, V_M) = T_x\mathbb{R}^N$ for all x in an open set $W \subseteq \mathbb{R}^N$, then, for every $c \in C^\infty(\mathbb{R}^N; \mathbb{R})$, $L^* + c$ is hypoelliptic on W. Moreover, if $\mathrm{Lie}_x([V_0, V_1], \ldots, [V_0, V_M], V_1, \ldots, V_M) = T_x\mathbb{R}^N$ for all $x \in W$, and if $f \in C^\infty(\mathbb{R}; \mathbb{R})$ never vanishes on the open interval J, then $L + c + f(t)\partial_t$ and $L^* + c + f(t)\partial_t$ are hypoelliptic on $J \times W$ for any $c \in C^\infty(\mathbb{R} \times \mathbb{R}^N; \mathbb{R})$.*

PROOF: All these assertions come from Theorem 7.4.3 combined with elementary observations about Lie algebras of vector fields. To prove the first, write $V_k^* = -V_k + \eta_k$ and observe that $L^* = \frac{1}{2}\sum_{k=1}^{M} V_k^2 + V_0' + c$, where

$$V_0' = -V_0 - \sum_{k=1}^{M} \eta_k V_k \quad \text{and} \quad c = \eta_0 + \frac{1}{2}\sum_{k=1}^{M}(\eta_k^2 - (V_k\eta_k)).$$

Hence, by the preceding discussion, it suffices to note that $\widetilde{\mathrm{Lie}}(V_0, \ldots, V_M) = \widetilde{\mathrm{Lie}}(V_0', V_1, \ldots, V_M)$.

To prove the second assertion, think of the V_k's as vector fields on $\mathbb{R} \times \mathbb{R}^N$ for which the coefficient of ∂_t vanishes. Then, to prove that $L + c + f(t)\partial_t$ is hypoelliptic on $J \times W$ under the stated hypothesis, one need only observe that $\mathrm{Lie}_{(t,x)}(V_0 + f(t)\partial_t, V_1, \ldots, V_M)$ equals the span of

$$\mathrm{Lie}_x([V_0, V_1], \ldots, [V_0, V_M], V_1, \ldots, V_M) \cup \{V_0 + f(t)\partial_t\}$$

and that the latter equals $T_{(t,x)}(\mathbb{R} \times \mathbb{R}^N)$ if

$$\mathrm{Lie}_x([V_0, V_1], \ldots, [V_0, V_M], V_1, \ldots, V_M) = T_x\mathbb{R}^N.$$

Finally, once one notes that

$$\widetilde{\mathrm{Lie}}([V_0, V_1], \ldots, [V_0, V_M], V_1, \ldots, V_M))$$

$$= \widetilde{\mathrm{Lie}}([V_0', V_1], \ldots, [V_0', V_M], V_1, \ldots, V_M),$$

the argument for $L^* + c + \partial_t$ is a repeat of the one for $L + c + f(t)\partial_t$. $\square$

We close by applying Theorem 7.4.18 to transition probability functions. Thus, let $\sigma : \mathbb{R}^N \longrightarrow \mathrm{Hom}(\mathbb{R}^M; \mathbb{R}^N)$ and $b : \mathbb{R}^N \longrightarrow \mathbb{R}^N$ be smooth functions with bounded derivatives of all orders, set $a = \sigma\sigma^\top$, and let $(t, x) \rightsquigarrow P(t, x)$ be the transition probability function determined (cf. Theorem 2.4.6) by the associated operator L given in (1.1.8). The next lemma allows us to apply Theorem 7.4.18 to $(t, x) \rightsquigarrow P(t, x)$. In its statement, the subscripts x and y on an operator are used to indicate on which variables the operator acts.

7.4.19 LEMMA. *Define the distribution* u *on* $C_c^\infty(\mathbb{R} \times \mathbb{R}^N \times \mathbb{R}^N; \mathbb{R})$ *by*

$$\langle \varphi, u \rangle = \int_0^\infty \left(\int \left(\int \varphi(t, x, y) P(t, x, dy) \right) dx \right) dt.$$

Then $(L_x + L_y^\top - 2\partial_t)u = 0$ *on* $(0, \infty) \times \mathbb{R}^N \times \mathbb{R}^N$.

PROOF: Let $\varphi \in C_c^\infty\big((0, \infty) \times \mathbb{R}^N \times \mathbb{R}^N; \mathbb{R}\big)$ be given. We must show that $\langle (L_x^\top + L_y + 2\partial_t)\varphi, u \rangle = 0$, and clearly this will follow if we show that $\langle (L_x^\top + \partial_t)\varphi, u \rangle = 0 = \langle (L_y + \partial_t)\varphi, u \rangle$. The second of these is easy. Namely, by Fubini's Theorem,

$$\langle (L_y + \partial_t)\varphi, u \rangle = \int \left(\int \left(\int [(L_y + \partial_t)\varphi](t, x, y)\, P(t, x, dy) \right) dt \right) dx$$

$$= \int \left(\int \partial_t \left(\int \varphi(t, x, y)\, P(t, x, dy) \right) dt \right) dx = 0$$

since φ is compactly supported in $(0, \infty) \times \mathbb{R}^N \times \mathbb{R}^N$. In order to handle the other term, recall the adjoint transition function $(t, y) \rightsquigarrow P^\top(t, y)$ introduced in § 2.5.1. Next, starting from $\big(f, \mathbf{P}_t g\big)_{L^2(\mathbb{R}^N; \mathbb{C})} = \big(\mathbf{P}_t^\top f, g\big)_{L^2(\mathbb{R}^N; \mathbb{C})}$ and applying Fubini's Theorem, we see (cf. (2.5.6)) first that

$$\int \left(\int [(L_x^\top + \partial_t)\varphi](t, x, y)\, P(t, x, dy) \right) dx$$

$$= \int \left(\int [(L_x^\top + \partial_t)\varphi](t, x, y)\, P^\top(t, y, dx) \right) dy$$

and then, just as before, that

$$\langle (L_x^\top + \partial_t)\varphi, u \rangle = \int \left(\int \partial_t \left(\int \varphi(t, x, y)\, P^\top(t, y, dx) \right) dt \right) dy = 0. \quad \square$$

7.4.20 THEOREM. *Define the vector fields* $V_0, \ldots, V_M$ *from* σ *and* b *as in* (7.4.2). *If* $W \subseteq \mathbb{R}^N$ *is an open subset such that*

$$\mathrm{Lie}_x\big([V_0, V_1], \ldots, [V_0, V_M], V_1, \ldots, V_M\big) = T_x\mathbb{R}^N \quad \text{for all } x \in W,$$

then there exists a smooth $p : (0, \infty) \times W \times W \longrightarrow [0, \infty)$ *such that* $P(t, s, \Gamma) = \int_\Gamma p(t, x, y)\, dy$ *for all* $\Gamma \in \mathcal{B}_{\mathbb{R}^N}$ *contained in* W.

Proof: In view of Theorem 7.4.18 and Lemma 7.4.19, all that we have to do is check that, for $(x, y) \in W \times W$,

$$\text{Lie}_{(x,y)}\big([V_0', V_1'], \ldots, [V_0', V_{2M}'], V_1', \ldots, V_{2M}'\big) = T_{(x,y)}(\mathbb{R}^N \times \mathbb{R}^N),$$

where $V_k' = (V_k)_x$ and $V_{M+k}' = (V_k)_y$ for $1 \leq k \leq M$, and

$$V_0' = (V_0)_x - (V_0)_y - \sum_{k=1}^{M}(V_k \eta_k)(y)(V_k)_y,$$

where the subscript is used to indicate on which variable the action is taken. But

$$\begin{aligned}
\text{Lie}_{(x,y)}&\big([V_0', V_1'], \ldots, [V_0', V_{2M}'], V_1', \ldots, V_{2M}'\big) \\
&= \text{Lie}_x\big([V_0, V_1], \ldots, [V_0, V_M], V_1, \ldots, V_M\big) \\
&\quad \oplus \text{Lie}_y\big([V_0, V_1], \ldots, [V_0, V_M], V_1, \ldots, V_M\big),
\end{aligned}$$

and so there is nothing more to do. □

Although Theorem 7.4.20 greatly extends the class of operators L whose transition probability functions we can say admit a smooth density, it would be a great mistake to assume that all these densities satisfy the sort of estimates which we proved in Chapter 4. To make it clear exactly how wrong such an assumption would be, consider the operator

$$L = \tfrac{1}{2}\partial_{x_1}^2 + \frac{x_1^2}{1 + x_1^2}\partial_{x_2}.$$

In this case $V_1 = \partial_{x_1}$ and $V_0 = \frac{x_1^2}{1+x_1^2}\partial_{x_2}$. Since

$$[V_1, V_0] = \frac{2x_1}{(1+x_1^2)^2}\partial_{x_2} \quad \text{and} \quad [V_1, [V_1, V_0]] = 2\frac{1 - 3x_1^2}{(1+x_1^2)^3}\partial_{x_2},$$

it is clear that $\text{Lie}_x\big([V_0, V_1], V_1\big) = T_x\mathbb{R}^2$ for all $x \in \mathbb{R}^2$, and therefore the associated transition probability $P(t, x)$ function admits a smooth density $p(t, x, y)$ on $(0, \infty) \times \mathbb{R}^2 \times \mathbb{R}^2$. Next, observe that $P(t, x)$ is the distribution of $\big(X_1(t, x), X_2(t, x)\big)$, where

$$X_1(t, x) = x_1 + B(t), \quad X_2(t, x) = x_2 + \int_0^t \frac{\big(x_1 + B(\tau)\big)^2}{1 + \big(x_1 + B(\tau)\big)^2}\, d\tau,$$

and $t \rightsquigarrow B(t)$ is a standard, $\mathbb{R}$-valued Brownian motion. In particular, $P\big(t, \mathbb{R} \times (-\infty, x_2)\big) = 0$, and so $p(t, x, y) = 0$ for all (x, y) with $y_2 \leq x_2$. Although it is somewhat out of place, one should remark that this conclusion rules out the possibility that $p(t, x, y)$ might be always real analytic whenever the vector fields are real analytic and satisfy Hörmander's condition.

7.5 Historical Notes and Commentary

Perhaps the first systematic treatment of what we now call pseudodifferential operators was given by H. Weyl in [58], where his motivation seems to have been a mathematically rigorous model of what we would now call quantization. A second early version of the theory was given by A.P. Calderón in [8], where it plays a crucial role in his proof of uniqueness for Cauchy initial value problems. The treatment given here is based on the paper [28] by J.J. Kohn and L. Nirenberg, which may be the first places where the theory was given the form which has become standard. On the other hand, by modern standards, the theory presented here is primitive. Indeed, in the years since [28] appeared, the theory has been vastly extended and is now seen as a special case of the much more general theory of Fourier integral operators. For the reader who wants to learn more and has lots of time to do so, Hörmander's multi-volume treatise [23] is indispensable.

Theorem 7.4.3 of Hörmander appears in [22]. Earlier, in [21], he had given a complete characterization in terms of their symbols of those constant coefficient, partial differential operators which are hypoelliptic. However, so far as I know, [22] is the first place in which a general theory for variable coefficient, non-parabolic operators appears. In my opinion, Hörmander's own proof of Theorem 7.4.3 is more revealing and sharper than the one we have given, which is taken from Kohn's [27]. However, Hörmander's proof is much more difficult than Kohn's and, as we learned since, his estimate is not optimal. In fact, starting with the paper [49] by L. Rothschild and E.M. Stein, when $V_0 = \sum_{k=1}^{M} c_k V_k$ for smooth functions c_k, a beautiful connection has been established between the δ in (7.4.4) and the volume growth of balls in the Cartheodory metric corresponding to the vector fields $\{V_1, \ldots, V_M\}$. Further, under the same condition, A. Sánchez-Calle proved that the associated heat kernel is bounded above and below in terms of a Gauss kernel in which $|y - x|^2$ is replaced in the exponential by the square of the Cartheodory distance and $t^{-\frac{N}{2}}$ is replaced by one over the volume of the Cartheodory ball around x of radius $t^{\frac{1}{2}}$. An excellent review of this material can be found in D. Jerison and Sánchez-Calle's article [26].

Probability theory has made a couple of contributions to our understanding of operators of the sort discussed in § 7.4. For example, as mentioned in § 6.4, Varadhan and my "support theorem" provides them with a strong minimum principle. More significant, in his groundbreaking paper [38], P. Malliavin initiated a program to prove regularity results using the ideas on which the results in Chapter 3 were based. As he realized, his ideas should handle solutions to equations involving operators of the sort dealt with in Hörmander's Theorem. His program was taken up and completed by various authors, but perhaps the most thorough treatment is the one given by Kusuoka and me in our papers [32], where we used probabilistic techniques to prove not only Hörmander's Theorem but also the estimates of Sánchez-Calle. I chose to not present these proofs mostly because I have become increasing skeptical about whether their virtues justify the difficulty of their details. In addition, as I said in the introduction to this chapter, I felt obliged to give at least a sample of the powerful techniques used in this chapter. Be that as it may, for those who, even after my warnings, would like to see what is involved in carrying out Malliavin's program, I suggest the book [6] by R. Bass, who made a valiant effort to present this material in a palatable way.

Finally, it should be recognized that, as stated, Hörmander's Theorem suffers from the same weakness as the results in § 2.2. Namely, in order to write an L given by (1.1.8) in the form in (7.4.1), the diffusion matrix a has to admit a smooth square root. Thus, it is important to know that there is an improved statement of Theorem 7.4.3 in which this weakness is no longer present. The precise statement is that if L is presented in the form given by (4.4.1) and

$$ V_k = \begin{cases} \sum_{j=1}^{N} b_j \partial_{x_j} & \text{if } k = 0 \\ \sum_{j=1}^{N} a_{ij} \partial_{x_j} & \text{if } 1 \le k \le N, \end{cases} $$

then the conclusion of Theorem 7.4.3 holds for L. Interestingly, just as was the case in Chapter 2, Oleinik, acting on a suggestion by Kohn, played a crucial role in making this improvement, which appeared for the first time in her work with E.V. Radekevich [47]. Refinements of their results appear in the work by C. Fefferman and D. Phong [18]. For further references, see [26].

Notation

===

Notation	Description
$\mathbf{1}_A$	The indicator function of the set A.
$a \wedge b \,\&\, a \vee b$	The minimum and the maximum of a, $b \in \mathbb{R}$.
$a^+ \,\&\, a^-$	The positive part $a \vee 0$ and negative part $(-a) \vee 0$ of $a \in \mathbb{R}$.
$\|\alpha\|$	Denotes $\sum_{i=1}^N \alpha_i$ for a multiindex $\alpha \in \mathbb{N}^N$.
$B(x, r)$	The ball of radius r centered at x.
$\mathcal{B}_E$	The σ-algebra of Borel measurable subsets of E.
$\mathcal{B}_t$	The σ-algebra generated by $\{\omega_\tau : \tau \in [0, t]\}$.
$C_{\mathrm{b}}(E; \mathbb{R})$	Space of bounded continuous functions from E into $\mathbb{R}$.
$C_{\mathrm{c}}(G; \mathbb{R})$	Space of continuous, $\mathbb{R}$-valued functions having compact support in the open set G.
$C^{1,2}([0, \infty) \times \mathbb{R}^N; \mathbb{C})$	Space of functions $(t, y) \in [0, \infty) \times \mathbb{R}^N \longrightarrow \mathbb{R}$ continuously differentiable once in t and twice in y.
$f \star g$	The convolution product of functions f and g.
$\mathbb{E}^\mu[X, A]$	To be read *the expectation value of X with respect to μ on A.* Equivalent to $\int_A X \, d\mu$. When A is unspecified, it is assumed to be the whole space.
$e_\xi(x)$	The imaginary exponential $e^{\sqrt{-1}\,(\xi, x)_{\mathbb{R}^N}}$.
$\Gamma(t)$	Euler's Gamma function.
γ	Standard Gauss measure on Euclidean space.
$H_s(\mathbb{R}^N; \mathbb{C}) \,\&\, H_s(\mathbb{R}^N; \mathbb{R})$	Sobolev space, values in $\mathbb{C}$ or $\mathbb{R}$.
$K \subset\subset E$	To be read: *K is a compact subset of E.*

$L^p(\mu; E)$	Lebesgue space of E-valued functions f with $\|f\|_E^p$ is μ-integrable.
$\mathbf{M}_1(E)$	Space of Borel probability measures on E.
$\hat{\mu}$	The characteristic function (Fourier transform) of μ.
$\mathbb{N}$	The non-negative integers: $\mathbb{N} = \{0\} \cup \mathbb{Z}^+$.
p'	The Hölder conjugate $\frac{p}{p-1}$ of $p \in [1, \infty]$.
S^m	The shift map on sequences.
$\mathbb{S}^{N-1}$	The unit sphere in $\mathbb{R}^N$.
$\mathscr{S}(\mathbb{R}^N; \mathbb{C})$ & $\mathscr{S}(\mathbb{R}^N; \mathbb{R})$	Schwartz test function, values in $\mathbb{C}$ or $\mathbb{R}$.
$\mathscr{S}'(\mathbb{R}^N; \mathbb{C})$ & $\mathscr{S}'(\mathbb{R}^N; \mathbb{R})$	Tempered distributions, values in $\mathbb{C}$ or $\mathbb{R}$.
Σ_s	The time shift transformation on pathspace.
$[t]_n$	The largest number $m2^{-n}$ dominated by t. Equivalently, $[t]_n = 2^{-n}[2^n t]$, where $[t] \equiv [t]_0$ is the integer part of t.
$\mathbb{Z}^+$	The strictly positive integers.
ω_{N-1} & Ω_N	The surface area and volume of the unit sphere and ball in $\mathbb{R}^N$.
$\langle \varphi, \mu \rangle$	Alternative notation for $\int \varphi \, d\mu$.
$\langle \varphi, u \rangle$	Action of distribution u on test function φ.
$\| \cdot \|_{\mathrm{u}}$	The uniform or "sup" norm on functions.
$\| \cdot \|_{\mathrm{H.S.}}$	The Hilbert–Schmidt norm.
$|\Gamma|$	Lebesgue measure of Γ.

References

[1] Adams, R.A., *Sobolev Spaces*, Academic Press, New York, 1975.

[2] Alexandrov, A.D., *Certain estimates concerning the Dirichlet problem*, Dokl. Akad. Nauk. SSSR **134** (1960), 1151–1154.

[3] Aronson, D., *The fundamental solution of a linear parabolic equation containing a small parameter*, Ill. J. Math. **3** (1959), 580–619.

[4] _____, *Bounds on the fundamental solution of a parabolic equation*, B.A.M.S. **73** (1967), 890–896.

[5] Azencott, R., *Methods of localization and diffusions on manifolds*, Publications of Istituto Matematico Ulisse Dini, Firenze, Italy (1971).

[6] Bass, R., *Probabilistic Techniques in Analysis*, Springer-Verlag, New York and Heidelberg, 1995.

[7] Bell, D., *The Malliavin Calculus*, Pitman Monographs and Surveys in Pure & Appl. Math., vol. 34, Longman, Wiley, Essex, 1987.

[8] Calderón, A.P., *Uniqueness in the Cauchy problem for partial differential equations*, Amer. J. Math. **80** (1958), 16-36.

[9] Chavel, I., *Riemannian Geometry: a Modern Introduction*, Cambridge Tracts in Math. #108, Cambridge Univ. Press, New York, 1993.

[10] Davies, E.B., *Heat Kernels and Spectral Theory*, Cambridge Tracts in Math. #92, Cambridge Univ. Press, New York, 1989.

[11] De Giorgi, E., *Sulle differentiabità e l'analiticità degli interali multipli regolari*, Mem. Accad. Sci. Torino Cl. Sci. Fis. Mat. Narur. (III) (1957).

[12] Doob, J.L., *A probability approach to the heat equation*, T.A.M.S. **80** (1955), 216–280.

[13] Eells, J. and Elworthy, D, *Stochastic dynamical systems*, Control Theory and Topics in Functional Analysis, Lectures in Semin. Course, Trieste 1974, vol. III, Atomic Energy Agency, Vienna, 1976, pp. 179–185.

[14] Evans, L.C., *Partial Differential Equations*, Graduate Studies in Math. #19, AMS, Providence, RI, 1998.

[15] Fabes, E. and Riviere, N., *Systems of parabolic equations with uniformly continuous coefficients*, Jour. D'Analyse Math. **XVII** (1966), 305–335.

[16] Fabes, E. and Stroock, D., *A new proof of Moser's parabolic Harnack inequality using the old ideas of Nash*, Arch. for Ratl. Mech. & Anal. **96** (1986), 327–338.

[17] ———, *The L^p-integrability of Green's functions and fundamental solutions for elliptic and parabolic equations*, Duke Math. J. **51** (1984), 997–1016.

[18] Fefferman, C., *The uncertainty principle*, B.A.M.S. **9** (1983), 129-206.

[19] Friedman, A., *Partial Differential Equations of the Parabolic Type*, Prentice Hall, Englewood Cliffs, N.J., 1964.

[20] Gikhman, I.I. and Skorohod, A.V., *Stochastic Differential Equations*, Springer-Verlag, New York and Heidelberg, 1972.

[21] Hörmander, L., *Linear Partial Differential Operators*, Springer-Verlag, New York and Heidelberg, 1967.

[22] ———, *Hypoelliptic second order differential operators*, Acta. Math. **121** (1967), 147–171.

[23] ———, *The Analysis of Partial Differential Operators, I–IV,*, Springer-Verlag, New York and Heidelberg, 1983–1987.

[24] Itô, K., *On Stochastic Differential Equations*, Mem. Amer. Math. Soc. #4, AMS, Providence, RI, 1951.

[25] Itô, K., *On stochastic differential equations on a differentiable manifold, 1 & 2*, Nagoya Math. J. #1 and Mem. Coll. Sci. Univ. Kyoto A. #28 (1950 and 1953), 35–47 and 82–85.

[26] Jerison, D. and Sanchez-Calle, A., *Subelliptic, Second Order Differential Operators*, Springer Lecture Notes in Math. #1277, Springer-Verlag, New York and Heidelberg, 1987, pp. 46–77.

[27] Kohn, J.J., *Pseudo-differential operators and non-elliptic problems*, Pseudo-Differential Operators (C.I.M.E., Stresa, 1968), Edizioni Cremonese, Rome, 1969, pp. 157-165; Pseudo-differential operators and hypoellipticity, Proc. Sympos. Pure Math. (1970), vol. 23, AMS, Providence, RI, 1973, pp. 61-69.

[28] Kohn, J.J. and Nirenberg, L., *An algebra of pseudo-differential operators*, Comm. Pure Appl. Math. **18** (1965), 269–305.

[29] Kolmogorov, A.N., *Uber der Analytischen Methoden in der Wahrscheinlictkeitsrechnung*, Math. Ann. **104** (1931), 415–458.

[30] Krylov, N.V. and Safonaov, M.V., *A certain property of solutions of parabolic equations with measurable coefficients*, English version: Mathematics of the USSR-Izvestiya **16**:1, 151–164 (1981), Izv. RAN. Ser. Mat. **44** #1 (1980), 161–175.

[31] Krylov, N.V., *Lectures on Elliptic and Parabolic Equations in Hlder Spaces*, Graduate Studies in Math. #12, A.M.S., Providence, RI, 1996.

[32] Kusuoka, S. and Stroock, D., *Application of the Malliavin calculus, II,* J. Fac. Sci. of Tokyo University, Sec. IA **32** (1987), 1–76; *Application of the Malliavin calculus, III,* J. Fac. Sci. of Tokyo University, Sec. IA **34** (1987), 391–442.

[33] ———, *Asymptotics of certain Wiener functionals with degenerate extrema,* Comm. Pure Appl. Math. **XLVII** (1994), 477–501.

[34] Ladyženskaja, O.A., Solonnikov, V.A., and Uralćeva, N.N., *Linear and Quasilinear Equations of Parabolic Type,* Translations of Math. Monographs #23, A.M.S., Providence, RI, 1968.

[35] Levi, E.E., *Sulle equazioni lineari totalmente ellittiche alle derivate parziali,* Rend. del. Circ. Mat. Palermo **24** (1907), 275–317.

[36] Li, P. and Yau, S.T., *On the parabolic kernel of a Schrödinger operator,* Acta. Math. **156** (1986), 153–201.

[37] Malliavin, P., *Geometrie Differentielle Stochastique,* Presses Univ. Montréal, Montréal, 1978.

[38] ———, *Stochastic calculus of variations and hypoelliptic operators, Proceedings of the International Symposium on Stochastic Differential Equations, Kyoto 1976,* Wiley, New York, 1978, pp. 195-263.

[39] McKean, H.P., *Stochastic Integrals,* Reprinted by AMS Chelsea, Academic Press, New York, 1969.

[40] Molchanov, S.A., *Diffusion processes and Riemannian geometry,* Russian Math. Survey **30** (1975), 1–63.

[41] Moser, J. A new proof of De Giorgi's theorem concerning the regularity problem for elliptic differential equations, Comm. Pure Appl. Math. **XIII** (1960), 457–468.

[42] ———, *A Harnack inequality for parabolic differential equations,* Comm. Pure Appl. Math. **XVII** (1964), 101–134.

[43] Nash, J., *Continuity of solutions of parabolic and elliptic equations,* Amer. J. of Math. **80** (1958), 931–954.

[44] Nirenberg, L., *A strong maximum principle for parabolic equations,* Comm. Pure Appl. Math. **VI** (1953), 167–177.

[45] Norris, J. and Stroock, D., *Estimates on the fundamental solution to heat flows with uniformly elliptic coefficients,* Proc. London Math. Soc. **62 #3** (1991), 373–402.

[46] Oleinik, O.A., *Alcuni risulati sulle equazioni lineari elliptico-paraboliche a derivate parziali del second order,* Rend. Casse Sci. Fis. Mat., Nat. Acad. Naz. Lineri, Ser. 8 **40** (1966), 775– 784.

[47] Oleinik, O.A. and Radekevich, E.V., *Second Order Equations with Non-negative Characteristic Form,* Plenum Press, New York, 1973.

[48] Pogorzelski, W., *Étude de la solution fundamental de l'equation parabolique,* Richerche di Mat. **5** (1956), 25–57.

[49] Rothschild, L.P. and Stein, E.M., *Hypoelliptic operators on nilpotent Lie groups*, Acta. Math. **137** (1977), 247–320.

[50] Saloff-Coste, L., *A note on Poincaré, Sobolev, and Harnack*, Internat. Math. Research Notices **2** (1992), 27–38.

[51] Stroock, D.W. and Varadhan, S.R.S., *On degenerate elliptic-parabolic operators of the second order and their associated diffusions*, Comm. Pure Appl. Math. **XXV** (1972), 651–713.

[52] ———, *Multidimensional Diffusion Processes*, Grundlehren Series #233, Springer-Verlag, New York and Heidelberg, 1979.

[53] Stroock, D., *Probability Theory, an Analytic View*, Cambridge U. Press, New York, 1993 & 1998.

[54] ———, *An Introduction to the Analysis of Paths on a Riemannian Manifold*, Mathematical Surveys and Monographs #74, AMS, Providence, RI, 2000.

[55] ———, *Markov Processes from K. Itô's Perspective*, Annals of Math. Studies #155, Princeton Univ. Press, Princeton, NJ, 2003.

[56] Varadhan, S.R.S., *On the behavior of the fundamental solution of the heat equation with variable coefficients*, Comm. Pure Appl. Math. **20** (1967), 431–455.

[57] Weyl, H., *The method of orthogonal projection in potential theory*, Duke Math. J. **7** (1940), 414–444.

[58] Weyl, H., *The Theory of Groups and Quantum Mechanics*, Dover Press, New York, 1950.

Index